STATIC TEST METHODS FOR COMPOSITES

Yu. M. Tarnopol'skii
T. Kincis

Translated from the third, revised and supplemented Soviet edition under the editorship of

George Lubin

VNR VAN NOSTRAND REINHOLD COMPANY

Library of Congress Catalog Card Number: 84-2254
ISBN: 0-442-28281-8

Manufactured in the United States of America

Published by Van Nostrand Reinhold Company Inc.
135 West 50th Street
New York, New York 10020

Van Nostrand Reinhold Company Limited
Molly Millars Lane
Wokingham, Berkshire RG11 2PY, England

Van Nostrand Reinhold
480 Latrobe Street
Melbourne, Victoria 3000, Australia

Macmillan of Canada
Division of Gage Publishing Limited
164 Commander Boulevard
Agincourt, Ontario M1S 3C7, Canada

15 14 13 12 11 10 9 8 7 6 5 4 3 2 1

Library of Congress Cataloging in Publication Data

Tarnopol'skiĭ, I͡U. M.
Static test methods for composites.

Includes bibliographical references and indexes.
1. Fibrous composites—Testing. I. Kincis, T. I͡A.
(Talivaldis I͡Anovich) II. Title.
TA418.9.C6T37 1984 620.1′18 84-2254
ISBN 0-442-28281-8

Foreword

The rapid growth in the development and application of composits has reached the stage where these materials are no longer considered to be exotic, but are treated as practical engineering structures. In aircraft alone, the use of composites has gone up from 2–3% of the total weight of the plane to 10–15% and some, more advanced structures may use up to 40–60%.

To aid in introducing and utilizing these materials, numerous design and manufacturing processes have been developed to the point that structurally effective and durable parts are being economically fabricated and are expected to perform as well or better than the previously used metallic materials.

In working with composites, it is necessary to develop statistical design allowables similar to those used in the metal industry. To obtain these design allowables, extensive testing has been required and it was found early in the composites history that the test data scatter is much greater than that of metals, and the resulting figures have had to be greatly reduced. As is explained in the text of this book, the scatter is basically caused by the large variations in properties of the components of the composite materials, their orientation, process parameters, relative proportion of constituents and processing techniques.

It is no exaggeration when we claim that a specimen cut from a sheet of metal represents fairly closely the strength of the metal. In the case of composites, an individual specimen usually represents only the strength of that particular specimen and has merely a general relation to the strength of the final article fabricated from the particular composite.

This state of affairs has necessitated the need for destructive testing of the final molded part to prove the effectiveness of the design. This in turn frequently has resulted in overdesigned structures and minimal weight savings.

Precise strength data is essential for the design and economic fabrication of production composites of the future.

To try to resolve the problem of data variations, a group of engineers and scientists at the Institute of Polymer Mechanics of the Latvian SSR Academy of Sciences in Riga conducted a survey of all available literature on the testing of polymers and amplificied it by extensive additional research. The results of the survey constitute the material in this book. Here, all identifiable variables have been investigated where possible, and formulas for the development of design data have been either selected from literature or derived and modified by the authors with the help of the Institute personnel. These formulas are offered to the readers for use in designing composites together with an extended set of references including papers, books, publications and standards from all over the world. The effects of all fabrication parameters are discussed and the numerous test methods and test specimens described, evaluated and rated for accuracy and effectiveness.

After reading this book, one realizes how inadequate and incomplete are the sample preparation and testing techniques used by most laboratories. The old metal methods and the newer fiberglass technology are too unreliable to apply to modern composites and many new procedures have had to be introduced and perfected.

In my own experience I have found that much of the early data on composites is unreliable and a comparison of values generated on identical materials by different laboratories shows great scatter and variability. While the authors of this book do not offer magic solutions to all our problems, they shed much light on the hidden parameters and offer many suggestions on the choice of tests and material control.

Most books on composites or anisotropic materials deal with basic design concepts and the theory of elasticity of anisotropic structures. In this book, information is presented for the planning of destructive tests, determination of applicable variables and the verification and control of test results.

In selecting specimen shapes and fabricating operations, it is required that all manufacturing conditions and processes for the material, product or structure be properly simulated. Suggestions are offered on the relative importance and effects of most production variables. Test results on specimens with different configurations are discussed and recommendations made for optimum reliability.

Other subjects covered include the scale effect, stress concentrations, specimen conditioning and types of loading. A final recommendation includes a list of all information which should be a part of the test report.

George Lubin

Preface

Over the last decades, the possibility of industrial production of materials with properties assigned in advance has first arisen in human history. The fact is closely bound up with the creation, development and improvement of composites. The advent of composites has been prompted by contemporary technological demands on materials. At the basis of composites lies the concept of reinforcement—an idea unique in its simplicity, involving the combination of polar opposites as to their material properties, a compliant matrix and a strong and stiff reinforcement.

The idea has originated in nature, so far from extremely simplified forms. These are the stalks and leaves of plants, human and animal bones which behave as fiber-reinforced anisotropic materials—composites. It is noteworthy that the concept of reinforcement is broader than simply strength or workability. It involves also an increase in material reliability. Apparently composites are the only materials for which an increase in strength is accompanied by an increase in fracture toughness.

The mechanical properties of composites, in contrast to metals, are characterized by a multitude of constants. Their experimental evaluation is fraught with essential methodical difficulties. It should be borne in mind that the level of standardization of test methods for composites leaves much to be desired. In practice, a variety of specimen shapes and sizes, manufacturing methods of specimens and experimental techniques are being employed. This has resulted in incomparable results and contradictory judgements about the structural potential of composites. A realistic approach of mechanical testing requires strict regulation of the number and methods of evaluation of strength and stiffness of composites, increasing the necessity of a critical analysis of the existing methods, their evaluation and generalization.

In an effort to select and evaluate the most promising static test methods for advanced composites in tension, compression, shear and bending on flat and ring specimens, the book is based on world experience, mainly that of the USSR and USA. Selection of the test methods and their areas of application is substantiated by voluminous factual material. The majority of tables and illustrations contain references to works on which they have been based; these works contain all the necessary additional information. The references are particularly to Soviet scientific publications, with which American readers are less familiar.

It is up to the reader to judge the pros and cons of the book; we would only like to mention that all three Soviet editions were highly appraised by specialists and the scientific community. On our part, the American edition has been prepared by Elga A. Ozolina, Vladimir L. Kulakov and Lilly L. Volgina; their efforts are gratefully acknowleged by the authors. Thanks are also due to Susan Munger, Senior Editor of Van Nostrand Reinhold Company, Inc., for her help. It is also a pleasure to note that the American edition is being recommended to readers by Dr. George Lubin, one of the pioneers in composite mechanics and technology. The problem of new materials is a global problem. The future advances of mankind greatly depend on its solution.

Yu. M. Tarnopol'skii

USSR, Riga

Introduction

In order to of estimate strength and stiffness, structural materials are subjected to mechanical testing. Historically, mechanical testing of structural materials, followed by practical application of test results, may be traced back to July 4, 1662 [77], when tows made from Riga and Holland yarns were compared [137]. Since then test procedures for materials, primarily metals, have attained a comparatively advanced level. The history of the development of test methods for technical materials is treated in detail in [235]. The advent and widely expanding use of composites in highly loaded primary structures have forced a revision of the subject of mechanical testing. New test methods are continuously being developed, already existing techniques verified and reexamined. The principal difficulties arising in the testing of composites have been analyzed in [198].

Research practice has far exceeded test methods specified by standards. So far there are only a few standards written in terms of anisotropy. Numerous investigations of composites by various techniques have resulted in ambivalent judgments about the potential capabilities of the materials. This fact still further emphasizes the need for a critical analysis of existing methods, their estimation and generalization.

Advanced fibrous composites with unidirectional, laminated, and multiply oriented reinforcement are nonhomogeneous, essentially anisotropic materials. For this class of materials the usual terms—tensile, compression, shear, and bending tests—become meaningless without specification of the direction between the load and the axes of elastic symmetry of the material. Material anisotropy and structural peculiarities cause a number of serious problems. First of all, the number of determinable strength and elastic characteristics necessary for complete description of the material depends on the type of material anisotropy. Selection of the loading scheme for which the material characteristics are most simply related to the experimentally determinable values, selection of the analytical apparatus for experimental data processing, and estimation of the range of their validity are of principal significance.

For fibrous composites it is difficult to establish a uniform state of stress in the characteristic volume of the specimen, even for the simplest types of tests. The difficulties increase with the anisotropy of the material, i.e., for materials with high-modulus and high-strength reinforcement (boron, carbon, and organic fiber reinforced composites). In composite testing, the measurable strain essentially depends on the boundary conditions, i.e., on the method of fastening and loading of specimens. This phenomenon is a characteristic of highly anisotropic materials and represents a specific case of Saint Venant's principle.

Anisotropy of elastic properties enhances the requirements imposed on specimen shape and size, elimination of end effect (proper choice of the distance from the grips to the reference section on a specimen), load transmission, fiber orientation, and angle of specimen cutting. Anisotropy of strength in the case of improper choice of the loading scheme and specimen fastening leads to alteration of the failure mode, for instance, to delamination or "shearing-off" of the specimen in grips under tension. A special problem is the selection of specimen width; it is important to avoid any edge effect, i.e., critical interlaminar stresses.

This book is devoted to methods of evaluation the elastic and strength characteristics of fibrous polymeric composites during short-term static tests under normal conditions. The authors have systematized and generalized worldwide experience in the field. Surveys published in recent years [27, 38, 39, 45, 194, 244, 266] and the experience accumulated at the Institute of Polymer Mechanics of the Latvian SSR Academy of Sciences, as well as numerous articles and reports presented at special conferences of the ASTM [10, 46–51], have been made good use of. The bibliography of this book does not claim to be comprehensive; however, it reflects sufficiently well the state of the art of mechanical testing of fibrous composites, advances made in the field, and the level of standardization achieved.

In spite of this progress, the degree of mastery and standardization is different for various methods. As before, shear stiffness and, especially, shear strength determination present difficulties. With the accumulation of experience a number of methods for estimation of strength and stiffness is in need of further corrections.

Test methods for annular specimens have been greatly extended during the last years. This fact allows us to consider test methods for flat and annular specimens in tension, compression, shear, and bending from a common standpoint—according to loading type. All the chapters include test methods and techniques for advanced composites with carbon, boron, and organic fiber reinforcement of various schemes. Attention has been focused on unidirectional materials which may be identified as laminae. Characteristics of a lamina must be determined not only for engineering certification of the material; they also form the basis for determining the properties of hybrid materials and composites with varying transverse reinforcement layups.

Among the composites which are discussed in this book, glass fiber composites with fibrous, laminated, and multidirectional reinforcement occupy the most important place. For substantiation of the test methods for glass fiber composites practically no experimental data are presented; these test methods are numerous and their accuracy is high. There is also no necessity to confirm and experimentally substantiate selection of the shape and size of a specimen, loading scheme, etc., for glass fiber composites with a traditional reinforcement scheme. For high-modulus materials the necessary experimental data have been presented, since the majority of problems are still under study. Test techniques for high-modulus and multidirectionally reinforced composites have not yet been refined.

This book is based on experience in testing of the "first generation" of fibrous composites with unidirectional, laminated, and multidirectional reinforcement, based on a polymeric matrix and reinforced with conventional and high-modulus fibers. Experience has shown that the methods evolved can be successfully applied to fibrous composites of the next generations—to carbon, metallic, and ceramic matrix based materials. However, experience in testing polymeric composites cannot directly be applied to modern materials without due regard for the specific properties of those matrices.

At the end of each chapter devoted to the analysis of an individual type of tests is a summary table. (See chapters 3, 4, and 5.) These tables are meant for fast visual orientation in the type of testing and for preliminary evaluation of the potential and selection of the optimum test method.

Knowledge of the properties of composites under uniaxial static loading is necessary but often insufficient for their rational use in primary structures. The development of methods for studying properties of fibrous composites under complex state of stress is a difficult task. There is a need for systematization and generalization of destructive as well as nondestructive methods for studying long-term static, dynamic, and fatigue properties of composites. The problems associated with peculiarities of testing at elevated and reduced temperatures, under conditions of increased humidity, under special environmental conditions, and in particular, prognostic methods of testing need special attention. Solution of all the above problems will form a scientific basis for future standardization of test methods for composites on an international scale.

In preparation of the first two editions significant assistance was lent to the authors by A.K. Mālmeisters. His advice and comments after having seen the outline of the book and the manuscript have greatly favoured the success of the work. The authors are truly grateful to I.G. Zhigun, who has analytically and experimentally evaluated a series of test methods for high-modulus and multidirectionally reinforced materials, and also to L.L. Volgina who has typed the manuscripts and prepared illustrations for all three editions.

Symbols

COORDINATES

x, y, z	structural axes, rectangular coordinate system
θ, z, r	structural axes, cylindrical coordinate system
1, 2, 3	material axes, rectangular or cylindrical coordinate system. In most cases treated in this book, they coincide with the structural axes in a rectangular or cylindrical coordinate system, which are preferred.

GEOMETRICAL CHARACTERISTICS

b, l, h	characteristic linear dimensions of the specimen, explained in the text
r	current radius
R, R_i, R_0	mean, inner, and outer radii, respectively, of ring or tube specimens
θ	fiber orientation or lamination angle

LOAD

P	concentrated load
P^u	concentrated load at specimen failure
p	uniform pressure
p^u	uniform pressure at the failure of a specimen
M	bending moment
M^u	bending moment at the failure of a specimen
M_T	the applied torque
M_T^u	the applied torque at the failure of a specimen

STRESSES AND STRENGTHS

σ	normal stresses: subscripts denote direction, superscripts denote type of loading
τ	tangential stresses: subscripts denote plane, super scripts denote type of loading
σ^u	strength due to normal stresses: subscripts denote direction, additional superscripts denote type of loading
τ^u	strength due to tangential stresses: subscripts denote plane, additional superscripts denote type of loading

DISPLACEMENTS AND STRAINS

u, v, w	displacements in the x, y, z or θ, z, r directions
ε	linear strains: subscripts denote direction, superscripts denote type of loading
γ	shear strains: subscripts denote plane
φ	angle of twist

MATERIAL CHANACTERISTICS

E	modulus of elasticity: subscripts denote direction or material constituents, superscripts denote type of loading
G	shear modulus: subscripts denote plane, superscripts denote type of loading
ν	Poisson's ratio: the first subscript denotes the loading direction, the second denotes transverse strain direction
V	volume content of material constituents, defined by subscripts

SUBSCRIPTS

x, y, z / θ, z, r	structural axes direction
1, 2, 3	material axes direction
xy, xz, yz / $\theta z, \theta r, zr$	shear planes in structural axes rectangular coordinate system

$0°, 45°, 60°, 90° \ldots$	orientation angles to the load direction
bear	bearing
break	breaking
cr	critical
exp	experiment
f	fiber; fictitious
h	horizontal
i	inner (inside)
m	matrix
mean	mean value
o	outer (outside)
shear	shearing
v	vertical; voids
L	longitudinal
T	transversal
τ	shear

SUPERSCRIPTS

b	bending
c	compression
t	tension

Contents

Chapter 1
Fibrous Composites

1.1. COMPOSITES

1.1.1. Types of Fibrous Composite

Synthetic composites hold a prominent place among advanced structural materials. They consist of two or more components having various, usually contrasting physical and mechanical properties [197]. Tailoring of materials to have prescribed properties may be carried out by properly selecting the properties of the components, their proportions, and the structure of the composite. At the present time, the most widely used composites consist of reinforced polymeric materials.

The attempt of combining light weight with strength and stiffness in principal directions has attracted attention to the idea of reinforcing a flexible matrix with various fibers. Only this class of materials ensures the possibility of combining increase in both strength and fracture toughness [111]. Intensive work on the development high-strength and high-modulus materials reinforced with filaments and fibers is going on the world over.*

At present, polymeric composites (glass fiber, carbon, boron, and organic

*The principle of reinforcement of synthetic resins with fibrous materials was first patented in 1908 (Backeland) [252, p. 15]. Then followed industrial production of molded laminates based on phenol and melamine resins. Reinforcement of synthetic resins with mineral fibers (glass fibers) was patented by the firm AEG in 1935. Introduction of the method was delayed because of the absence of suitable resins; industrial output of fibrous polymeric composites was mastered only after World War II. Subsequently, the choice of synthetic resins and reinforcement materials for use in the production of plymeric composites expanded, and new technological processes were developed, in particular filament winding. However, the principle of producing these materials remained unchanged.

A history of composites, primarily based on experience in the USA, is given in a book edited by G. Lubin [89, p. 1].

composites) are widely used in various fields of technology, and their fraction in the total output of synthetic materials is increasing continuously. The idea of reinforcing a yielding matrix with fibers has so many structural and technological advantages that plainly it will soon lead to the production of higher-strength engineering materials. Production of polymeric composites has grown into a powerful branch of industry. One prospect is the establishment of industrial production of metals reinforced with fibers. (The first advances in production of boron-aluminium and graphite-aluminium are impressive.) However, practical production of metallic composites still requires solution of complicated technological problems.

The idea of polymeric composite production is extremely simple [108, p. 15; 231, p. 7]. Nature itself continuously creates composites [197], for example, the trunks and stalks of plants, or the bones of animals. It is worth noting that natural composites have an ordered structure. In polymeric composites, the fiber reinforcement provides strength and stiffness, while the polymeric binder provides integrity and moldability of the material. The principle of production of the material allows the high strength and stiffness characteristic of the reinforcing fibers to be combined with technological properties which are valuable in a polymer. By changing the arrangement of the fibers, anisotropy of properties can be successfully controlled.

The first practical fibrous polymeric composites were glass fiber composites. The structural advantages of oriented glass fiber composites are well known, primarily high specific strength in the reinforcement direction. However, this basic advantage is countered by a series of problems associated with comparatively low stiffness even in the direction of reinforcement, pronounced shear creep, and sensitivity to fiber waviness and misalignment.

The difficulty usually lies in the fact that composite properties are judged on the basis of one or perhaps two parameters. The fact is that composites inherit not only the favorable, but also unfavorable properties of the components. However, in describing the structural capabilities of a material it is often customary to emphasize only its high specific strength and stiffness and to say nothing of its unfavorable properties. This is mainly due to insufficient testing, in which only some of the parameters which characterize the material and its behavior in a structure are evaluated, and to incorrect assessment of the test results (because apparatus has been used which does not take into account the properties of materials being tested). To permit correct use of fibrous composites, new tests must be developed to determine transverse and shear characteristics, which were not considered previously, and methods of determination of elastic and strength characteristics in the fiber direction, in particular for unidirectional materials (for example,

Table 1.1.1. Typical Ratios Characterizing Interlaminar Shear and Transverse Tension and Compression resistance of composites [271].

PARAMETERS ANISOTROPY	FIBERGLASS COMPOSITE C 0°	FIBERGLASS COMPOSITE 0/90°	CARBON COMPOSITE 0°	CARBON COMPOSITE 0/90°	BORON COMPOSITE, 0°	ORGANIC COMPOSITE, 0°
E_x/G_{xz}	20–35	10–20	40–80	20–4	30–60	25–40
σ_x^u/τ_{xz}^u	30–40	10–15	20–40	5–15	20–50	10
E_x/E_z	5–8	3	20–30	15	8–12	12–18
$\sigma_x^{tu}\sigma_z^{tu}$	25	—	25–50	20	15–30	50
$\sigma_x^{tu}\sigma_z^{cu}$	6–10	—	6–10	—	10	15–20

tension of boron composites, compression of organic composites) must be improved.

The first rush of publications on composites was connected with determination and estimation of specific structural properties. Reliable numerical data were obtained, which permitted estimation not only of properties in the reinforcement direction, but also of unfavorable characteristics of fibrous composites. It turned out that the majority of fibrous composites have poor resistance to interlaminar shear and transverse tension. Shear resistance is characterized by ratios E_x/G_{xz} and σ_x^u/τ_{xz}^u; transverse tension and compression resistance by the ratios E_x/E_z, $\sigma_x^{tu}/\sigma_z^{tu}$, $\sigma_x^{tu}/\sigma_z^{cu}$. Here, E_x and E_z are the moduli of elasticity in the x and z directions; G_{xz} is the interlaminar shear modulus; σ_x^u and σ_z^u are the strengths in the x and z directions; τ_{xz}^u is the shear strength in the xz plane; the x and y axes are located in the plane of reinforcement layup, but the z axis is perpendicular to this plane; the superscript t designates tension and c compression. Typical values of the parameters for fibrous composites with various layups are presented in Table 1.1.1.

The following basic approaches have been taken in order to overcome the mentioned disadvantages: passing over to high-modulus reinforcement,* increase in fiber-matrix interface strength at the expense of various surface treatments, and a search for multidirectional reinforcement schemes with the aim of generating improved interlaminar bonds. The advent of hybrid (multi-fibrous) materials, permitting the combination of reinforcement materials with different mechanical characteristics and the solution of

*It should be noted that this does not mean contrasting high-modulus materials to traditional glass fiber composites, the capabilities of which are still far from being fully utilized, but rather estimation of the structural properties of new composites and the areas of their efficient use.

problems pertaining to the introduction of relatively expensive high-modulus and high-strength reinforcement materials, was marked in [112] as one of the most important advances in the field of composites in the last decade. Successful development of hybrid composites was predetermined by the advances made in the field of composite mechanics, in particular, the theory of laminated media [32].

The most widely used high-modulus fibers are boron and carbon fibers; lately certain organic fibers have been added (aramids). The great interest shown in these materials may be attributed not only to their structural advantages, but also to the possibility of multi-tonnage production. Though the technology of the majority of high-modulus fibers is in an early stage of development at present, it does not differ significantly from the technology for glass fiber composites. This fact makes it possible to consider boron, carbon, and organic fibers as the most promising reinforcing materials. It has been predicted that the production of carbon composites and particularly organic composites will exceed the production output of glass fiber composites in the immediate future.

A specific feature of a number of high-modulus composites is significant anisotropy of the reinforcing fibers. For example, for carbon composites the ratio of the moduli along and across the fibers, E_{fx}/E_{fz}, can reach 40–50; a number of organic composites possess the same degree of anisotropy. In addition to other, well-known features of composites (low interlaminar shear and transverse tension strengths) a new factor appears—a significant difference in elastic properties along and across the fibers. This fact influences the analysis of test results for composites. In particular, testing of organic composites in compression becomes complicated.

The use of high-modulus fibers in composites according to traditional fiber layups, when structures are formed by packing of laminae, does not result in proportional increase in all elastic characteristics. The main distinctive feature of high-modulus composites is the anisotropy of their elastic properties, which is still higher than that of glass fiber composites. For organic and carbon composites, the anisotropy of elastic properties is due to the high anisotropy of the reinforcing fibers themselves. This specific feature is illustrated in Table 1.1.1. The strength anisotropy of glass fiber, boron, and carbon composites is practically the same. The main peculiarities are associated with the behavior of organic and boron composites under tension-compression, with low interlaminar shear and transverse tension strengths for carbon composites and with low compression strength along the fibers for organic composites.

In reviewing the published data on mechanical properties of boron, carbon and organic composites [37, 232] it became apparent that the numerical estimates in many cases vary over a wide range. This is evident from Table 1.1.1, which presents test data obtained by various authors on materials of similar structure. A wide scatter is ascribed not only to non-uniform production technology of composites and variability of component properties, but also to a number of specific features of testing essentially anisotropic materials. The main difficulties lie in building up a uniform state of stress in the reference section, even in the case of the simplest tests. In selection of specimen size for high-modulus composites [45, Ch. 9; 59, 229] attention must be focused on the elimination of end and edge effects; this concerns the choice of the distance from the grips to reference section and the width of the gauge section of the specimen. For high-modulus composites, it is necessary to specify a rigorous tolerance for reinforcement layup as well as specimen cutting angle.

The specific features of fibrous composites have shown that there is a need to refine methods of design and testing in order to estimate phenomena connected with the low shear and transverse properties of these materials. Numerous investigations carried out in recent years have revealed the necessity of taking account of the structural features of the materials. Along with the design of composites specialists have sought methods and procedures to eliminate the unfavorable characteristics of composites with traditional reinforcement schemes. Efforts in recent years have led to the creation of materials which are free of such deficiencies. Advances in the development and study of multi-directionally reinforced composites, consisting of two or three filament systems are discussed in detail in [271].

1.1.2. Testing Criteria

Mechanical testing techniques and data processing are different for various types of composites. The properties of materials are so varied that a common approach is difficult to find. Thus, testing techniques and data processing for materials reinforced with discrete particles and with continuous fibers differ greatly in many respects, since the former are quasi-isotropic, but the latter are essentially anisotropic materials. Hence one must discuss testing of fibrous composites in terms of anisotropy. It has been stated in the introduction that the usual terms—tension, compression, shear, and bending tests—are meaningless unless the angle between the load and the axes of elastic symmetry of the material is indicated. Therefore description of material properties should be based on the theory of elasticity of an anisotropic body [91, 126]. In this manner the structural peculiarities of fibrous composites, the possibility of transition to continuum for a lamina, and the ways of determining integral characteristics of multidirectional or hybrid materials based on experiments on a lamina should be taken into account.

The first difficulty in testing composites is connected with the determination of the number of measurable strength and elastic characteristics. Details and precision of the information obtained depend on choice of the material model. For fibrous composites, even the simplest description, based on Hooke's model, involves measurement of a large number of parameters. Determination of the type of anisotropy of various types of materials with fibrous, laminated, and multidirectional structures and the number of measurable strength and elastic characteristics are of special interest.

It is necessary to emphasize that the precision of the elastic approximation depends on the loading direction relative to the principal axes of symmetry of the composite. Under loads parallel to the direction of reinforcement,* fibrous composites follow Hooke's law with high precision (many materials right up to failure). If the load acts at an angle to the reinforcement direction or perpendicular to the reinforcement plane, the stress-strain curve becomes essentially nonlinear.

For fibrous composites, the majority of the measurable values are not constants, regardless of test conditions. At the same time, in an ideal case the results obtained must characterize only the properties of the materials and must not depend on numerous secondary factors. For this purpose, in mechanical testing of composites the following criteria must be resolved: selection of specimen shape and size and its preparation for the test; selection of testing machine power and apparatus for measurement of the forces and deformation; determination of the limits within which the conditions of the experiment can change (loading rate, specimen geometry, inaccuracy of loading) so that these deviations can be disregarded; possible scatter of results and the number of specimens used; processing of the experimental results and evaluation of their reliability.

In this book an effort is made to answer these questions. Test methods regulated by state standards (ASTM, DIN, GOST, ISO)* or by departmental standards and the methods used in research practice are dealt with. The basic methods, standards, characteristics determinable in each kind of testing, experimentally measurable values, suggested specimen shape, typical

*Structures of composites are designed so that the main loads coincide with the reinforcement direction, i.e., the strength field coincide with the effective stress field.

*ASTM—American Society for Testing and Materials, Philadelphia, Pa., USA. DIN—Deutsche Industrie Normen, the obsolete but retained designation of standards issued by the German Committee on Industrial Norms and Standards (Deutscher Normen Ausschuss, FRG). GOST—the state standard of the USSR. ISO—International Organization for Standardization, Geneva, Switzerland. Several ASTM standards form the basis of the ISO and American National Standards. British Standards, as well as the national standards of a whole series of other countries, were not used in writing this book.

test equipment, and shortcomings and limitations of the various methods treated in this book are listed in summary tables which conclude the chapters. The tables give an idea of the state of the art and capabilities of each of the test methods and they can serve as starting points for planning of the experiment.

1.2. ANISOTROPY OF FIBROUS POLYMERIC COMPOSITES

1.2.1. Classification

There is no generally accepted classification of fibrous polymeric composites. The mechanics of materials has its specific requirements for division of fibrous composites into separate classes. The general principles by which materials obtain their names are specified in classification of fibrous composites. Fibrous composites are then divided into quasi-isotropic and anisotropic materials, the type of anisotropy is established, and the transition to a uniform medium is accomplished. The following general principles may underlie classification of fibrous composites: materials, i.e., classification by reinforcement or matrix material; structure, i.e., by type of reinforcement and its layup; technology, i.e., by the method of conversion into end products.

More often the fibrous composites obtain their names from the reinforcing fibers: glass fiber, boron, carbon, graphite, organic (aramid), glass fiber/boron, etc., composites. Classification by matrix material is less widespread. It is difficult because of the broad selection of resins and their combinations. This principle is used for more precise definition of the name of the material only: for example, unidirectional boron-epoxy composite.

Classification by method of conversion or the technological classification divides fibrous composites into cast, molded, and wound. Randomly reinforced materials are processed by casting and molding, and oriented ones by winding, press molding, and contact molding. The processing method has a strong effect on the properties of the material. The specimen technology and shape must correspond to the production method and intended use of the material. This explains the different approaches to testing of wound and molded materials.

Even the most successful material mechanics or technological nomenclature for a material does not address the details of mechanical tests of fibrous composites. Type of reinforcement and its layup in the polymeric matrix are the most important classification criteria in this regard. The main requirement for classification, from the point of view of the mechanics of materials, is to determine the law of deformation and the relationship of the properties to the angular coordinate. By assuming, in the first approximation, that fibrous composites follow Hooke's law, the entire variety of fibrous composites can be divided into isotropic and anisotropic materials.

1.2.2. Isotropic and Anisotropic Materials

Fibrous polymeric composites are nonhomogeneous compound materials. Their base is a polymeric matrix, reinforced by fibers or particles. For a sound selection of the experimentally determinable characteristics and mathematical apparatus for processing the test results, it must be determined to which class the composites belong—to isotropic or anisotropic materials. Further, to determine the calculated relationships connecting the experimentally determinable values (forces, deformations, displacements) with the macroscopic characteristics of the material it is necessary to accomplish the transition to a continuous medium and thus employ the well-developed apparatus of the theory of elasticity of continuous anisotropic media.

Depending on the size relationships of reinforcing elements and their arrangement in the polymeric matrix, two large groups of reinforced materials can be identified: randomly reinforced (matrix + particles) and regularly reinforced or oriented (matrix + continuous fibers) materials. The first group of materials incorporates reinforcement consisting of particles the sizes of which are commensurate in all directions, or of discrete fibers, for example, short fiber lengths, whiskers, etc. These particles are randomly arranged in the polymeric matrix, so that the material is quasi-isotropic, i.e., anisotropic in microvolumes but isotropic in macrovolumes (in the products). However, it should be remembered that in the conversion process, for example, in cast-molding, the product as a whole can become anisotropic, similar to metals in after pressure processing.

Materials of the matrix + continuous fiber class have an oriented structure. Anisotropy of fibrous composites is generated in production of the material or product by appropriate layup of the fibers. Such anisotropy is called *structural anisotropy*, as distinct from physical (inherent in the crystals), natural (wood, human and animal bones, etc.), technological (emerging in the conversion process, for example, of rolling steel, extrusion of polymeric pipes, etc.), or deformational (emerging in the process of loading the initially isotropic materials) anisotropies.

1.2.3. Types of Reinforcement Layup

The basic structural element of polymeric composites is the *lamina*. This is either a flat or curved (in wound products) arrangement of unidirectional fibers (unidirectional lamina) or woven fibers (woven lamina) in a polymeric matrix. Prepreg tapes used in the production of polymeric composites are woven laminae.

Depending on the mutual arrangement of reinforcing elements, three basic groups of composites can be distinguished: uni-, bi-, and multidimensional; these are also called unidirectional, laminated, and multidirectional composites.

Unidirectional materials are produced by placing all the fibers or filaments parallel to each other. They are also called the materials with 0° layup, indicating by this the absence of cross-laid fibers. If the fibers in such a material are arranged uniformly, the material is transversely isotropic or monotropic* in the planes perpendicular to the reinforcement direction. However, in a number of cases (for example, in tape winding) some layering of materials with 0° layup should be taken into account. This fact forces the material to be classified as orthotropic.

Laminated composites are formed by a stack of laminae with various prescribed orientations.

In the case of reinforcement with ribbons the material is transversely isotropic in the planes tangential to the surfaces of the ribbons.

Nonwoven laminated materials are formed of regularly alternating unidirectional laminae. Depending on the number of different directions along which the fibers are laid, the materials are called two-directional, three-directional, etc. In two-directional materials the fibers most often are arranged in mutually perpendicular layers—this is orthogonally reinforced material. The ratio between the number of longitudinal and transverse layers can vary (1 : 1, 1 : 3, 2 : 5, etc.). Materials with 1 : 1 layup are assumed to be balanced or equally strong (in principal axes only).

Orthogonally reinforced materials are orthotropic on the axes which coincide with the reinforcement directions. Materials reinforced in two nonorthogonal directions, with the same number of laminae in both directions, are orthotropic on axes directed along the bisector of the angle between fibers in neighboring laminae. Three-directional materials are formed by the same number of unidirectional laminae in three directions with angles of 60° between them, i.e., in the 0°, 60°, 120°. Such materials are isotropic in the planes parallel to the layup of the laminae. n-directional materials in which the same number of laminae is laid in the 0, π/n, $2\pi/n$, . . . , π directions are also transversely isotropic.

For more complicated reinforcement layups as described above, it is reasonable to employ the laminate code suggested by the Wright-Patterson Air Force Base, explained in detail in the appendix to [42].

* For terminology of the elastic theory of anisotropic bodies, see Section 1.3.

The fabrics used as reinforcement can be divided into flat and bulk types. In flat fabrics, the warp and weft filaments are interwoven within a single layer. Therefore, a material made of such fabrics will be laminated (in distinction from reinforcement with ribbons); in this case, the material can be anisotropic in the reinforcement planes. Fabrics can be formed so that the direction of the warp in all laminae coincides or forms a certain given angle between the directions of adjacent laminae. For these materials, the number of reinforcing fibers in different directions is determined by both the type of fabric, and a given scheme of laminae layup.

Bulk fabrics are used to produce multidimensionally reinforced materials. Multidimensionally reinforced materials are classified into three groups, depending on the way spatial bonds are formed. Materials in which spatial bonds are formed by curved fibers (all or only a part) in one and the same direction, belong to the first group. These materials are formed in the framework of the traditional system of two filaments: cross-linking is achieved by means of interweaving the fibers of the warp with the straight weft fibers. Depending on the type of weft filament binding (sewing), these materials fall into several subgroups [271, p. 26]. Materials in which the spatial bonds are formed by the fibers in the third direction, i.e., by forming a system of three filaments, belong to the second group. The fiber layups may be orthogonal in three directions or they can be placed at an angle on one of the reinforcement planes. Materials in which the spatial bonds are formed by whiskers belong to the third group. Due to the presence of spatial bonds, multidimensionally reinforced materials themselves do not present a stack of laminae, but they consist of repeating elements. Their division into layers accessible for analysis in terms of the theory of reinforced media is described in [271, p. 42–49].

Depending on the reinforcement layup with respect to the longitudinal axis of the specimen or another product, two types of reinforcement patterns in the specimen plane must be distinguished—symmetrical and asymmetrical. If all the reinforcement across the material thickness is placed uniformly, the material is homogeneous through the thickness. If the reinforcement pattern through the thickness varies, the material is nonhomogeneous through the thickness. Besides, the reinforcement through the material thickness can be placed symmetrically or asymmetrically with respect to the material midplane. In actual materials, the midplane concept is a mere geometric term; it is technologically unrealizable. In practice, symmetry through the material thickness is ensured by a sufficiently large number of laminae. The available theoretical apparatus allows correct processing only for test data on materials having a symmetrical and homogeneous structure.

1.2.4. Transition to Continuous Medium

It has been emphasized that polymeric composites are nonhomogeneous compound materials. In order to apply the existing theory, developed for homogeneous anisotropic materials, to processing of test results full transition to continuous medium must be accomplished. For homogeneous materials, introduction of the continuous medium model involves rejection of the molecular structure. The validity of this assumption in an engineering approach to the study of macroscopic properties of homogeneous materials is obvious.

For composites, transition to the continuous homogeneous medium model is considerably more complicated [71, p. 19]. One feature of the structure of all types of fibrous composites permits procedures to be found for overcoming structural inhomogeneity. Materials reinforced by fibers have a regular structure, and they contain a large number of a single type of structural element (fibers, filaments, strands, cords, roving or fabric layers, etc.), which are impossible and even inadvisable to consider separately. This opens up a new step in creation of a continuous medium model, called the energetic leveling method [29, p. 72]—the reinforcing elements are distributed through the bulk of the body, and the medium is considered to be homogeneous, but to have some new properties which depend on the properties of the components of the system. The reinforcement plays the main role in the reinforced directions, and the polymeric matrix in the transverse planes. Therefore, the idealized medium turns out to be anisotropic, as a rule.

To employ the energetic leveling method three levels of description of the model* are introduced which differ in gage length from each other (Fig. 1.2.1). The lower level h is the level of structural nonhomogeneity. Its scale equals the characteristic size of the reinforcing elements—the particle or fiber diameter or lamina thickness. The following level, the level of inspection H, is one in which it is possible to replace the nonhomogeneous material with a locally homogeneous one of equivalent strength and stiffness, and to calculate the macroscopic characteristics of the composite—characteristics which are determined from macroscopic tests. Finally, the highest level, the scale of which is designated Λ, is equal to the characteristic size of the product and/or the distance at which the leveling stresses and strains experience a noticeable change. At this level, the effective loads and average characteristics may be successfully connected by analytical relations, i.e.,

*The principal difficulties in the treatment of structural nonhomogeneity of solid deformable bodies are discussed in [133].

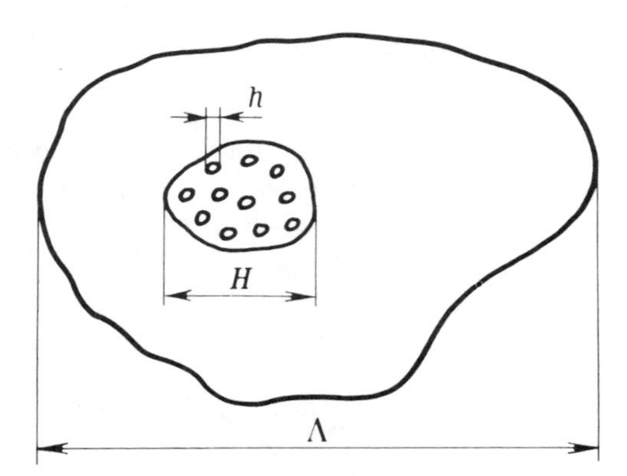

Fig. 1.2.1. Model of a composite [29].

expressions are obtained which connect the applied load and the forces, displacements, and deformations determined in the test. On the basis of a solution of the boundary problem for a body with the desired macroscopic elastic characteristics, the leveling stress, deformation, and displacement fields are determined, and a connection is established between the averaged elastic characteristics, effective load, and the resulting displacements (or deformations): deflection of a beam, angle of twist of a rod, deflection of a plate, etc.

In this manner, with a large number of reinforcing elements fibrous composites can be considered to be homogeneous, anisotropic materials. The number of fibers for which the error from replacement of the non-homogeneous medium by a homogeneous, anisotropic one, does not exceed the assigned value depends on the ratio of reinforcing element size to structure (or specimen) size and on the nature of the effective load (the smoother the load, the smaller the error). For the majority of structural elements of fibrous composites, this substitution is completely permissible if the zones of attachment and the zones of concentrated load are eliminated from the analysis. This permits the use of the methods of the theory of elasticity of an anisotropic body.

With the seeming obviousness of this approach, it cannot be forgotten that a nonhomogeneous material is being studied and that the error of transition to a continuous medium must be estimated. Numerous errors in processing the results of bending tests of thin specimens, in which the number of laminae is insufficient for complete transition, are connected with just these characteristics of the material. This is the cause of error in study of compression strength of thin specimens, when the surface laminae have lost stability. It should be noted that transition to a continuous medium is possible only for homogeneous through the thickness materials (layup angle $\theta = \text{const.}$). For material of nonhomogeneous through the thickness layup $[\theta = \theta(h)]$, a step-by-step consideration is necessary, since upon loading at an angle to the axes of elastic symmetry, the average material characteristics depend on coordinates.

In transition to the continuous medium model an apparent loss of information on redistribution of the force and physical fields between components of the material, takes place, which is especially important in analysis of failure of composites. However, this method does not mean that the actual structure of the material is ignored. If it is necessary to find the structural stresses and strains, there is a need to turn to the level h again. The properties of fibrous composites can be expressed through the properties of the components and the stresses and strains in the material components can be determined from these values, calculated for a quasi-homogeneous material. In turn, the latter properties can be compared with various criteria.

1.2.5. St. Venant's Principle

In testing of fibrous composites the measurable strain depends on the boundary conditions, i.e., on the method of fastening and loading the specimen. This phenomenon, which is typical of structures of highly anisotropic materials, is a specific manifestation of St. Venant's principle. According to St. Venant's principle, for an isotropic, elastic medium disturbances introduced by a self-balanced system of forces, rapidly fade out at a distance from the source of the disturbance greater than the characteristic size of the specimen H [Fig. 1.2.2(a)].

For an anisotropic medium, the disturbances are damped differently in different directions, i.e., the manifestation of St. Venant's principle is also anisotropic. Disturbances are damped more slowly in the direction of the

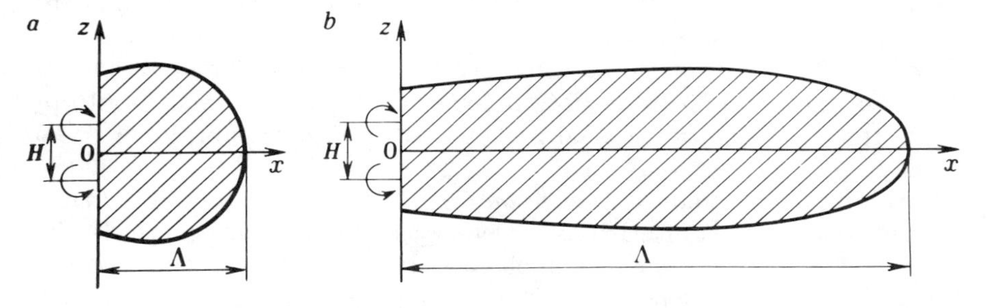

Fig. 1.2.2. Damping of disturbances in isotropic (a) and anisotropic (b) materials [29].

greatest stiffness and more rapidly in the direction of the least stiffness. As a result, the region of noticeable disturbances is elongated in the direction of the greatest stiffness [Fig. 1.2.2(b)]. The characteristic magnitude of the disturbances region in this direction is on the order of

$$\Lambda \sim H(E_i/G_{ij})^{1/2}.$$

This phenomenon is understandable from an analysis of the structure of fibrous composites: in view of the low shear and transverse tension resistance of the binder, redistribution of the forces between the reinforcing elements is hampered. This peculiarity should be taken into account in selection of the specimen size.

1.3. DETERMINABLE CHARACTERISTICS

1.3.1. Elastic Constants

The principal attention in this book is focused on fibrous polymeric composites, i.e., anisotropic materials. Therefore, in order to conduct qualitative experiment on fibrous composites, familiarity with the basic concepts and relationship of the theory of elasticity of an anisotropic body is required. Besides, special monographs (see, for example, [7, 16, 91, 104, 124, 125, 126, 135]), such knowledge is presented in any book on composites. In this book, the information is presented as far as the planning of the experiment (determination of the number of characteristics) and verification of the results require that.

In the general case of anisotropy, deformability of an elastic body is characterized by 21 elastic constants, 18 of which are independent [126, p. 27]. Experimental evaluation of all the 18 elastic constants is a time-consuming procedure. However, fibrous composites are characterized by some structural symmetry. It permits the number of determinable elastic characteristics to be reduced. The theory of elasticity of an anisotropic body considers five cases of elastic symmetry. The most frequently encountered types of elastic symmetry of fibrous composites are orthogonal anisotropy (or orthotropy) and transversal isotropy (also called monotropy or transtropy).

An anisotropic body is *orthotropic* when three mutually perpendicular (orthogonal) planes of elastic symmetry pass through each point. In this case, deformability of the body is characterized by twelve elastic constants, nine of which are independent: three moduli of elasticity E_x, E_y, E_z; three shear moduli G_{xy}, G_{xz}, G_{yz}, and three Poisson's ratios ν_{xy}, ν_{yz}, and ν_{zx}. Here the coordinate axes x, y, and z are the principal axes of orthotropy, which are parallel to the lines of intersection of elastic symmetry plane xy and xz, yx and yz, zx and zy, respectively. The remaining three elastic constants—Poisson's ratios ν_{yx}, ν_{xz}, and ν_{zy} are determined according to the relationships

$$E_x \nu_{xy} = E_y \nu_{yx}; \qquad E_x \nu_{xz} = E_z \nu_{zx}; \qquad E_y \nu_{yz} = E_z \nu_{zy}. \tag{1.3.1}$$

Orthotropic materials are orthogonally reinforced fibrous composites, the reinforcing directions of which coincide with the specimen axes. Non-orthogonally reinforced fibrous composites have axes, directed along the bisectors of the angle between the fibers of adjacent laminae and all the real unidirectional fibrous composites.

In the case of the plane problem, an orthotropic body is characterized by four elastic constants: moduli of elasticity E_x and E_y, shear modulus G_{xy}, and Poisson's ratio ν_{xy}. Poisson's ratio ν_{yx} is determined according to the relationship $E_x \nu_{xy} = E_y \nu_{yx}$. Often, instead of G_{xy}, the much easier obtainable elastic modulus E_{45° (at an angle of 45° to the axes x and y) is assumed to be the independent constant. In the case G_{xy} is determined according to the formula

$$\frac{1}{G_{xy}} = \frac{4}{E_{45^\circ}} - \frac{1}{E_x} - \frac{1}{E_y} + \frac{2\nu_{yx}}{E_x}. \tag{1.3.2}$$

An anisotropic body is *transversely isotropic* if there is an axis of elastic symmetry in each point of the body, in relation to which the elastic properties in all the directions of the plane perpendicular to the axis are equivalent. In this case, there are five elastic constants: moduli of elasticity E_x and E_y, shear modulus G_{xz} and Poisson's ratios ν_{xy} and ν_{xz}. The remaining elastic constants are determined according to the relationships

$$E_y = E_x; \qquad G_{yz} = G_{xz}; \qquad G_{xy} = E_x/[2(1 + \nu_{xy})];$$

$$E_x \nu_{xz} = E_z \nu_{zx}; \qquad \nu_{yz} = \nu_{xy}; \qquad \nu_{zy} = \nu_{zx}; \qquad \nu_{yz} = \nu_{xz}. \tag{1.3.3}$$

A characteristic example of a transversely isotropic material is a qualitative unidirectional composite.

1.3.2. Relationships Between Elastic Constants

Certain relationships between the elastic constants of an anisotropic body should be noted which can successfully be used for controlling of test data.

It follows from the equality of compliance coefficients that

$$E_1 \nu_{12} = E_2 \nu_{21}; \qquad E_1 \nu_{13} = E_3 \nu_{31}; \qquad E_2 \nu_{23} = E_3 \nu_{32} \tag{1.3.4}$$

or in a general form

$$E_i \eta_{ij} = E_j \eta_{ji} \qquad (i, j = 1, 2, \ldots, 6).$$

All moduli of elasticity (E_1, E_2, E_3) and all shear moduli (G_{12}, G_{13}, G_{23}) are positive. Analogously, all the stiffness constants $(C_{11}, C_{22}, \ldots, C_{66})$ are positive also.

Besides,

$$0 \leqslant \nu_{12}\nu_{21} < 1, \qquad 0 \leqslant \nu_{13}\nu_{31} < 1, \qquad 0 \leqslant \nu_{23}\nu_{32} < 1 \tag{1.3.5}$$

or in a general form

$$0 \leqslant \eta_{ij}\eta_{ji} < 1 \qquad (i, j = 1, 2, \ldots, 6).$$

For any anisotropic material, the following inequality must be satisfied

$$\nu_{32}\nu_{21}\nu_{13} = \nu_{23}\nu_{12}\nu_{31} < \tfrac{1}{2}(1 - \nu_{12}\nu_{21} - \nu_{13}\nu_{31} - \nu_{23}\nu_{32}) < \tfrac{1}{2}. \tag{1.3.6}$$

Let us assume that $E_1 > E_2 > E_3$. In hydrostatic compression, the volume of the body should decrease; therefore, it should be equal to

$$\nu_{23} + \nu_{31} + \nu_{12} < \frac{3}{2} \quad 1 + \frac{E_1}{E_3} \quad . \tag{1.3.7}$$

Moreover, it follows from relationship (1.3.5) that

$$-1 < \nu_{13} < 1, \qquad -1 < \nu_{12} < 1, \qquad -1 < \nu_{23} < 1. \tag{1.3.8}$$

Further, the numerical values of Poisson's ratios must satisfy the condition of increase in volume Δ^+ (a decrease in strength) under tension or a decrease in volume Δ^- (increase in strength) under compression. In triaxial uniform tension, the inequality designating the increase in volume has the form [14, p. 41]

$$\frac{1}{E_x} + \frac{1}{E_y} + \frac{1}{E_z} - \frac{2\nu_{xy}}{E_x} - \frac{2\nu_{yz}}{E_y} - \frac{2\nu_{zx}}{E_z} \geqslant 0. \tag{1.3.9}$$

By introducing the symbol of reduced Poisson's ratio of an orthotropic material ν_0, the inequality can be expressed in the following form:

$$\nu_0 = \frac{\dfrac{\nu_{xy}}{E_x} + \dfrac{\nu_{yz}}{E_y} + \dfrac{\nu_{zx}}{E_z}}{\dfrac{1}{E_x} + \dfrac{1}{E_y} + \dfrac{1}{E_z}} \leqslant 0.5. \qquad (1.3.10)$$

This quantity, as for isotropic materials, must not exceed 0.5.

The restrictions imposed on Poisson's ratios are treated in a general form in [1].

1.3.3. Strength

Like the elastic constants, the strength of anisotropic materials depends on the direction of action of the load. Normal (or longitudinal; or transverse) strength σ_1^u (σ_2^u; σ_3^u) is found during uniaxial tension (compression) along axis 1 (2; 3). In a similar manner, the shear strength τ_{23}^u (τ_{31}^u; τ_{12}^u) is determined as the ultimate value of stress τ_{23} (τ_{31}; τ_{12}).

As a rule, the normal strengths σ_x^u, σ_y^u, σ_z^u and shear strengths τ_{xy}^u, τ_{yz}^u, τ_{xz}^u are used as the strength characteristics of an orthotropic material on the x, y, and z axes. However, for prediction of the time of failure during the simultaneous action of several components of stress, knowledge of these six strengths is insufficient. The number of independent characteristics is determined by the strength criteria. Development and analysis of such criteria is the subject of special studies (see, for example, [18, 25, 31, 75, 107, 134, 181, 182, 193, 212, 240, 241]).

The strength of a fibrous composite depends on both the direction and the sign of the effective stresses. In the general case, both the tensile and compression strengths ($\sigma_1^{tu} \neq \sigma_1^{cu}$, $\sigma_2^{tu} \neq \sigma_2^{cu}$, etc.) and the shear strength, under positive and negative tangential stresses ($\tau_{12}^{u(+)} \neq \tau_{12}^{u(-)}$, etc.) are different.

1.3.4. Effects of Temperature and Time

The temperature and time dependence of the fibrous polymeric composite mechanical characteristics is primarily determined by the characteristics of deformation of the polymer binders, since inelastic properties of the most widespread reinforcing materials begin to appear at considerably higher stresses and temperatures than they do in the matrix materials [195]. Consequently, temperature fluctuations in a comparatively narrow range, which practically do not affect the properties of metals, can show up in the mechanical properties of fibrous composites.

The characteristics determined by the reinforcing fibers depend to a lesser extent on temperature and time. The most widespread reinforcing materials —glass, boron, carbon—can be considered to be linearly elastic right up to failure. Therefore, in the types of tests in which basic load is carried by the reinforcement (tension-compression in the reinforcement direction), a linear connection is found between the strain and stress. This makes it possible to use Hooke's law within the framework of the problems treated in this book.

1.4. TEST SPECIMENS

1.4.1. General Requirements

Selection of the specimen shape and manufacturing technology requires that all the manufacturing conditions and processes of the material, product or structure should be properly simulated. There are two essentially different manufacturing processes for oriented fibrous composites: laminating with subsequent pressing or contact molding and filament (wet) or prepreg (dry) winding. The specimen shape in mechanical testing must allow easy simulation of the technology of the material; an evident example of a successful selection of the specimen shape is the development of ring specimens for studying the mechanical properties of wound materials.

The method of fabrication of the material, the appropriate specimen shape and size and test methods are summarized in Fig. 1.4.1. As seen in the figure, specimens for mechanical testing are classified into flat specimens and bodies of revolution (rings, tubes), depending on the production technique. Flat specimens, in turn, are subdivided into bars and plates. Ring specimens may be split into segments. Flat specimens are fabricated by pressing packages of correctly oriented unidirectional laminae. Ring and tubular specimens are fabricated by winding.

The specimens and test methods presented schematically in Fig. 1.4.1 are especially intended for specification of a lamina. It can be seen from the figure that for specification of flat laminae it is sufficient to have specimens of the same shape, but with a different reinforcement layup. For specification of wound laminae it is necessary to have specimens of various shapes—ring as well as tubular specimens. Ring specimens of a unidirectional material are used for experimental assessment of the characteristics in the direction of fiber layup, tubular specimens—with a 90° angle of winding—to determine the characteristics perpendicular to the direction of fiber layup. Tubular specimens have other uses as well. Tubular specimens with different fiber

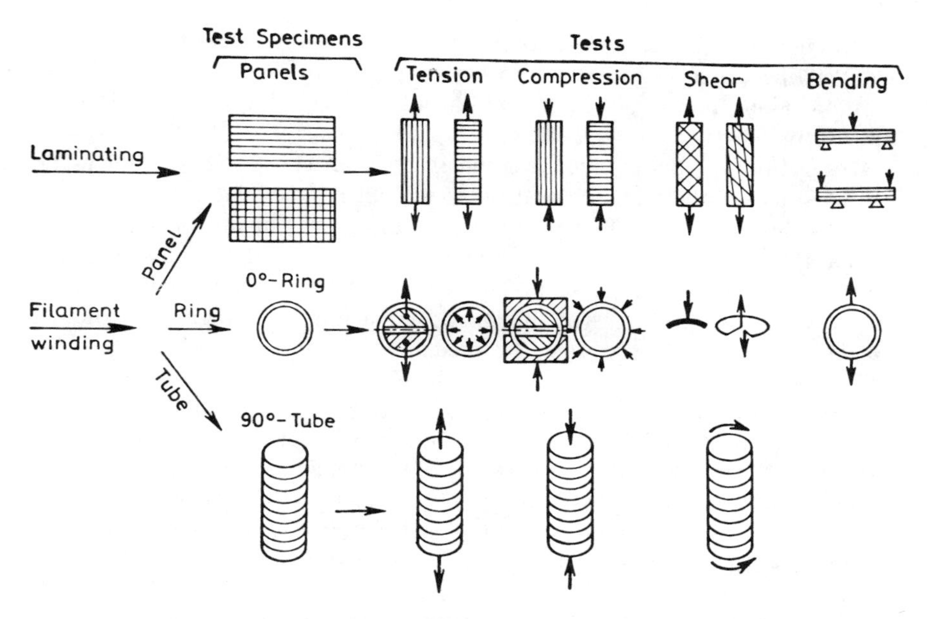

Fig. 1.4.1. Methods of material fabrication and respective specimens and test methods.

layups (always symmetrical about the longitudinal axis) are employed to assess shear characteristics and to study complex states of stress. (This type of loading is not shown in Fig. 1.4.1.)

Besides the technological requirements, the optimal specimen design for mechanical tests must satisfy the following requirements [39]: it must be appropriate for all types of mechanical tests; the test fixture must be as simple and cheap as possible; methods of specimen fastening in the testing machine and the test itself must be easily performed; the failure mode, location of failure, and the numerical value of strength must be reproducible; it should be applicable to the assessment of elastic characteristics and to the study of environmental effects; and, finally, the test must be insensitive to the method of fastening. From the following chapters it will be obvious that satisfaction of all these requirements is not an easy task; for example, for determination of elastic properties and strength, specimens of varied shapes and sizes are usually employed.

The specimen shape, to a great extent, depends on the objective of the test: verification of scientific hypotheses, engineering specification of the material, or quality control of the material. In scientific investigations unique specimens and loading types are employed for evaluation of the specific phenomenon or effect, which are inappropriate for serial testing (specimens with a fiber layup at an angle to the loading direction or having a nonhomogeneous reinforcement layup through the thickness also belong to this group [39]).

Tests aimed evaluating the mechanical characteristics of fibrous polymeric composites are the very foundation of technical specification of materials and of design efforts. For this reason, simple specimens of orthotropic material must be fabricated, which may be characterized by a uniform state of stress within the gage section. To facilitate the experiment and data processing, the specimens of this group must have symmetry of structure in the plane as well as across the specimen thickness. In cases where the latter requirement is technically unrealizable (for instance, in wound structures), all the detrimental side phenomena associated with the lack of structural symmetry can be eliminated by a proper selection of a sufficiently large number of laminae. In processing of test results for nonuniform across the thickness materials [$\theta = \theta(h)$], the peculiarities of their deformability must be taken into account; otherwise serious errors are unavoidable. And finally, specimens used in mechanical testing for quality control of fibrous polymeric composites and structures should not necessarily be applicable for numerical evaluation of the mechanical characteristics of the material.

1.4.2. Scale effect and Stress Concentration

The statistical nature of the strength of fibrous composites is due to the effect of size on the strength of the materials in structures. In distinction from classical structural materials, sensitivity of fiber composites to the scale effect in static loading under uniform stress is always present. The short-term static tensile strength decreases with increase in specimen cross-sectional area; a decrease in strength is observed even with small specimen thickness [233].

The decrease in strength of the thicker specimens is explained by the fact that an increase in cross-sectional area increases the probability of appearance of defects which lead to decrease in strength. The decrease in strength of thin specimens is due to the relatively large effect of surface damage (whence the higher requirements on production and mechanical processing technology of small-size specimens) and disruption of the conditions of transition to a continuous medium (insufficient number of laminae).

The scale effect of fibrous composites is clearly anisotropic. The effect of absolute dimension on strength depends on the dimension at the expense of which change in volume takes place. Experimental data show, for example, that change in specimen thickness of glass fiber composites has a considerably greater effect than change in width and length.

The scale effect in other types of loading and for different fibrous composites has plainly been insufficiently studied. The same remark applies to the effect of stress concentration on the measured strength under various loading conditions. However, it is well known that the presence of comparatively small sources of stress concentration (only mechanical disturbances of material structure—cracks, scratches on the surface, notches—not stress concentration due to a sharp change in product shape, are borne in mind) leads to an appreciable decrease in fibrous composite strength. For instance, in [64] it has been established that two scratches 0.2 mm ($= 0.1h$) in depth on the surface of the specimen cause a decrease in strength of carbon composite in three-point bending of one-third. For ring specimens, the presence of cracks or notches on the inner or outer surfaces can lead to a specific type of failure: "unwinding" [55, p. 152; 213; 231, p. 242]. Therefore, there must be stringent standards for the outer surfaces of specimens and specimen treatment; for example, care should be taken not to produce marks on the specimen by a sharp instrument.

1.4.3. Specimen Conditioning and Test Conditions

The behavior of a specimen of fibrous composites in testing can be determined by its past history, to a considerable extent. In this case, the specimen not only preserves a "memory" of production technology and the storage time, but also of the conditions directly before the tests. Therefore, the reproducibility of the manufacturing conditions, of the conditioning before tests and the test conditions must be ensured. In brief, conditioning and test conditions (after the capital letter T) are briefly recorded by the expression Time/temperature/relative humidity : T-temperature/relative humidity. For example, the record 40/23/50 : T-23/50 means that the specimens are held at 23°C and a relative air humidity of 50% for 40 hours and tested at 23°C and a relative air humidity of 50%.

In ASTM standard D 618-61 (revised in 1971) six typical cases (A–F) of specimen conditioning for rigid plastics are described. Case A is common to all mechanical tests, if the specimen conditioning is not specified. Cases B–F regulate the conditioning at elevated temperatures and relative air humidity. In case A, two conditions are recommended: 40/23/50 for samples up to 7 mm thick, and 88/23/50 for samples thicker than 7 mm. Specimen conditioning is carried out in a thermostat with forced air circulation.

The following encironmental requirements are established by ASTM standard D 618-61 (71): air temperature 23 ± 2°C (or ± 1°C), relative humidity 50 ± 5% (or ± 2%). These quantities are measured directly at the testing machine, but not further than 60 cm from the place of installation of the specimen. A temperature of 20–30°C is considered to be room temperature.

If the tests are conducted at elevated or reduced temperatures, after preliminary conditioning for a period of 0.5 hour the samples are transferred to the test conditions and are held under these conditions for not more than 5 hours, but not less than necessary to achieve thermal equilibrium.

1.4.4. Static Loading

Depending on the nature of loads, mechanical tests of materials are classified into static and dynamic. Static tests are characterized by a smooth and relatively slow change in load on the specimen during the test; such a slow acceleration of moving parts of the testing machine that the inertial forces developed in them can be disregarded; the ability to determine the forces applied to the specimen at any time in the test by the simple static equilibrium method; and simplicity of measurement of the strain of the specimen, which can be done at any time during the test.

Static tests of materials are characterized as either short term or long term. While long-term static testing does not require special explanation, "short-term static testing must be precisely defined and substantiated. Some authors consider strength to be short term with a loading time less than 1 minute [168]. It is sometimes assumed [113, p. 231] that the time of a specimen under the maximum load should not be more than 10^{-4} hour. It is shown in [194] that in shear testing the time of loading to failure must not exceed 3–6 sec. The discrepancies in the definition of short-term static testing must be eliminated, since the effect of polymeric matrix creep on the characteristics of the material depends mainly on the loading rate and loading conditions (stepwise, continuous). With this aim, the effect of strain rate $\dot{\varepsilon}$ and the range of values of $\dot{\varepsilon}$ which allow the effect of time to be neglected while giving comparable results, has been evaluated in the description of each kind of test.

The absence of valid recommendations on selection of loading conditions forces preliminary testing of the behavior of a specimen under load: whether the load remains constant over time and how deformation of the specimen changes after the first and several successive loadings. A sharp decrease in load indicates failure of the specimen or the unreliability of the equipment (specimen slippage). "Training" of the specimen—repeated loading with a small, brief load—usually decreases the scatter of the measured values. Stepwise and continuous loading of a specimen can give sharply differing experimental results. Therefore, all features of behavior of a specimen and loading conditions should be fixed in the test report.

1.4.5. The Test Report

A test report must contain the following information, which is common to all types of loading and is reflected in many standards:

1. Complete identification of the material tested—its type (percentage content of components, reinforcement scheme, production criteria, manufacturer, grade of material); history (whether it was loaded and how, type of preloading, how long held in storage or under other conditions besides environmental); shape and dimensions of products (plate, bar, tube)
2. Method of fabrication of specimens
3. Shape and dimensions of specimens
4. Number of specimens
5. Specimen conditioning specifications
6. Temperature and relative humidity of air in test room
7. Method of fastening or supporting specimens
8. Testing machine and apparatus characteristics
9. Loading conditions (application of a load, deformation rate, change in the loading conditions in time)
10. Formulas for calculation
11. Values measured in the experiment
12. Calculated values, their statistical characteristics
13. Estimation of data scatter
14. Rejected results and their evaluation
15. Test data
16. Name and signature of the investigator(s) and test data processor(s)

For serial tests of a single type (tensile, compression, etc.) it is advisable to use standard service lists for recording and processing of experimental data; this contributes to the systematic performing of experimental data analysis.

In processing results, the average value, standard deviation, coefficient of variation, and confidence interval with a given confidence probability should be calculated for each property.

1.5. EFFECT OF TECHNOLOGICAL FACTORS

1.5.1. Molding Conditions

The express dependence of strength and elastic properties on molding method and parameters is well known [61, 233]. This dependence is attributed to the extreme sensitivity of a polymeric matrix to temperature and loading history of the product, or simply, to fabrication technology. Therefore, for correct estimation of the load carrying capacity of a material in construction, it is necessary to know the history of the part being tested. More than that, specimens frequently are made by "hothouse" technology, i.e., under considerably more favorable conditions than in the subsequent processing of the material into a product. This results in a sharp difference in the properties of the specimens and the finished products. Production technology of fibrous polymeric composites, with only a short history of development, is continally being refined and the quality of materials is gradually improving with time. Items being studied do not always possess the best possible properties. Frequently test results reflect not so much the material properties as some defect in a still insufficiently developed technology.

The development of traditional materials is an independent process, preceding the process of fabrication of some structures from them; as delivered, these materials have distinctive, previously specified properties. The mechanical properties of fibrous polymeric composite parts are established in the molding process and are determined to a great extent by the parameters of the molding process. For example, the properties of molded parts depend on the shape of the channel (i.e., the path of the material flow in forming process), molding temperature T_0 and pressure p_0, curing time τ_0, the method of placement of the material in the mould, etc. The same parameters are of significance for other methods of processing as well—winding, contact molding. The principal effects on the properties of the material are caused by three basic parameters: $F_0 = f(T_0, p_0, \tau_0)$. The dependence is a complex function, which can reach maximum and minimum values at various combinations of the parameters T_0, p_0, τ_0 [233, p. 179]. Molding pressure (especially when low) has the greatest effect on strength perpendicular to the reinforcement plane. In the working range of p_0, the effect of this pressure on the properties in the fiber direction is small; molding pressure and temperature have little effect on the modulus of elasticity in the fiber direction. Forming parameters affect the properties determined by the polymeric matrix especially strongly.

Frequently, the molding and winding technology provides insufficient interlayer contact pressure for removal of irregularities of the monolithic structure of the composite—pores and voids, i.e., places where there is no adhesion between the fibers and the matrix. Such structural imperfections of the composite, which have little effect on the tensile strength in the fiber direction, can show up significantly in compression strength and stiffness, interlaminar shear and transverse tension strengths. In testing of fabric reinforced composite an increase in porosity from 2 to 15% led to almost a fourfold decrease in compression and interlaminar shear strengths [275, 276].

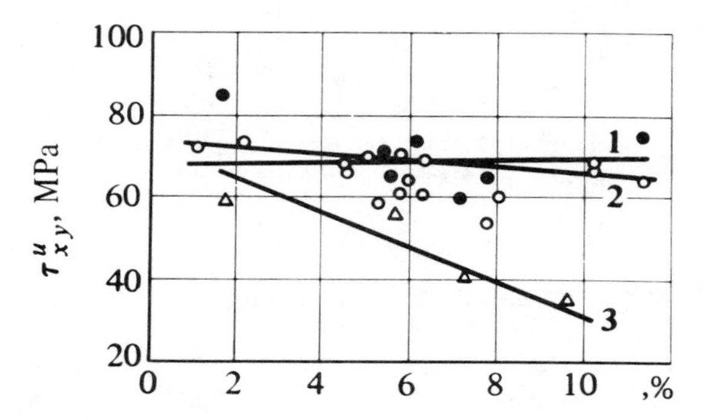

Fig. 1.5.1. Shear strength of unidirectional carbon composite versus void content V_v and test method [194]. (1) tension of a strip (○); (2) transverse compression (●); (3) three-point bending (△).

Analogously, similar results have been obtained in tests of boron and carbon composites [179]. Investigations [194] have revealed an essential effect of voids on shear strength of unidirectional carbon composites. As a rule, the shear strength decreases with an increase in void content. The change in strength and data scatter depend on the material being tested and the test method. It can be seen from Fig. 1.5.1 that three-point bending is the method most sensitive to void content, while tension of an anisotropic strip is the least sensitive, i.e., the methods of determining interlaminar shear strength are more sensitive to void content.

The presence of technological defects in filament wound items of high-modulus composites, in particular of materials having low fiber-matrix adhesion (for example, fiberglass composites), affects primarily the transverse tension strength σ_r^{tu}. In practice, however, winding tension N_0 shows no effect on the transverse tension strength.

Pores play an insignificant role in the study of the strength and stiffness of materials reinforced with whiskerized fibers: the unjustified increase in contact pressure can damage the crystals and reinforcing fibers. Pore elimination involves additional difficulties. However, elimination of porosity increases the interlaminar shear strength [226]. Porosity has little effect on the properties of multidirectional materials; in this case we have to deal with the problem of reinforcement waviness.

1.5.2. Fiber Misalignment

As a consequence of significant differences between the mechanical properties of the components, fibrous polymeric composites are extremely sensitive to fiber misalignment and waviness. Misalignment means deviation of the fiber direction in the laminae from the planned direction. Such deviations are due to imperfections in the technology, and they can be one of the causes of the large scatter of the test results, especially for unidirectional materials reinforced with high-modulus fibers. For more details on the effect of misalignment, see [271, p. 106].

1.5.3. Fiber Waviness and Tensioning

Reinforcement waviness plays a role analogous to that of misalignment. Woven materials have a unique feature—regular fiber waviness. Deviation of the reinforcing fibers from rectilinearity in some materials is of random nature and it is caused by imperfections in processing these materials into finished parts. As a result, fiber waviness is especially noticeable in parts produced by contact molding and molding in closed molds. In production, shrinkage during polymerization will increase fiber waviness in woven fabrics still further and change the pattern of fiber waviness of a given weave, if this is not prevented by an increase in the pretensioning.

In fabrication of a structure from unidirectional materials little waviness of the reinforcement takes place, as a rule. Such waviness can be a consequence of irregular spacing of the reinforcing fibers in the mould, a drop in tensioning below the critical level, or technological (chemical or thermal) shrinkage of the binder. In winding, waviness is a consequence of insufficient tensioning and "settlement" of the windings over the thickness of the product being wound. Fiber waviness is especially noticeable in materials reinforced with roving in two mutually perpendicular directions. Sometimes, very significant local waviness is caused by placing stock in a mold smaller than the fiber length of the material. The probability of considerable local fiber waviness increases with increase in dimensions of the part, which causes additional difficulties in placement of the reinforcing fibers into the mould.

Even minor fiber waviness is a cause of significant changes in the elastic and strength properties in the reinforcing directions, but at the same time, it will have practically no effect on properties peculiar to the polymeric matrix. This fact has often been demonstrated experimentally [231, p. 47] and it follows from the analysis of the simplest model, which assumes sinusoidal fiber waviness. The effect of minor waviness on the modulus of elasticity under tension for the sinusoidal case can be estimated as follows:

$$E_\theta^* = \frac{E_\theta}{1 + \dfrac{E_\theta}{G_{\theta r}} \cdot \dfrac{f^2}{2}}, \tag{1.5.1}$$

where E_θ is the modulus of elasticity under tension along the "straight" fibers; $G_{\theta r}$ is the interlaminar shear modulus; $f = \pi k A/l$ is a parameter, which characterizes fiber waviness (k is the number of halfwaves in base l; A is the amplitude of the sinusoid).

Even with minor waviness f, the modulus of elasticity in actual materials can be significantly lower than in a material with ideally straight fibers. As can be seen from equation (1.5.1), this is a consequence of the low interlaminar shear strength of the composites.

The modulus of elasticity perpendicular to the reinforcing fibers is

$$E_r^* = \frac{E_r}{1 + \frac{E_r}{G_{\theta r}} \cdot \frac{f^2}{2}} \tag{1.5.2}$$

from which it is seen that the curvature of the layers only slightly affects deformability in the direction perpendicular to the reinforcement, since E_r and $G_{\theta r}$ are quantities of the same order. Similarly, it can be shown that the shear modulus $G_{\theta r}^*$ of a medium with small initial irregularities is equal to

$$G_{\theta r}^* = \frac{G_{\theta r}}{1 + 2\left(\frac{G_{\theta r}}{E_r} - 1\right) f^2} \tag{1.5.3}$$

and depends little on the waviness of the reinforcing layers. Similar results have been obtained [247, p. 110] for thick-walled rings with regular waviness in the case where the equation of the wavy surface is given in the polar form $r = r_0(1 + \lambda \cos N\theta)$, where $\lambda \ll 1$ is the amplitude of the wave, N is the number of waves in the ring, and θ is the angular coordinate.

The effect of waviness on the elastic properties in the reinforcement direction depends significantly on the anisotropy. Consequently, the effect increases for materials reinforced with high-modulus fibers. In some materials, for example, carbon composites, a new factor appears—anisotropy of the reinforcing fibers. The formula (1.5.1) has been confirmed experimentally in tests of fiberglass composites. This formula requires some refinement for materials made with anisotropic fibers.

Fiber waviness makes the field of surrounding strains which develop under internal or external pressure non-axisymmetric. In estimation of E_θ in individual sections of the outer (or inner) surface of a ring, for example, from strain gage readings the value of E_θ turns out to be dependent on the angular coordinate. A change in loading method does not eliminate this error. In loading with concentrated forces, the results obtained depend on the angle between the direction of the applied load and the radius of the greatest fiber waviness.

Fiber tensioning during molding allows the elimination of or at least a decrease in the fiber waviness. Of course, complete straightness of all fibers can be achieved only for nonwoven materials; for woven materials, fiber tensioning leads only to a decrease in waviness of the reinforcement in the direction of the stress. Under certain conditions, it is accompanied by an increase in waviness of the fibers arranged in the perpendicular direction. A graph of the increase in strength versus the amount of fiber tensioning shows a peak, after which a drop in strength is observed. In this case, the type of resin and the fiber volume content have a considerable effect. With a high reinforcement content (over 70% by weight), the tensioning has practically no strengthening effect.

Selection of the tensioning deserves special attention in study of the properties of high-modulus materials on ring specimens. In winding brittle fibers (for example, carbon) the range of possible regulation of the tensioning is small. However, even in this range it is possible to eliminate the error introduced by fiber waviness. The effect obtained is dependent on both the fiber straightening and the compaction of the material.

The causes of local fiber waviness, the nature of the waviness, the magnitude and role of tensioning in various methods of forming fiberglass and carbon composites have been treated in detail in [63, 230]. In evaluation of the effect of technological parameters there is a need to study the characteristics which are most sensitive to temperature, pressure, and exposure time; stability of the technological process must be evaluated according to these parameters, not according to the properties in the reinforcing direction.

1.5.4. Initial Stresses and Their Experimental Determination

Fibrous polymeric composites are characterized by initial (or manufacturing) stresses, induced in the fabrication process. There are initial stresses of two kinds: microstresses and macrostresses. Initial microstresses are caused mainly by differences in the coefficients of thermal expansion of the material components. They are nearly completely retained in small specimens cut from the material. Therefore mechanical characteristics obtained during testing of such specimens incorporate the effects of initial microstresses.

Initial macrostresses are built up as thermoelastic stresses in the post-cure cooling process or as a result of stresses due to fiber tensioning or material nonhomogeneity. Upon cutting the specimens from the material, initial macrostresses are partly or completely obliterated. Therefore, mechanical tests provide strength characteristics devoid of the effect of initial macrostresses. Consequently, the initial macrostresses must be determined separately and in the process of structural design they must be combined with the stresses under external load.

The most widely used experimental technique of determination the initial macrostresses are the generalized Sachs' and Davidenkov's methods; both methods are exclusively meant for ring specimens. To study the process of initial macrostress generation, in order to permit evaluation of the effects of various technological factors, tensiometric mandrels and embedded strain gages are used.

A Generalized Sachs' Method. The method involves subsequent removal of thin inner or outer cylindrical layers and recording of circumferential strain on the opposite surface of the ring. For wound cylinders and rings from preimpregnated tape or fabric, Sachs' method is accomplished by layer-by-layer unwinding of the inner or outer layers of the specimen. For specimens fabricated by filament winding, layer unwinding is inapplicable and the cylindrical layers are removed by machining of specimens. Practically, outer layer unwinding or machining is more convenient. Sachs' method allows one to determine not only the absolute values of radial (σ_r^0) and circumferential (σ_θ^0) initial stresses, but also distribution of the initial stresses along the specimen radii $\sigma_r^0 = f(r)$ and $\sigma_\theta^0 = \varphi(r)$. By applying a tensiometer on the inner surface of rings from homogeneous material, the radial and circumferential normal stresses are calculated according to the following formulas:

$$\sigma_r^0 = -\frac{kE_r}{2}\left[\left(\frac{r}{R_i}\right)^{k-1} - \left(\frac{r}{R_i}\right)^{-k-1}\right]\varepsilon_{\theta i}(r), \tag{1.5.4}$$

$$\sigma_\theta^0(r) = r\frac{d\sigma_r^0}{dr} + \sigma_r^0(r) \tag{1.5.5}$$

where E_r and E_θ are the moduli of elasticity of the material in the radial and circumferential directions, respectively; $k = \sqrt{E_\theta/E_r}$; r is the actual radius; $\varepsilon_{\theta i}$ is the relative circumferential strain, measured on the inner surface of the ring ($r = R_i$).

In the testing of cylinders, the moduli of elasticity E_r and E_θ are replaced by the moduli for plane strain

$$E_r^* = \frac{E_r}{1 - \nu_{r\theta}\nu_{\theta r}}, \qquad E_\theta^* = \frac{E_\theta}{1 - \nu_{r\theta}\nu_{\theta r}}.$$

According to the formula (1.5.4) and the moduli of elasticity E_r and E_θ, radii R_i and r, and experimentally evaluated circumferential strain $\varepsilon_{\theta i}$, the curve $\sigma_r^0 = f(r)$ is plotted and subsequently used for calculation of the stresses $\sigma_\theta^0 = f(r)$ from the formula (1.5.5).

Calculation of the relations for thick-walled elements ($R/h < 4$) of non-homogeneous materials is more complex and requires a computer [27].

A Generalized Davidenkov's Method. Davidenkov's method differs from Sachs' method in that before removal of cylindrical layers the ring or cylinder is cut through the radius. In other words, tension (compression) of a circular ring according to Sachs is substituted by studying a curved bar in flexure.

The circumferential modulus of elasticity of thin rings can be assumed to be constant through the ring thickness, and the formulas for calculation the initial stresses take the form

$$\sigma_\theta^0(a) = E_\theta\left[\left(1 - \frac{2a}{h}\right)\varepsilon_{\theta i} - \frac{1}{3}(h-a)\frac{d\varepsilon_{\theta i}}{da} + \frac{2}{3}\int_0^a \frac{2h - 3a + \eta}{h - \eta}\cdot\frac{d\varepsilon_{\theta i}}{d\eta}d\eta\right] \quad (1.5.6)$$

$$\sigma_r^0(a) = \frac{1}{r}\int_{R_i}^{r} \sigma_\theta^0(a)\, dr \quad (1.5.7)$$

where a is the layer coordinate, for which the stress is calculated; h is the thickness of the specimen; and η is the thickness of the removed layer.

The more complex cases of distribution of the circumferential modulus of elasticity E_θ through the thickness of the ring (linear, hyperbolic) as well as the general case of testing thick-walled elements have been treated in [27].

When evaluating and making choice between Sachs' and Davidenkov's methods it should be taken into account that the circumferential strain $\varepsilon_{\theta i}$ according to Sachs' method is due to extension of an annular specimen, whereas according to Davidenkov's method it is due to bending of a curved bar. Therefore, when employing Sachs' method, there is a need for high precision in recording the strain $\varepsilon_{\theta i}$. This is of particular importance in the testing of thin-walled specimens.

The formulas of Sachs' method assume that the stresses σ_r^0 are proportional to the radial modulus of elasticity E_r or its function, the accuracy of their determination is considerably lower than that of measuring the circumferential modulus of elasticity E_θ, which is the basis of the formulas of Davidenkov's method. Circumferential stresses σ_θ^0 are calculated according to Sachs' method by differentiation of the relation $\sigma_r^0 = f(r)$, resulting in additional errors. According to Davidenkov's method, the circumferential stress σ_θ^0 is first determined in the processing of experimental data, i.e., this is the value determined with more accuracy, and the radial stresses are estimated by integrating the relation $\sigma_\theta^0 = \varphi(r)$.

In comparison, Davidenkov's method appears to be more accurate, but also more time-consuming. Besides, the scatter of experimental data is unlikely to justify the gain in accuracy of initial data estimates.

The process of initial macrostresses generation may be investigated by means of tensiometric mandrels and by application of the embedded strain gages. Strain gaging of the mandrel is an oblique method by which the information about initiation and building up of the initial stresses in the process of winding and subsequent heat treatment due to various technological factors is obtained by measuring the pressure on the mandrel from specimen's side. For this sake, heat resistant strain gages are glued on the inner surface of the metallic mandrel: to measure stresses in the circumferential direction, and to compensate in the axial direction. Simultaneously, mandrel temperature and various technological parameters (for instance, tensioning of the tape being wound) are measured.

The method of embedded strain gauges is used for investigation of the generation of circumferential initial strains and stresses in the process of heat treatment of thick-walled cylinders wound from preimpregnated fabric. For this reason, in the winding of a cylinder, the tensiometric tapes, made of the same fabric as the cylinder, with heat-resistant microgages glued on them are placed between fabric layers. In the vicinity of operating strain gages the compensatory strain gages and thermocouples are placed. The method of embedded strain gages is treated in detail in [58, 248].

Chapter 2
Tensile Testing

2.1. TENSION OF FLAT SPECIMENS

2.1.1. Introductory Remarks

Uniaxial tension is the most widespread and the most studied mechanical test for composites. This type of test of fibrous polymeric composites has been standardized in the USSR (GOST 25.60I-80), the USA (ASTMD 3039-76), West Germany (DIN 53392), Great Britain (BS 2782, part 10, method 1003), by the ISO, etc. The popularity of uniaxial tension as a test method is explained mainly by simplicity of accomplishment and ease of processing and analysis of the test results. The characteristics obtained in uniaxial tension are used both for material specifications and for estimation of load-carrying capacity. Practically all strength criteria include tensile strength. Only bending tests on simply supported bars can compete with uniaxial tension in simplicity of accomplishment, but not in processing of the test results.

Despite its apparent simplicity, the tension test is subject to a series of problems due to the structure and properties of fibrous polymeric composites. The main difficulty of tensile testing of fibrous polymeric composites is in establishing a uniform state of stress over the entire gage length. In the determination of elastic constants and of strength, the requirements for a uniform state of stress are different; elastic and strength properties are often studied on specimens of various shapes. For anisotropic materials Saint Venant's principle is satisfied more poorly than for isotropic materials. Compared with traditional materials, the end effect zones increase sharply. In this connection, the attempt to obtain reliable data on stiffness with a given gage length, i.e., of that section of the specimen in which the measurement of deformations takes place, results in an increase in specimen length.

This, in turn, causes the possibility of a transition from one type of failure to another.

The most frequent error in determination of strength is that the mathematical apparatus used in processing the test results does not correspond to the type of failure during experiment. The formula for the calculation of tensile strength as the maximum load per unit of cross-sectional area of the gage section assumes one failure mode—breaking of the specimen perpendicular to its longitudinal axis. However, in practice the specimen often fails by longitudinal delamination (i.e., peeling of a number of layers), shearing, or breaking outside the gage section in the test machine grips. Such errors are often encountered in tension at an angle to the reinforcement direction. They should be eliminated by a proper selection of specimen size and by proper clamping. The type of specimen clamping must ensure reliable alignment of high-strength composite specimens, in particular unidirectional ones, which are highly compliant in directions perpendicular to the direction of reinforcement and often slide in the grips.

2.1.2. The Stress-Strain Curve

It has been experimentally established that the stress-strain curve σ-ε of fibrous polymeric composites under tension depends on the reinforcement material and layup, the interaction of the reinforcement with the matrix, and the direction of the load relative to the principal axes of elastic symmetry and environmental conditions.

Stress-strain curves of fiberglass composites are shown in Fig. 2.1.1. For unidirectional fiberglass composites, the stress-strain curve is practically linear right up to failure of the material. The stress-strain curves of orthogonally reinforcement, nonwoven glass fiber composites and of glass cloth composites consist of two or more rectilinear sections with different slopes, i.e., with different values of the moduli of elasticity E_x^t. The stress-strain curve, registered by recorders of standard testing machines for metals, does not often show such peculiarities [251]. There are several causes of appearance of characteristic breaks ("knees") in the stress-strain curve, but all of them can be reduced to one phenomenon—partial failure of the material. Nonlinearity of the stress-strain curve can be also a consequence of imperfect experimental techniques, for example, instrument lag.

In loading fiberglass composites with multidirectional reinforcement in the directions of the principal axes of elastic symmetry, there is a linear section on the curve σ-ε the length of which depends on structural composition of the material. For materials based on whiskerized fibers, the σ-ε relation is practically linear right up to failure (Fig. 2.1.2, curve 1). The

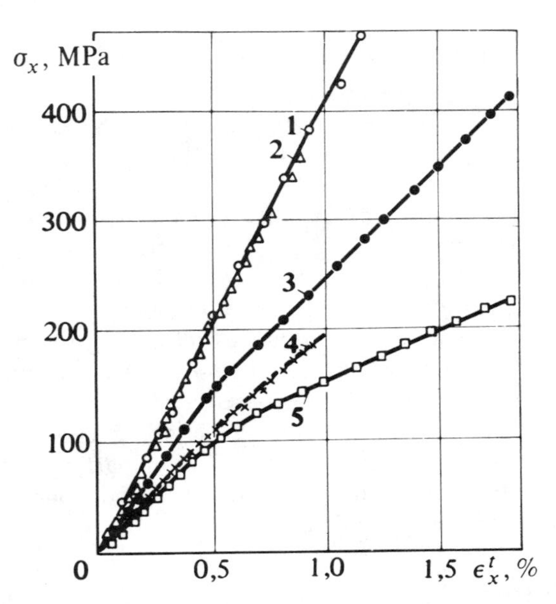

Fig. 2.1.1. Tension stress-strain curves of glass fiber composites at room temperature [214]. (1) 27-63S (0°); (2) AG-4S (0°); (3) 27-63S (0/90°); (4) AG-4S (0/90°); (5) glass cloth composite.

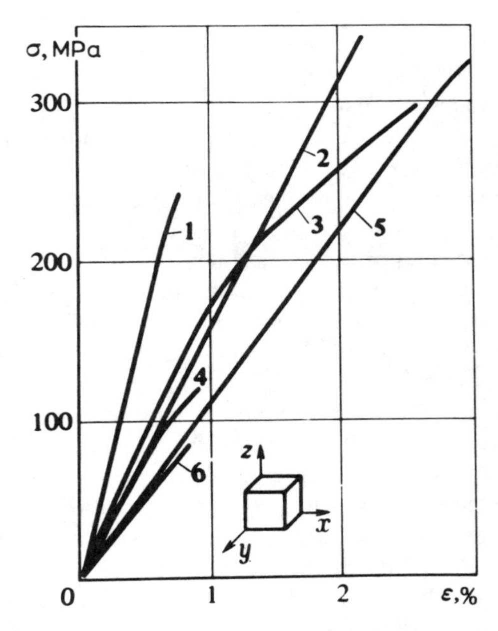

Fig. 2.1.2. Tension stress-strain curves under loading in the directions of principal axes of elastic symmetry of three-dimensionally reinforced materials based on whiskerized fibers (1), a system of two (2, 3) and three (4–6) filaments [271]. Loading direction: (1, 3, 5) over the x axis; (2, 4) over the y axis; (6) over the z axis.

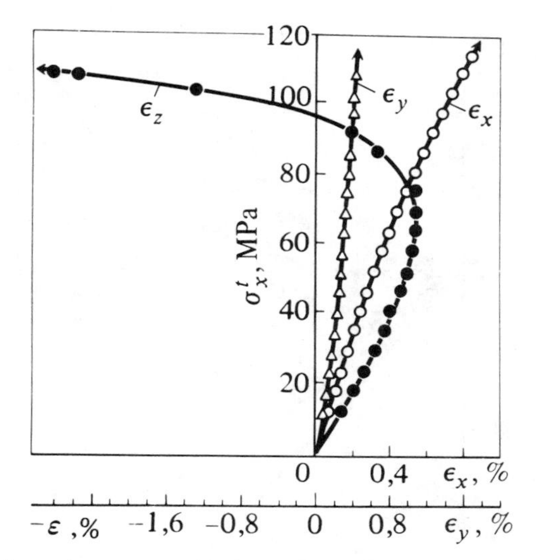

Fig. 2.1.3. Change in strains ε_x, ε_y, and ε_z in tension of glass cloth composite in the direction of the warp, along the weft, and perpendicular to the reinforcement plane (subscripts x, y, and z, respectively. Fiber content $V_f = 59.3\%$ by weight [116].

stress-strain curve for three-filament materials show some linearity at stresses, very near to the breaking stresses (Fig. 2.1.2, curves 4–6); the character of stress-strain curves in all directions of reinforcement is the same. Bi-directional materials have different stress-strain curves under loading in the 0° and 90° (x and y) directions (Fig. 2.1.2, curves 2 and 3): in the 90° direction, the σ-ε relation is practically linear right up to failure; in 0° direction linearity is preserved at stresses which do not exceed 50–70% of the breaking stresses.

Nonlinearity of stress-strain curves has been observed also in tension of boron and fiberglass composites, thereby the degree of nonlinearity depends to a great extent on the direction of the load [249].

Poisson's ratio for fibrous polymeric composites under tension is not a constant value, but decreases somewhat with increasing load. Poisson's ratio for fibrous polymeric composites the stress-strain curves of which have several linear sections changes numerical value after a stress level which corresponds to a breakpoint in the stress-strain curve is reached in the first loading. The phenomenon is expressed particularly distinctly in the reinforcement plane. This is seen in Fig. 2.1.3, in which the change of relative strains in principal directions of the materials is shown.

After repeated loading above the breakpoint, the numerical values of

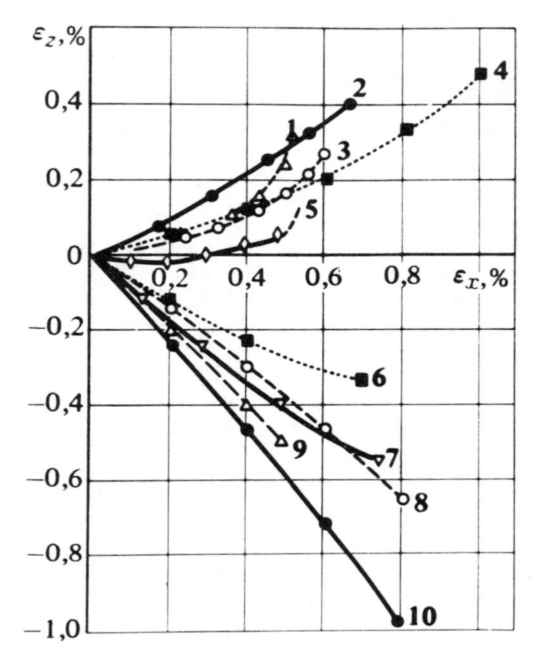

Fig. 2.1.4. Curves of variation in transverse strain ε_z over the specimen thickness with tensile strain ε_x of carbon composites T300/934 [56]. Layups:

1. $[45/-45/0/90]_s$
2. $[(\pm 45)/(\pm 45)/(0/90)/(90/0)]_s$
3. $[45/-45/90/0]_s$
4. $[(\pm 45)/(0/90)]_s$
5. $[0/45/-45/90]_s$
6. $[0/90/45/-45]_s$
7. $[90/45/-45/0]_s$
8. $[(0/90)/(\pm 45)]_s$
9. $[90/0/45/-45]_s$
10. $[(0/90)/(90/0)/(\pm 45)/(\pm 45)]_s$

Poisson's ratio stabilize, but they differ from the values obtained in the first loading.

It has been experimentally established that the sign of Poisson's ratio depends on the stacking sequence of the reinforcement. It can be seen from Fig. 2.1.4 (in which the function $\varepsilon_z = f(\varepsilon_x)$ is shown) that, depending on the stacking sequence of reinforcement layers of one and the same material, Poisson's ratio can turn out to be positive or negative due to edge effect.

Taking into account the mentioned peculiarities and difficulties of measuring the relative strains across the reinforcement direction (see

Section 2.1.3), it is reasonable to conduct experimental determination of Poisson's ratio only on specimens with uniform layup through the thickness and width—strictly speaking, only on specimens made of unidirectional materials. Hence it is necessary to indicate the loading level of the material.

2.1.3. Measurable Values

In testing of fibrous polymeric composites for uniaxial tensile strength σ_x^{tu}, elastic modulus E_x^t, and Poisson's ratios ν_{yx}^t and ν_{zx}^t are determined (directions of axes: x is the longitudinal axis of the specimen; the y axis is directed through the width of the specimen; and the z axis is perpendicular to the specimen plane). To calculate these characteristics, stresses and strains must be determined in the gage section of the specimen. The gage section of the specimen is assumed to be the section, in which by a proper selection of the specimen shape and size a uniform state of stress is ensured. In the gage section of the specimen the measurement base is distinguished, over which the strains are measured and the failure of the specimen must take place.

If a uniform state of stress is ensured in the gage section of the specimen, the stress in the direction of the longitudinal axis of the specimen is determined according to the formula:

$$\sigma_x^t = \frac{\rho^t}{F} \tag{2.1.1}$$

where σ_x^t is the normal stress in the cross section of the specimen, ρ^t is the load, and F is the cross-sectional area in the gage length of the specimen before testing.

Upon failure of the specimen, when $\rho^t = \rho^{tu}$, the short-term static tensile strength is being determined:

$$\sigma_x^{tu} = \frac{\rho^{tu}}{F}. \tag{2.1.2}$$

The breaking load ρ^{tu} is determined from the readings of the testing machine or the stress-strain curve.

The longitudinal strain is equal to:

$$\varepsilon_x^t = \frac{\Delta l}{l} \tag{2.1.3}$$

where $\Delta l = l_1 - l$ is the increase in gage length l, and l_1 is the gage length before and after loading.

The modulus of elasticity is determined according to the formula:

$$E_x^t = \frac{\sigma_x^t}{\varepsilon_x^t} = \frac{\Delta P^t}{F}\frac{l}{\Delta l} = \frac{\Delta P^t}{F}\cdot\frac{1}{\varepsilon_x^t} \tag{2.1.4}$$

where ΔP^t is the load increment; F is the cross-sectional area of the gage length of the specimen (only specimens having gage length with a constant cross section are applicable); l is a base of the tensiometer; Δl is the deformation over the base l at the load increment ΔP^t; ε_x^t is the strain measured by resistance strain gages in the gage length at the load increment ΔP^t.

In practice, the modulus of elasticity E_x^t is often determined directly from the stress-strain curve. If the stress-strain curve does not have linear sections, only determination of the tangent or secant modulus of elasticity is possible.

The Poisson's ratio is determined by

$$\nu_{yx}^t = -\frac{\varepsilon_y^t}{\varepsilon_x^t}, \qquad \nu_{zx} = -\frac{\varepsilon_z^t}{\varepsilon_x^t} \tag{2.1.5}$$

or

$$\nu = -\frac{\Delta l_T}{l_T}\frac{l_L}{\Delta l_L} = -\frac{\varepsilon_T^t}{\varepsilon_L^t} \tag{2.1.6}$$

where Δl_T is the transverse deformation of the specimen, over the base l_T; Δl_L is the longitudinal deformation of the specimen over the base l_L.

The values $\Delta l_L/l_L = \varepsilon_L^t$ and $-\Delta l_T/l_T = \varepsilon_T^t$, entering the formulas (2.1.4) and (2.1.6) can be measured directly by resistance strain gages.

Longitudinal and transverse deformation of the specimen are measured by means of mechanical, optico-mechanical tensiometers, or resistance strain gages.

Mechanical strain gages are the most accurate, but they are applicable only for specimens with a sufficiently large gage length. The mechanical strain gages sometimes slip on the surfaces of carbon composite specimens. Therefore where they are used, backup with resistance strain gages [203] is obligatory. The use of resistance strain gages (wire, foil) requires specific skills (in gluing them on and application of amplifying and recording apparatus), but they are usable for specimens of practically all sizes, they directly measure the strain and do not apply additional loads to the specimen. However, it must be taken into account that in cases where the load acts perpendicular to the reinforcement direction in the outer layers of the material, the resistance strain gage readings due to cracking of the matrix of the material can turn out to be incorrect. Besides, such resistance strain gages must be used, the coefficient of transverse tensosensitivity of which is equal or near to zero.

For measurement of large deformations which exceed the working range of wire and foil resistance strain gages, gages of special design or deformation converters (transducers) are used. In deformation converters, the measured displacement (elongation, deflection) is transmitted to an intermediate link (usually a thin strip or a ring loaded in bending), deformation of which can be measured by regular wire resistance strain gages. The deformation converters must be carefully calibrated beforehand.

2.1.4. Failure Modes

Macroscopic failure modes of fibrous polymeric composites under tension are determined mainly by the type of reinforcement layups, but the mechanical characteristics of the components of the material and their interaction, technological defects (voids, fiber waviness, etc.) and specimen dimensions (edge and end effects) also exert an essential effect. Failure modes, caused by imperfect experimental technique, have not been treated here.

Unidirectional composites loaded in the direction of reinforcement fail, as a rule, by breaking of the reinforcing fibers. At low fiber volume content, the failure of reinforcing fibers is preceded by failure of the polymeric matrix, so that the crack propagates in a direction perpendicular to the direction of action of the external load. At high volume content of the reinforcing fibers, the short sections of fracture front are perpendicular to the direction of action of the external load, but they are connected among themselves by relatively long cracks, directed parallel to tensile stresses and caused by shear or delamination as a result of partial failure of the reinforcing fibers or of technological defects (voids). As a result of the combination of perpendicular and longitudinal cracks the fracture surface becomes rough; this phenomenon can be reinforced by partial pull-out of the reinforcing fibers from the polymeric matrix.

In the case of loading of unidirectional composites at an angle to the reinforcement direction, the failure mode of the material changes, depending on the magnitude of the reinforcement angle. At small angles, the failure of the materials starts by shear and by splitting of the polymeric matrix parallel to the direction of reinforcement. With increasing angles, the tensile stresses exert still greater effect and, in the extreme case—in the case of tension perpendicular to the reinforcing fibers—the material fails in transverse tension.

The failure mode of composites with symmetric (balanced) angle-ply reinforcement (the angle of fiber layup to the loading direction $= \pm\theta$) depends on the angle of the reinforcement layup [122]. At small angles of fiber layup ($\theta < 30°$), the material fails as a result of fast propagation of the crack due to delamination (shear between fibers) and, besides, a V-shaped fracture surface is observed. In this case, interlaminar stresses play an essential role. At large fiber layup angles ($\theta > 60°$), the crack propagates along the fibers as a result of rupture of the polymeric matrix, without any preceding delamination. Upon failure of the material with a reinforcement layup at $\pm 45°$, the combination of both failure modes is observed. At the angles of reinforcement layup $\theta = \pm 15°$, the strength of a balanced crossply composite is twice the strength of the respective unidirectional composite (with a fiber layup $+15°$ or $-15°$). At fiber layup angles $\theta \approx \pm 45°$, the strength of a balanced composite does not essentially differ from the strength of the respective unidirectional composite, and at large fiber layup angles a balanced composite is weaker than a unidirectional composite.

As a result of the above-mentioned failure variations the tensile strength of fibrous polymeric composites, determined as a maximum load per unit of cross-sectional area of the specimen, is a conventional characteristic of the material.

The state of stress in a specimen of fibrous polymeric composite under uniaxial tension is characterized by appearance of interlaminar stresses: normal tensile or compressive $\sigma_z^{t(c)}$ or tangential τ_{xz} and τ_{yz}. Three cases are possible, depending on the fiber layup:

1. In polymeric interlayers of symmetric composites with a fiber layup $\pm\theta$, only interlaminar tangential stresses τ_{xz} act.
2. In polymeric interlayers of crossply composites (at angles 0 and 90°), only interlaminar normal stresses σ_z and tangential stresses τ_{yz} act.
3. In the case of a combination of two preceding cases, for example, in composites with a fiber layup at angles $\pm\theta_1$ and $\pm\theta_2$, interlaminar stresses of all three kinds (σ_z, τ_{xz}, and τ_{yz}) act in the polymeric interlayers.

The sign of interlaminar normal stress σ_z (σ_z^t or σ_z^c) depends on the stacking sequence of reinforcement layers and thus essentially affects the material strength. Distribution of interlaminar stresses through the width of the specimen is uneven—there is a certain zone of concentration of interlaminar stresses on specimen edges free from external load (the so-called "edge effect"). During the last decade, increased interest has been focused on the edge effect, since concentration of interlaminar stresses can be the cause of failure of the material.

The experimental technique depends greatly on another peculiarity of fibrous polymeric composites—anisotropy of strength, characterized mainly by large ratios of strengths $\sigma_x^{tu}/\sigma_z^{cu}$, which for high-strength composites is in the range 6–40. Relatively poor resistance to transverse compression σ_z^{cu} does not allow for large side pressure on the specimen and interferes with safe specimen fixation and often leads to failure of the specimen in grips.

2.1.5. Loading Conditions

A specimen is loaded in accordance with the purpose of the test: in determination of the elastic constants, up to a given load; in determination of strength, up to failure of the specimen. In testing of fibrous polymeric composites, as a rule, machine loading with continuous or stepwise recording of the loads and deformations is used. In determination of the elastic constants of the material, a specimen installed in the machine is initially loaded with a force corresponding to 10–20% of the expected short-term static strength of the material σ_x^{tu}, the load is then decreased to $0.05\sigma_x^{tu}$, and this state is adopted as the initial load. Establishment of a certain constant initial load on a specimen decreases the scatter of the measured values. In determination the strength of the material, there is no necessity of preliminary loading.

In selecting the loading conditions (single or multiple), it should be taken into account that the stress-strain curves σ-ε at the first and subsequent loadings are different. This difference depends on the level of the first load, i.e., on whether or not the breakpoint level on the stress-strain curve is exceeded. It has been noted in [82] that in testing boron composites preliminary loading to 0.75 σ_x^{tu} results in elevation of the breakpoint ("proportionality limit") and a decrease and stabilization of the numerical values of the elastic modulus during repeated loading.

In the determination of short-term static strength, a specimen is steadily loaded at a fixed constant strain rate until failure. In this case, a stress-strain curve is plotted, and the breaking strength is recorded. For this purpose, it is better not to use the standard testing machine recorders, but independent registering apparatus. The short-term static strength is determined by formula (2.1.2).

In determination of the modulus of elasticity and the Poisson's ratio, the specimen is loaded several times (at least three times). When the scatter of the measured values is large, it has to be loaded 6–10 times with a force at which the stress does not exceed the level of the first break in the stress-strain curve.

The effect of the strain rate on the mechanical properties of metals has been studied in sufficient detail. An increase in the strain rate of metals contributes to a growth in the strength characteristics, especially the tensile strength of the material. However, significant changes in the mechanical properties of metals are observed only at strain rates corresponding to impact loading, i.e., exceeding 10^{-2} s^{-1}.

A change in strain rate in the 10^{-4}–10^{-2} s^{-1} interval affects the mechanical properties of fibrous polymeric composites differently. Thus, an increase in the short term static strength and relative elongation of the material, which is particularly intensive in the 0.00015–0.0008 s^{-1} interval, is observed with increase in the strain rate of glass cloth composites. If the strain rate is higher than 0.0008 s^{-1}, the increase in the characteristics is linear, but less intensive. The elastic modulus remains practically unchanged in this range of strain rates. The scatter of the measured values increases with decrease in strain rate.

Unfortunately, results of similar studies are unknown for composites with other reinforcing materials and other schemes of reinforcement layup as well as of composites based on polymer binders of various types.

The strain rate shows up also in the nature of the strain diagram [139]. The complete failure of a specimen is always preceded by a series of micro-failures, accompanied by small drops in the load, which are recorded at low strain rates (a sawtooth stress-strain curve is observed). At high strain rates, the strain diagram up to failure of the specimen turns out to be smooth.

Loading rate, i.e., the cross-head speed of the testing machine for materials whose deformation is described by Hooke's law, is estimated according to the formula

$$V_l = V_\sigma \frac{l}{E_x^t} \tag{2.1.7}$$

where V_σ is the rate of stress increase, l is the length of the free segment of the specimen (between grips), and E_x^t is the modulus of elasticity under tension of the specimen.

The mean strain rate (without taking account of nonuniformity in strain distribution over the length of the specimen) at the calibration of the testing machine equals (in s^{-1})

$$\dot{\varepsilon} = \frac{\Delta P^t}{\Delta t} \cdot \frac{1}{E_x^t F} = \frac{V_\sigma}{E_x^t} = \frac{V_l}{\sigma_0 l} \tag{2.1.8}$$

where ΔP^t is the load increment (fixed upon calibration) Δt is the time, necessary for the load increment ΔP^t (measured in calibration), F is the average cross-sectional area of the gage section of the specimen, V_l is the cross-head speed of the testing machine, and l is the free length of the specimen.

In the determination of the strength and elastic constants of glass fiber composites, the strain rate is chosen within $\dot{\varepsilon} = 0.0008–0.0025\ s^{-1}$ or the stress rate should be $V_\sigma > 2500$ MPa/min.

The recommendations in various standards, pertaining to selection of the strain rate in tension are considerably different:

STANDARD	V_l, mm/min	$\dot{\varepsilon} \times 10^3, s^{-1}$
ASTM D 3939-76	–	0.167–0.333
ISO DIS 3268 (1972, for cloth-reinforced glassfiber composites):		
qualification tests		
specimens with variable cross section	4	0.64
specimens with constant cross section	2	0.20
routine tests		
specimens with variable cross section	10	1.45
specimens with constant cross section	5	0.49

A selection of a low loading rate in all types of mechanical tests is usually explained by the necessity of taking account the instrument lag. However, the other side of the problem has not been completely discussed—the effect of the polymer binder creep.

2.2. SPECIMEN SHAPE AND DIMENSIONS

2.2.1. General Requirements

Specimen shape for tensile testing should conform to generally accepted requirements: in the gage section of the specimen, where strains and stresses are measured, there should be a uniform state of stress; failure must take place in the gage length of the specimen. The measured short-term static strength of a material, the scatter of the test results, material consumption and cost of fabrication of specimens are evidence of the correct selection of the specimen shape.

Despite many international and national standards, in practice there is no generally accepted approach to selection of specimen shapes and dimensions, nor are there universal methods of specimen fastening in the testing machine. The variety of specimen shapes used, a few of which are shown in Fig. 2.2.1, indicates the difficulties of establishment of a single method of tensile testing of fibrous polymeric composites. The data presented in this figure—specimen shape, the strength of the same fiberglass composite, and the

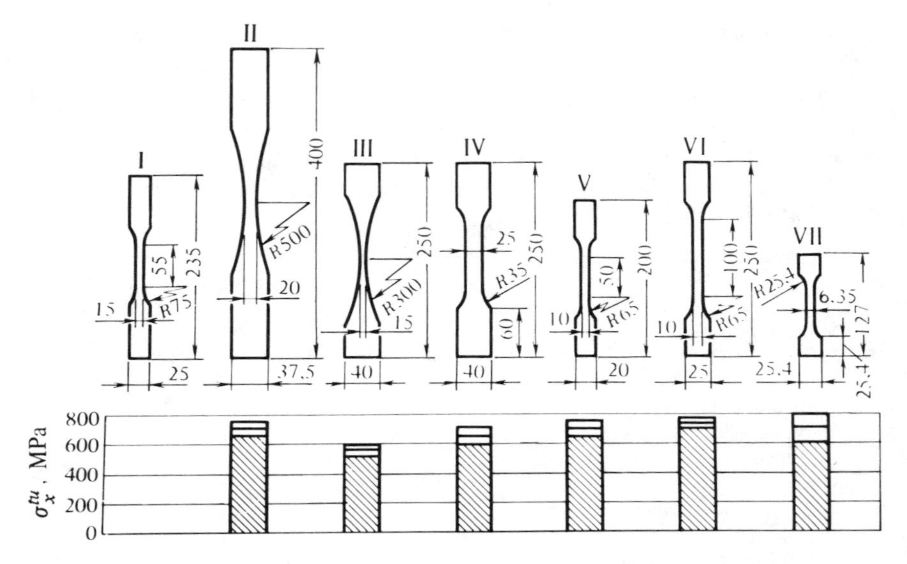

Fig. 2.2.1. Types of specimens for tensile testing of glass fiber composites and minimum, mean, and maximum strengths σ_x^u, obtained with them [233]: (I) according to GOST 4649-55 (there are no strength data); (II) according to Erickson and Norris; (III) a short specimen according to Erickson and Norris; (IV) an elongated specimen according to DIN 53455/1; (V) according to DIN 53455/1; (VI) a modified specimen according to DIN 53455/2; (VII) according to Dietz and McGarry.

scatter of the results—concern an early stage of development of methods of mechanical testing of fibrous polymeric composites, and are only of historical interest.

All the specimens shown in the figure are of variable cross section (dumbbell shape). This type of specimen is well established for testing isotropic homogeneous materials, and such specimens were initially used for testing fibrous polymeric composites. However, very soon the peculiarities of fibrous polymeric composites—splitting and longitudinal delamination tendencies—led to the revision of the approach to selection of the shape and dimensions of specimens with variable cross section. The fact is that, in contrast to testing of isotropic homogeneous materials, the elastic constants and strength of fibrous polymeric composites cannot both always be determined on specimens of the same shape and dimensions.

In determination of the elastic constants a uniform state of stress must be ensured over the entire gage length of the specimen. Consequently, for determination of the elastic constants specimens of variable cross section over the gage length (Fig. 2.2.1, types II and III) are unusable. In determination of the elastic constants, the load on the specimen is as a rule considerably lower than the breaking load, and therefore in the selection of the specimen shape evaluating the effect of stress concentration outside the gage length of the specimen is a task of secondary importance.

In the determination of strength, in contrast, the effect of stress concentration in any part of the specimen must be reduced by means of selection of the appropriate shape and dimensions so as to ensure failure of the material in the gage section. Therefore, in strength determination, particularly in the case of high-strength fibrous polymeric composites, specimens with variable cross section over the gage length are preferred. This facilitates the fastening of specimens of high-strength unidirectional materials, for which it is extremely difficult to ensure reliable transmission of high tensile load (see Section 2.2.3) in the testing machine.

Another type of specimen, Fig. 2.2.1(a) [59] has been successfully used as a replacement for the standard D638 ASTM Tensile specimen. It is now used in numerous U.S. Government Specifications and in AMS Specifications for testing fiberglass fabric laminates. This specimen usually shows strength values 10% or more higher than the ASTM specimen, and all breaks occur in the gage section.

Attempts at unification of specimen shapes, case of specimen fabrication, and elimination of negative phenomena characteristic of variable cross section specimens, have resulted in the development and widespread introduction of the simplest type of specimen with constant cross section, the strip. In correct practice, this type of specimen is universal, i.e., used for the determination of strength and elastic constants. An attempt to eliminate the difficulties connected with fastening specimens of high-strength materials has also led to the appearance of a special type of specimen—sandwich beams. A more detailed treatment of the problems pertaining to all types of specimens is given in later in this chapter.

2.2.2. Specimens with Variable Cross Section

Specimens with variable cross section are used for determination of elastic constants and tensile strength. On the whole, a nonhomogeneous state of stress is characteristic of specimens of this shape. However, with correct selection of specimen dimensions a uniform state of stress can be provided in the gage length. In the determination of the tensile strength of high-strength composites, specimens with variable cross section demonstrate higher and more stable results than those with constant cross section. This can be explained by the lower flexural stiffness of the gage length and, consequently, the lesser effect of bending as a consequence of inaccurate installation of the specimens in the testing machine.

Variable cross-section specimens can be made by three methods of

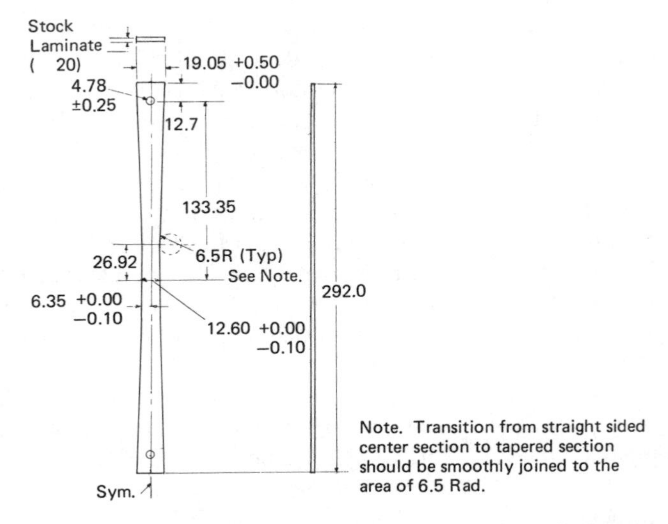

Fig. 2.2.1(a). Elongated Bow Tie tension specimen. [59]

longitudinal profiling: by varying the thickness, width, or thickness and width simultaneously of the specimen (Fig. 2.2.2). In the latter case, in order to ensure symmetry of specimen shape high accuracy of mechanical treatment is required. The most effective means of decreasing the net section is to change its width; however, consideration of the transmission of the tensile load, in this case, results in too great a specimen length. A change in the specimen thickness is admissible only for materials with uniform fiber layup through the thickness; in the nonuniform case, the structure of the material in the gage section may be different. A decrease in the thickness of the net section (i.e., flexural stiffness of the specimen in the plane perpendicular to the reinforcement plane) has been preferred because inaccuracy in installation of the specimen in the testing machine and, consequently, the effect of bending is the greatest in this plane. Because of relatively large splitting area, specimens with variable cross section are especially usable in testing unidirectional composites, i.e., materials with the highest strength ratios $\sigma_x^{tu}/\tau_{xz}^u$.

The longitudinal profile of the specimen must be chosen so that the overall tensile load may be transmitted to the reference section of the specimen without disturbing its integrity. The main limiting factor is the interlaminar shear strength; exceeding this value leads to splitting and delamination of the specimen. In the determination of the minimum section of a specimen it is necessary to assume a compromise, satisfying two opposite requirements:

assurance that a sufficient volume of the material is tested (with allowance of scale effect) and consideration of an actually assured tensile force, which depends on the power of the testing machine and on the reliability of power transmission to the specimen.

The largest cross section of a specimen is established by multiplying the minimum cross section by a stress concentration factor which depends on the assumed narrowing of the cross section. In [69] it has been established that for specimens of unidirectional carbon composites the stress concentration factor is within the range of 1.25–1.5 under the load in the reinforcement direction and 1.1–1.2 under the load perpendicular to the reinforcement direction. In this work, for the case of variable specimen thickness the following relationships have been used to establish the longitudinal profile of the specimen. A change in the cross section of a specimen is determined by the following relationships:

(a) Provided in all sections of a specimen the tangential stresses are lower than the interlaminar shear strength ($\tau_{xz} < \tau_{xz}^{u}$), allowing the ultimate tensile stress to be achieved at the minimum section,

$$\frac{dz}{dx} = \frac{\tau_{xz}^{u}}{\sigma_{x}^{tu}} \cdot \frac{z}{z_0} \tag{2.2.1}$$

whence

$$z = z_0 e^{\frac{1}{z_0} \frac{\tau_{xz}^{u}}{\sigma_{x}^{tu}} x} \tag{2.2.2}$$

and the length of a profiled section of a specimen is as follows:

$$l = \frac{\tau_{xz}^{u}}{\sigma_{x}^{tu}} h \ln \frac{h_1}{h} \tag{2.2.3}$$

(for designations of the coordinate axes, see Fig. 2.2.2(c); z is an ordinate at the abscissa, equal to x; $z_0 = h/2$).

(b) On the assumption of interaction between tensile and shear strengths,

$$\frac{dz}{dx} = \frac{\tau_{xz}^{u}}{\sigma_{x}^{tu}} \left(\frac{z^2 - z_0^2}{z_0^2} \right)^{1/2} \tag{2.2.4}$$

whence

$$z = \frac{z_0}{2} \left(e^{\frac{1}{z_0} \frac{\tau_{xz}^{u}}{\sigma_{x}^{tu}} x} - \frac{1}{e^{\frac{1}{z} \frac{\tau_{xz}^{u}}{\sigma_{x}^{tu}} x}} \right) \tag{2.2.5}$$

and

$$l = \frac{\tau_{xz}^{u}}{\sigma_{x}^{tu}} h \ln\left[\frac{h_1}{h} + \sqrt{\left(\frac{h_1}{h}\right)^2 - 1}\right]. \tag{2.2.6}$$

From relationships (2.2.1) and (2.2.4) the latter is considerably more rigorous, i.e., a change in the specimen thickness effected must be considerably smoother (the radii of curvature differ approximately by a factor of three).

The longitudinal profile of the specimen described by relationships (2.22) and (2.2.5) appears as an exponent. From a practical point of view, it is reasonable to form this profile by some radius of revolution R, corresponding to a ring segment with a chord of length l and a height, equal to $\left(\frac{h_1}{2} - \frac{h}{2}\right)$:

$$R = \frac{l^2}{4(h_1 - h)}. \tag{2.2.7}$$

By using (2.2.4), we obtain

$$R = \left(\frac{\sigma_x^{tu}}{\tau_{xz}^{u}}\right)^2 \frac{h^2}{h + h_1} \tag{2.2.8}$$

This simplification leads to an increase in gradient dz/dx by 5–10%, so that the radius calculated according to the formula (2.2.8) must be accordingly increased. For unidirectional high-strength composites having a $\sigma_x^{tu}/\tau_{xz}^{u}$ ratio up to 50 the necessary radius of the narrowed region is appreciable (> 1000 mm). In tension perpendicular to the plane of fiber layup, when $\sigma_x^{tu}/\tau_{xz}^{u}$ ratio is not large (~ 2), the specimen length turns out to be insufficient for elimination of bending stresses resulting from inaccurate installation in the testing machine; in such cases, soft extension pieces made of light alloys may be used (Fig. 2.2.3).

In order to improve transmission of tensile forces from the testing machine grips to the specimen, tabs are used which are glued o the end sections of the specimen (on choice of the material and dimensions of tabs see Section 2.3.3). To improve precision of assembly of the specimens in the grips it is reasonable to use holes drilled in the end sections of the specimens. Pins located in specimen grips fit through these holes, ensuring accurate alignment.

2.2.3. Specimens with Constant Cross Section

For determination of the short term static strength, modulus of elasticity and Poisson's ratio, specimens of the simplest shape—in the form of a

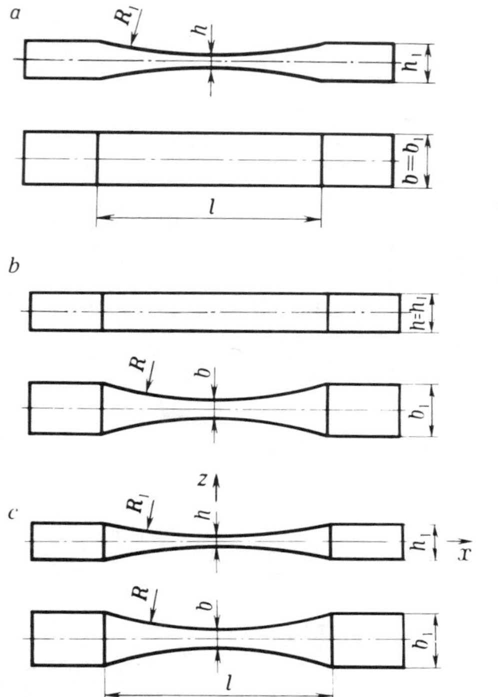

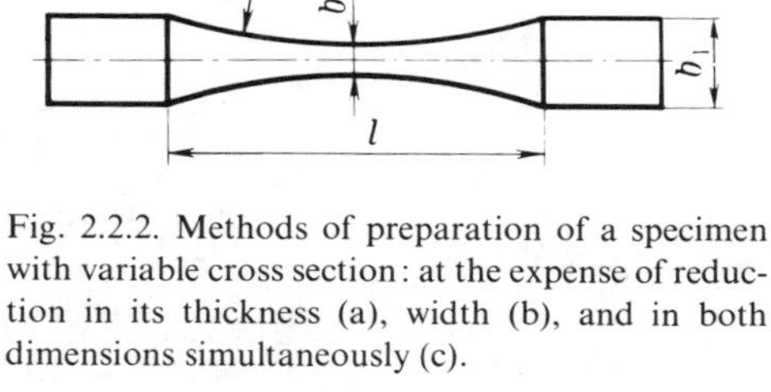

Fig. 2.2.2. Methods of preparation of a specimen with variable cross section: at the expense of reduction in its thickness (a), width (b), and in both dimensions simultaneously (c).

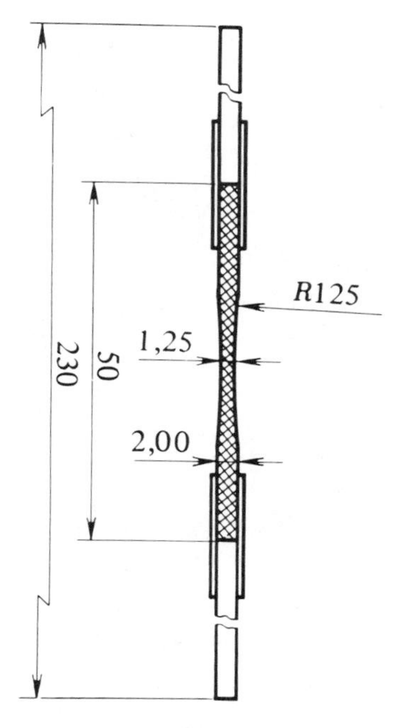

Fig. 2.2.3. A specimen with an elongation for testing unidirectional composites under tension perpendicular to the reinforcement direction (the specimen width 10 mm) [69].

strip (Fig. 2.2.4)—have fully justified themselves in practice. Although such specimens provide the most stable measurements, they also are not devoid of shortcomings. The principal one is the difficulty of ensuring reliable fastening in the testing machine. Since strips do not have a gage section with a reduced cross section, it is necessary to have the tensile force considerably higher than that of specimens with variable cross section at equal strength of the material being tested. In order to increase the reliability of fastening and improve transmission of the tensile forces, tabs have been used which are glued to the specimen (For recommendations for transmission of the tensile forces, selection of the material and dimensions of tabs in more detail, see Section 2.3).

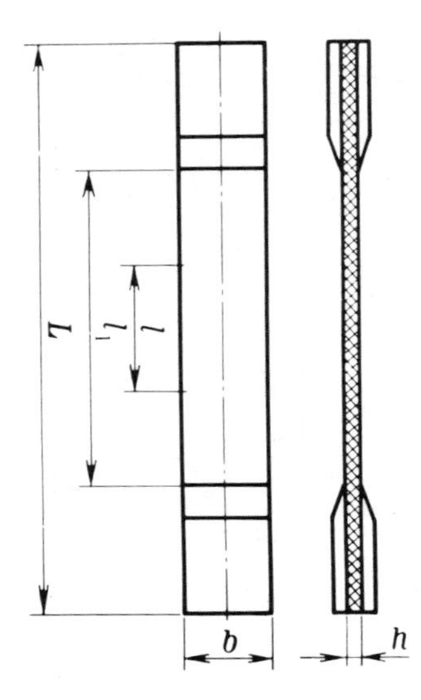

Fig. 2.2.4. A specimen of constant cross section with tabs.

2.2.4. Sandwich Beams

Sandwich beams are used for determination of elastic constants and tensile strengths of fibrous polymeric composites, mainly those of high modulus and high strength. For this purpose, sandwich beams are loaded in four-point bending. The principal advantage of sandwich beams is the absence of all phenomena connected with attachment and loading of specimens in tensile testing. For more detail on sandwich beams, see Section 5.1.

2.3. LOADING OF FLAT SPECIMENS

2.3.1. Deformational Features of an Anisotropic Bar

In a general case of anisotropy, a bar under uniaxial tension is not only extended in the direction of loading and contracting in the transverse directions, but it undergoes shears in all planes parallel to coordinate planes [126, p. 79; 104, p. 12]. In contrast to isotropic materials, the shear moduli are independent of elastic constants.

Let us compare behavior of specimens under tension, having two principal types of fiber layup—symmetrical (layup angle $\pm\theta$) and asymmetrical (layup angle $+\theta$ or $-\theta$) (Fig. 2.3.1). In the first case, we have a material which is orthotropic in the fiber layup plane, and for a regular fiber layup

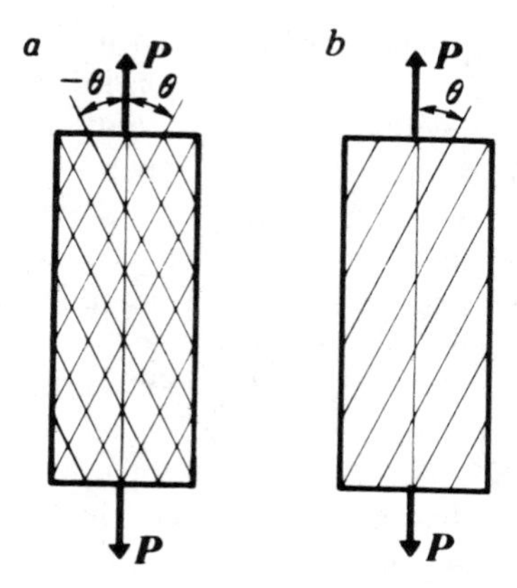

Fig. 2.3.1. A symmetrical (a) and asymmetrical (b) reinforcement.

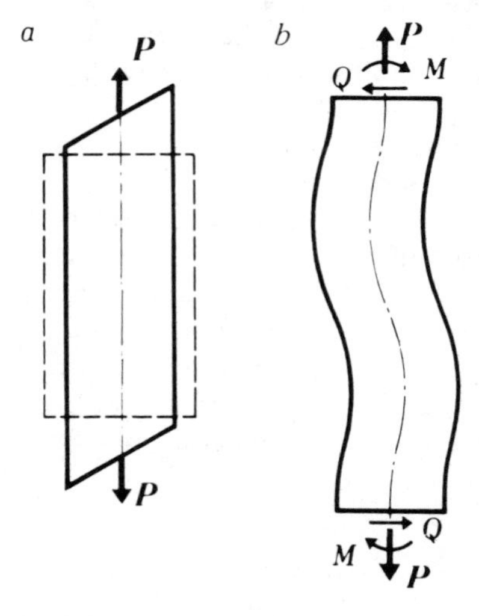

Fig. 2.3.2. Free (a) and restricted (b) deformation of a specimen with symmetrical reinforcement [164].

and precise load application the longitudinal axis of the specimen remains straight during deformation, but all additional effects develop only as the result of interaction of the reinforced layers of the material and restriction of their deformation in the testing machine.

In the case of free tension—without restriction of deformation—on a specimen with an asymmetrical fiber layup, the axis remains straight but the specimen itself distorts into a parallelogram in the reinforcement plane [Fig. 2.3.2(a)].

If the end sections of the specimen are fixed in the testing machine grips, then, besides tensile forces, bending and shearing forces also develop and the deformation of the specimen becomes nonuniform [Fig. 2.3.2(b)]. In this case, the effect of bending and shear depends not only on the ratio of elastic constants of the material being tested, but also on the specimen dimensions, mainly on the ratio of the length to its width l/b. In this case, it is not possible to use the formula (2.1.4) for determination of the modulus of elasticity E_x^t. It is also impossible to evaluate the strength correctly, since restriction of the free deformation of a bar with an asymmetrical layup leads to a nonuniform distribution of stresses and deformations through the specimen thickness, these irregularities being most evident next to the clamps. For example, it has been established in testing of highly flexible carbon composite bars [34] that the greatest nonuniformity in distribution of stresses is exhibited at layup angles $\theta = 5$–$30°$; at $\theta = 10°$ the stresses through the specimen width varied from 462 to 296 MPa next to the grips and from 386 to 289 MPa in the middle of the specimen length.

If the specimen is nonuniform through the thickness or asymmetrical relative to its midplane layup of laminae, then analogous phenomena will be observed in the plane perpendicular to the reinforcement plane when there is some restriction of deformation.

The peculiarities of deformation of an anisotropic material under tension forced a reexamination and refinement of the technique of testing under tension and of methods of data processing. Since consideration of the effect of tensile, bending, and shear forces in two planes of the specimen is not possible in processing of the test results, in planning an experiment all the effects connected with bending and shear should be reduced to a minimum or one must refrain from testing a material with fiber layup at an angle to the direction of action of the load [39].

2.3.2. Transmission of Tensile Forces

The most widespread means of transmission of the tensile forces to load the specimen in self-closing wedge-type grips (Fig. 2.3.3). Reliable fastening of a specimen in the grips is ensured by the condition

$$F \geqslant \tfrac{1}{2}P \tag{2.3.1}$$

where P is the load and F is the friction force on one of the side faces of the fixed specimen.

For self-closing wedge-type grips, the friction force is determined by the formula

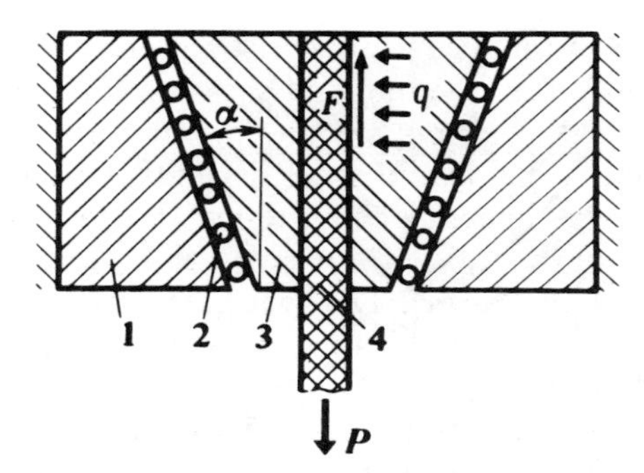

Fig. 2.3.3. Loading scheme of a specimen in self-closing wedge grips [270]: (1) grip body; (2) roller bearings; (3) movable grip jaws; (4) a specimen.

$$F = \frac{P}{2} \frac{f}{\operatorname{tg}(\alpha + \varphi)} = Pf\beta \tag{2.3.2}$$

where f is the friction coefficient of the material on the wedge surfaces, α is the angle of slope of grip jaws, φ is the reduced angle of rolling friction on the inclined surface of grips, and β is the coefficient of intensity of transmission of normal pressure q.

The condition (2.3.1) with allowance for (2.3.2) takes the form:

$$\beta f \geqslant 1. \tag{2.3.3}$$

The value β is determined by the angle of slope of the grip jaws α and by the reduced angle of rolling friction φ (see Fig. 2.3.4). The numerical value of the friction coefficient f depends on the material of the specimen and the surface of the grip jaws. For advanced unidirectional composites and steel surface of jaws which are cut at an angle $\pm 45°$ and a pitch of 1 mm, the friction coefficient varies within the range $f = 0.26$–0.32; if the steel surface of the grip jaws is smooth, then $f = 0.17$ [270].

In selecting the value of α it is necessary to take into account not only the necessity of satisfying condition (2.3.3), but also the strength of the material under transverse compression σ_z^{cu}. At small angles of α the normal pressure q is very high and it usually leads to longitudinal delamination of the specimen. In order to eliminate the longitudinal delamination, it is necessary to have $q < \sigma_z^{cu}$ or

$$\beta \sigma_x^{tu} \frac{S}{S_1} < \sigma_z^{cu} \tag{2.3.4}$$

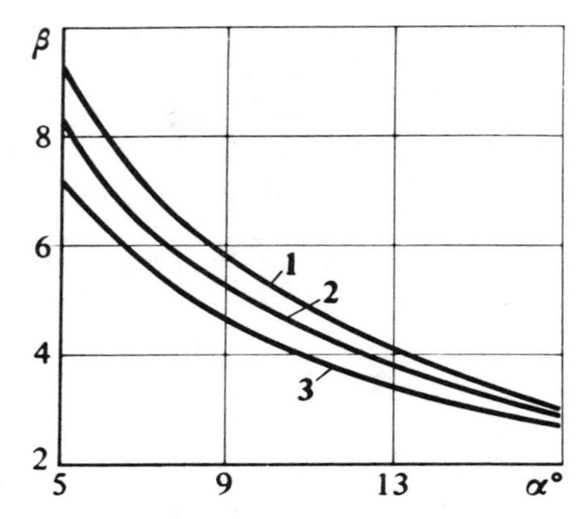

Fig. 2.3.4. Dependence of the parameter β on the angle α at φ, equal to [270]: (1) 10; (2) 2°; (3) 3°.

where σ_x^{tu} and σ_z^{cu} are the strengths of the material of the specimen under tension and transverse compression, respectively; $s = bh$ is the cross-sectional area of the specimen in its gage section; $s_1 = b_1 l_1$ is the area of one of the side surfaces of the specimen bearing normal pressure q in the grips; b and b_1 are the widths of a specimen in the gage section and at the sites of load application, respectively; h is the specimen thickness; l_1 is the length of the specimen section bearing normal pressure q.

It follows from the condition (2.3.4) that

$$\frac{s_1}{s} \geqslant \beta \frac{\sigma_x^{tu}}{\sigma_z^{cu}} \tag{2.3.5}$$

or

$$l \geqslant \beta \frac{\sigma_x^{tu}}{\sigma_z^{cu}} \frac{b}{b_1} h. \tag{2.3.6}$$

By knowing the strength ratio $\sigma_x^{tu}/\sigma_z^{cu}$ (for advanced high-modulus fibrous polymeric composites, this lies within the range of 6–40) and the value of β, determined by condition (2.3.3), it is possible to establish the specimen dimensions for which longitudinal delamination and slip in the grips are excluded.

For specimens with constant cross section, having $b = b_1$, satisfaction of condition (2.3.6) unfortunately leads to a considerable length of the loaded section l_1, particularly at large specimen thicknesses h. The length l_1 can be reduced by using specimens with variable cross section. Their successful use, however, is possible only with exclusion of a phenomenon connected with low shear strength of fibrous polymeric composites—splitting of the specimen head through the width of its gage length b. In order to exclude splitting of specimen heads, the following condition must be satisfied (for $h = \text{const.}$):

$$F_1 < \tau_{xz}^{u} h l_1 \tag{2.3.7}$$

where the splitting force F_1 is

$$F_1 = (b_1 - b) l_1 q = (b_1 - b) l_1 \frac{F}{s_1} = (b_1 - b) l_1 \frac{\sigma_x^{tu}}{2} \frac{s}{s_1}.$$

It follows from the relationship (2.3.7) that

$$(b_1 - b) < 2h \frac{\tau_{xz}^{u}}{\sigma_x^{tu}} \frac{s_1}{s}$$

or

$$(b_1 - b) \frac{b}{b_1} < 2 l_1 \frac{\tau_{xz}^{u}}{\sigma_x^{tu}}. \tag{2.3.8}$$

In establishing the geometrical dimensions of specimens with variable cross section it is necessary to correctly select the radius of curvature in the transition area from the head to the gage section of the specimen (see Section 2.2.2). An insufficient radius of curvature leads to reduced values of the measured strength σ_x^{tu} [270], as can be seen from the data for specimens of a unidirectional carbon composite presented below:

	σ_x^1, MPa		
	CONSTANT CROSS SECTION	VARIABLE CROSS SECTION	
THICKNESS h, mm		$R \leqslant 80$ mm	$R \geqslant 125$ mm
1.2	950	650	820
1.25	1250	1020	1150
1.9	1150	1020	1080

[a] Constant cross-sectional area of the gage section.
[b] Variable cross-sectional area of the gage section.

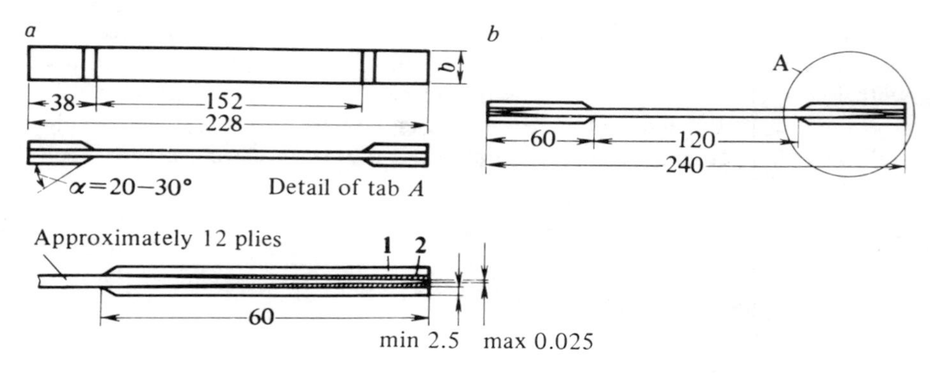

Fig. 2.3.5. Specimens with constant cross section for testing boron composites [130]: (a) with glass cloth composite tabs; (b) with combined tabs (1—glass cloth composite; 2—aluminium).

2.3.3. Tabs

Tensile load is frequently applied to the specimen by means of bonded tabs (Fig. 2.3.5). Tabs protect the surface of end sections of the specimen against damage in the grips and improve transmission of tensile forces to the specimen. In [130] it has been shown how the type of load application affects distribution of the tensile forces in the layers of reinforcement. 16-layer boron composites with titanium tabs were studied. In loading a specimen with tangential forces only [Fig. 2.3.6(a)], the maximum tangential stresses in the outer reinforcement layers exceeded the mean stresses in a specimen by a factor of 1.5, but the inner material layers turned out to be understressed. Additional loading of specimens perpendicular to the plane of tabs led to reduction in maximum stresses in the outer reinforcement layers (by ~20%) and the increase in stresses in the inner layers [Fig. 2.3.6(b)], i.e., to more uniform distribution of stresses.

Tabs are glued to composite sheets which are then cut into specimens. The adhesive is selected depending on the material of specimen and on the test temperature. Usually metal adhesives are used. However, more important than selection of the adhesive is assurance of a uniformly glued seam [131].

The modulus of elasticity of the tab material must be lower (in testing of boron composites, $E_{tob} \approx E_{spec}/4$ is suggested [89, p. 710]), and elongation at rupture higher than the respective characteristics of the specimen material. In testing of boron composites, tabs are usually made of glass cloth composites, with the principal axes of the tab material directed at an angle of $\pm 45°$ to the axis of the specimen. ASTM Standard D 3039–74 for specimens of unidirectional and orthogonally reinforced composites recommends

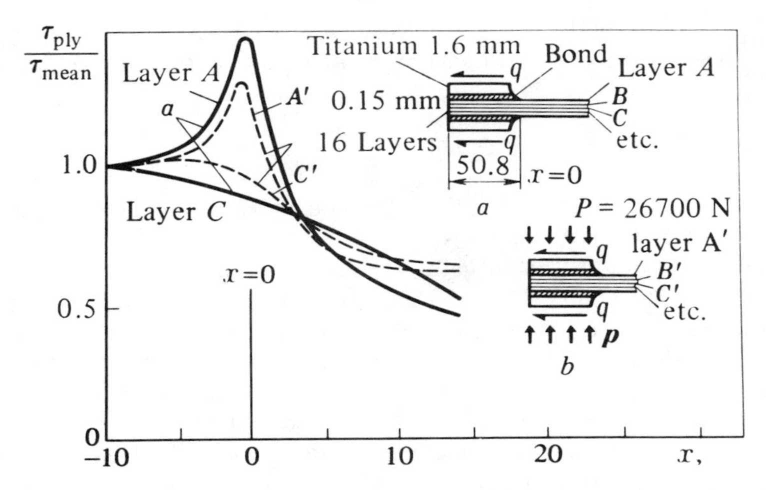

Fig. 2.3.6. Distribution of tangential stresses in layers of boron composite specimen [130] (τ_{ply} is the stress in the layer being considered, τ_{mean} is the mean stress): (a) in loading with tangential forces (q) only; (b) in loading with normal (p) and tangential (q) forces.

tabs of balanced, unwoven, orthogonally reinforced glass fiber composites; the directions of fiber layup in adjacent layers of the specimen and tabs must coincide. The gluing of tabs to specimens of carbon composites is a very tricky and unreliable procedure. In practice, it has been established that an effective means of improving the transmission of tensile forces to the specimen is the gluing of veneer tabs; devices with wooden inserts have also been used [13]. In loading perpendicular to the fiber layup direction, alumina (aluminum oxide) covered fabric tabs are sufficiently effective.

Comparison of tabs made of various materials and of different dimensions is given in [270]. In testing unidirectional carbon composite (a strip specimen 1.2 mm thick and 10 mm wide), the following strength values have been obtained:

	σ_x^{tu}, MPa
Without tabs	950
With tabs	
of pressed veneer	880
of duraluminium	840
of glass cloth composite of length	
60 mm	1180/1150*
90 mm	1240/1170
120 mm	1310/1190

[a] Left-hand values of σ_x^{tu} are for a specimen with 10 mm gage length; right-hand values are for 50 mm.

The dimensions of the tabs were selected in such a manner that the breaking load taken up by them was higher than the breaking load in the specimen gage section:

$$\sigma_x^{tu} bh < 2\tau_{xz}^u b_{tcb} l_{tcb} \tag{2.3.9}$$

where τ_{xz}^u is the lowest among the interlaminar shear strengths of the tab, specimen, and adhesive materials.

Dimensions of tabs are usually selected experimentally. According to ASTM Standard D 3039–76, the thickness of tabs must be equal to $h_{tcb} = (1.5–4)h$, where h is the specimen thickness. For specimens of advanced composites, the ratio $\sigma_x^{tu}/\tau_{xz}^u$ is high (up to 50) as a result of which the load carrying capacity of glass cloth composite tabs is rather limited and thin specimens (2–2.5 mm) must be used. Therefore, attempts have been made [130, 131] to use combined (fiberglass-metal) or metallic tabs. However, these attempts did not yield positive results. Aluminium tabs create high stress concentration in a specimen and failure takes place near the ends of tabs, but titanium tabs give the same results as fiberglass tabs [131]. Sometimes wedge-shape metallic inserts [Fig. 2.3.5(b)] are used to improve the transmission of the tensile forces; however, detailed information on their effectiveness is still missing.

Data on the transition sections of tabs are highly controversial. The tabs described in several works (for example, [203]) and in ISO recommendations do not have a transition section (i.e., $\alpha = 90°$; for notation, see Fig. 2.3.5). In [98], in a study of graphite composites the slope of the transition section of tabs was selected depending on the direction of fiber layup: in unidirectional composites, $\alpha = 10°$; for a fiber layup at angles of $+22.5°$, $\pm 45°$, and $90°$, the slope is $\alpha = 25°$. In references [89, 130] it is suggested that the slope of the transition section of tabs for boron composites be selected equal to 20–30°. ASTM D 3039-76 recommends that the slope of the transition section of tabs be 25°.

Reliability of specimen installation in the testing machine grips and load carrying capacity of tabs can be improved by openings drilled (to a template) in the end parts of specimens, in which fixing pins are inserted to hold the specimen in position. This type of fastening is treated in more detail in Section 2.3.4.

2.3.4. Specimen Fastening

Tensile tests are carried out in testing machines with a sufficient tensile capacity. Load measurement error on the machine scale should not exceed

$\pm 1\%$. It is desirable that the working stresses be within the 10–90% range on the machine scale selected. After careful measurement of its gage length (detailed instructions of tensile testing are included in all standards) a specimen of the material being studied is installed in the testing machine grips, centered, and gripped.

In tensile tests of any material, especially of high-strength, high-modulus, strongly anisotropic composites, one of the most difficult operations is precise installation of specimens in the testing machine grips. Installation of specimens by eye and closing of wedge grips with a blow, occasionally observed in mechanical testing of metals, is inadmissible for fibrous polymeric composites, since this results in inaccurate, arbitrary fixing of the specimens in the grips and damage of the material being tested.

Testing machine grips usually are self-aligning, with cylindrical or ball hinges. However, because of friction in the grip joints there is always some deviation of the axis of the specimen from the direction of action of the load. This results in bending of the specimen, irregular distribution of strains over the gage length, and premature failure of the specimen, especially in the case of brittle materials, to which the majority of fibrous polymeric composites belong. Moreover, misalignment during installation of a specimen in the testing machine grips often results in slipping of the specimen. Therefore, in practice, attempts are constantly made to improve the centering and fastening of specimens. All known methods can be reduced, in principle, to the following two types:

1. Specimens have accurately located openings in their end sections, into which lock pins are inserted.
2. The position of the specimen is fixed under a small load, with the testing machine grips open.

The fixing pins are used primarily for accurate and reproducible installation of specimens, but at the same time they prevent, to a certain degree, slippage of specimens in the grips. Structurally, the fastening units for specimens can be diverse. In the simplest case, the fixing pins rest in V-shaped grooves in the grips (Fig. 2.3.7). A more complicated, but, judging by the test results, successful design of a gripping device has been described in [171] (Fig. 2.3.8). Specimens with end section dimensions (together with the tabs) of $57.2 \times 12.7 \times 3.9$ mm, are inserted into the grooves of the wedge grips; the groove dimensions are $57.2 \times 15.9 \times 3.8$ mm. This is the first stage of centering, which eliminates gross errors. The grips are kept in the semi-open state by means of two screws. Then the fixing pins (4 in the figure) are inserted, and the specimen is loaded with a small force (the stress in the

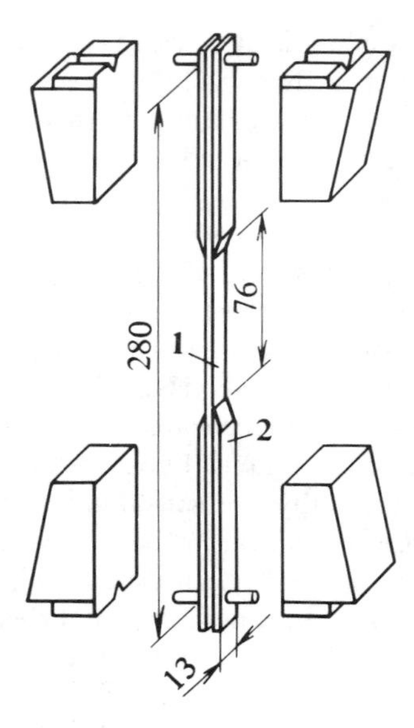

Fig. 2.3.7. Fastening of specimen by means of fixing pins, seated in V-grooves on the grips [82]: (1) 6-ply boron composite ($h = 0.76$ mm); (2) 4-ply glass cloth composite tab.

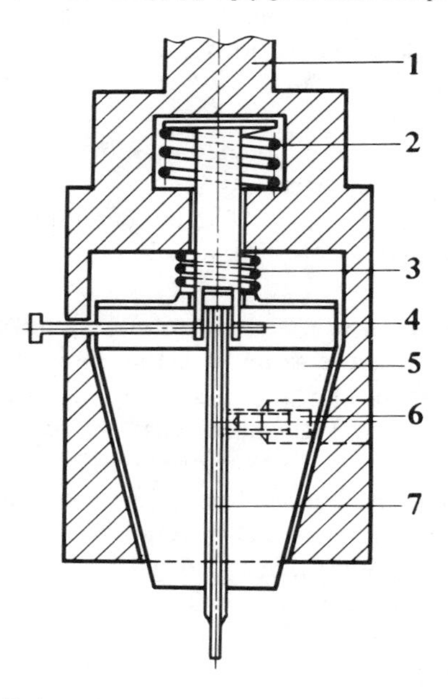

Fig. 2.3.8. Device for installation of specimen with testing machine grips open [171]: (1) device body; (2) outer spring; (3) inner spring; (4) fixing pin; (5) collet grips; (6) screw; (7) specimen.

gage length of the specimen is 40 MPa), which is balanced by spring 2. After this, screws 6 are tightened and spring 3 clamps the grips to the specimen. With further loading, the grips are drawn still further into the device and they firmly hold the specimen.

The results obtained in this device (Table 2.3.1) differ from the results of standard devices, especially, for composites with a rigid matrix. The measured strength and modulus of elasticity turn out to be considerably higher and approach the theoretical limiting values. Moreover, scatter of measured values with this device is less than with standard devices (in tests of Celanese/Epi-Rez 508, 6% versus 12%).

Of course, the device described is not the only possible solution to the problem of fastening specimens in the testing machine grips. However, the results achieved again confirm the importance of this problem. Where auxiliary centering devices are lacking, the position of a specimen in the testing machine grips is carefully selected experimentally, so that the bending stress in the specimen plane and perpendicular to it does not exceed a fixed percentage of the tensile stress. For this purpose, ASTM D 3039-76 recommends gluing three resistance strain gages to the specimen (see Fig. 2.3.9). The specimen should be fixed so as to satisfy the following relationships between strains measured with resistance strain gauges [11]:

$$\frac{\varepsilon_3 - \dfrac{\varepsilon_1 + \varepsilon_2}{2}}{\varepsilon_3} \leqslant 0.02; \qquad \frac{\varepsilon_1 - \varepsilon_2}{\varepsilon_1} \leqslant 0.02.$$

2.3.5. Specimen Dimensions

A specimen for testing of materials in uniaxial tension has three functionally different parts: a gage length and two transition and two loading sections. In the gage length of a specimen, deformations are measured and stresses are calculated according to the geometrical dimensions and the external load. The transition sections serve to attenuate stress perturbations associated with fastening and loading of the specimen (end effect). Loaded areas serve for fastening of a specimen in the testing machine; they take up and transmit the external tensile load to the gage length of a specimen. For selection of the dimensions of the loaded area of the specimen, see Section 2.3.2.

The dimensions of the gage length of specimens are selected with allowance for the following requirements: in the gage length, there must be a uniform state of stress; the values measured must not depend on the cross-sectional dimensions of the specimen; reliable attachment of measuring instruments must be assured.

Table 2.3.1. Results of Tensile Tests of Unidirectional Graphite Composites [171].

CHARACTERISTICS	CELANESE/EPI-REZ 508			MORGANITE II/EPI-REZ 508, variable cross-section specimens	
	STANDARD GRIPS	MODIFIED GRIPS			
	Variable cross-section specimens	Variable cross-section specimens	Straight bars	STANDARD GRIPS	MODIFIED GRIPS
σ_x^{tu}, MPa	605	1062	917	855	1000
	475	1131	924	841	1041
	590	1137	896	917	1020
	458	1254	820		
	564	1062			
	457	1081			
		1096			
	525	1117	889	869	1020
Average strength σ_x^{tu}, MPa	—	2.13	1.69	—	1.17
$\sigma_{x\ \text{modif.}}^{tu}/\sigma_{x\ \text{stand.}}^{tu}$	0.42	0.90	0.72		
$\sigma_x^{tu}/(V_f\sigma_f^{tu})$	297	365	324	124	131
$E_x^t \times 10^{-4}$, MPa	—	1.23	1.09	—	1.06
$E_x^t/(V_f E_f^t)$	0.81	0.99	0.88	—	
δ_x^t, %	0.18	0.33	0.27	0.75	0.77

Crosshead speed $v = 1.3$ mm/min.
Fiber content $V_f = 60$ % (Celanese), $V_f = 50$–55% (Morganite II).
Celanese fibers: strength $\sigma_f^{tu} = 2069$ MPa; modulus of elasticity $E_f^t = 613$ GPa; elongation at failure $\delta_f^t = 0.34$%.

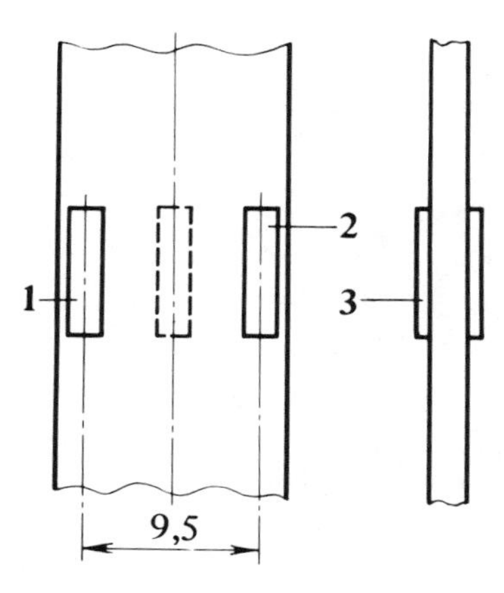

Fig. 2.3.9. Scheme of bonding strain gages (1, 2, 3), used for alignment of the specimen in test machine grips.

In connection with the first requirement, the specimen length with a given measurement base length must be selected with allowance for the end effect. For strongly anisotropic materials, the end effect is highly significant. It is sufficient to note that it requires a change in the gage length of the specimen, according to DIN 53455/2, from 50 to 100 mm, to produce a 10% increase in the measured short-term static strength of glass cloth reinforced composite; this is a consequence not of an increase in gage length proper, but of an increase in the total length of the specimen, which leads to a more uniform state of stress in the gage length of the specimen.

Uniformity of the state of stress in the gage length of the specimen depends also on the length-to-width ratio: the greater l/b, the easier it is to assure a uniform state of stress in the gage length of the specimen.

The most complete investigation of the end effect zone in bars of rectangular cross section is contained in [184]. It has been shown that stress distribution in the end effect zone and the length of the zone depend on the ratios of elastic constants ($\alpha = E_x/E_z$ and $\beta = E_x/G_{kz} - 2\nu_{zx}$), on the ratios of geometrical dimensions ($m_1 = c/h, m_2 = a/h, n = b/a$), and on the distribution of external load ($t = q/p$) (for notation, see Fig. 2.3.10; the width of a specimen $b =$ const.). The highest stresses are observed in the section where $x = a$, i.e., in the transition section from the head of a specimen to its gage length. Thereby, the highest normal stresses σ_x are reached at the sides of the section ($z = \pm h$); however, at $\alpha > 1$ and $\alpha/\beta < 2$ the maximum value of σ_x may be reached away from the side of the section. The extremal values of normal stresses $\bar{\sigma}_x = \sigma_x/p$ and the length of end effect zone $\Delta l/(2h)$ are

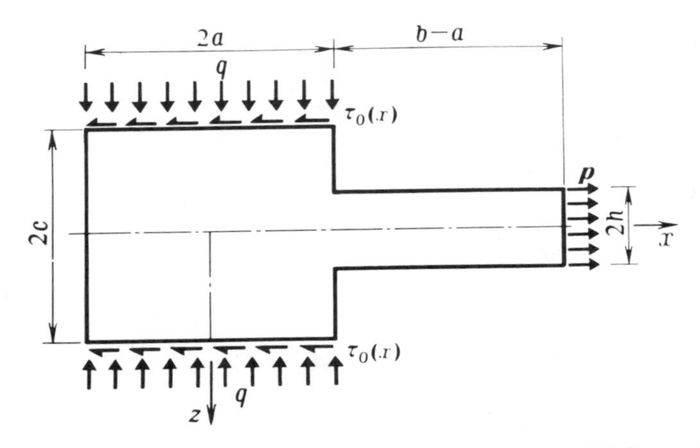

Fig. 2.3.10. Loading scheme for specimen under tension [184].

presented below versus the coefficients α and β, calculated at $m_1 = 2$ and $m_2 = 5$ [184]:

	$\alpha = 40$ $\beta = 150$	$\alpha = 10$ $\beta = 25$	$\alpha = 1$ $\beta = 2$
$\bar{\sigma}_x$ at $x = a$, $z = \pm h$	2.03	1.15	0.88
$\bar{\sigma}$ at $x = a$, $z = 0$	0.80	0.96	1.10
$\Delta l/(2h)$	4.40	3.73	1.90

It is clear from the data presented that the outer layers of the specimen ($z = \pm h$) are overloaded, but the inner layers ($z = 0$) underloaded. For materials with strong anisotropy of the elastic properties ($\alpha = 40$, $\beta = 150$), the values of extreme stresses in the section $x = a$ differ by a factor of 2.5, but for isotropic materials ($\alpha = 1$, $\beta = 2$) only by 25%. The length of the end effect zone for strongly anisotropic materials does not exceed 4.4 thicknesses of the gage length of the specimen. For specimens with constant cross section specified by GOST, ASTM, and ISO standards, the length of the transition section (between tabs or grips and the gage length) is equal to 2.5–31.75 times the thickness of the gage length of the specimen, i.e., on the whole, it is in agreement with requirements.

The effect of geometrical parameters m_1 and m_2 on the relative stresses $\bar{\sigma}_x$ is obvious from the following table [185], in which $\bar{\sigma}_x$ was calculated at $\alpha = 20$ and $\beta = 150$:

	$\bar{\sigma}_x$ AT $x = a$, $z = \pm h$, AND m_1 EQUAL TO					
m_2	1.0	1.2	1.4	1.6	1.8	2.0
3	2.55	2.47	2.42	2.49	2.52	2.53
5	1.92	1.73	1.71	1.74	1.76	1.78
7	1.64	1.32	1.36	1.39	1.42	1.44
9	1.47	1.13	1.14	1.19	1.23	1.26
11	1.37	0.97	1.00	0.98	1.09	1.14

It follows from the table that the increase in the length of loaded section, i.e., a parameter m_2, results in a decrease in normal stresses on the edge of the section, thereby, the decrease in $\bar{\sigma}_x$ of specimens with constant cross section ($m_1 = 1.0$) occurs significantly move slowly than for specimens with variable cross section ($m_1 > 1$). A variation in the parameter $t = q/p$, i.e., a degree of transversal compression of a specimen in the grips, affects $\bar{\sigma}_x$ inessentially. Experimental investigation of the distribution of deformations along the length and height of a specimen [185] confirms the main conclusions of a theoretical analysis, described above. Let us note in the conclusion that the values of stresses σ_z and τ_{xz} in the end effect zone are two to three orders lower than σ_x.

Purposeful studies of the effect of cross-sectional dimensions of a specimen on the strength and elastic constants of all types of fibrous polymeric composites have not been carried out and general regularities have not been established. There are no clearly formulated conclusions in the literature on the minimum necessary number of reinforcement layers. It has been noted [97] that 5- and 10-ply specimens of carbon composites with constant cross section fail at statistically indistinguishable stresses, but 3-ply specimens give understated strength values. In the latter case, the conditions of load transmission from layer to layer are poorer than in multilayer specimens, and the measured strength values correspond to those predicted theoretically by bundle theory. In the determination of the strength of boron composites [82], 6-ply specimens with constant cross section gave more nearly constant results than did 10-ply ones, in which the strength of specimens with long (89 mm) tabs was greater than that of specimens with short tabs (38 mm). Apparently, the optimum number of reinforcement layers needs more precise definition with allowance for conditions of load transmission, flexural stiffness of the specimen, scale effect, and reinforcing material (its compliance, hardness, and anisotropy of mechanical properties). It is of value to note that in ASTM Standard D 3552-77 for tensile testing of fiber-reinforced metal matrix composites, there is the additional requirement that in the gage

Table 2.3.2. Dimensions of Specimens with Constant Cross Section.

DIMENSION	GOST II262-76, FIBERGLASS 0°	ASTM D 3039-76, FIBERGLASS, CARBON, AND BORON COMPOSITE 0°	90°	0/90°	ISO 3268, GLASS CLOTH COMPOSITE
Gage length l, mm	—	127	38	127	100
Distance between end plates l_1, mm	150	152	89	178	150
Length of each end plate, mm, no less than	70	38	38	38	50
Entire length L, mm, no less than	250	228	165	254	250
Width b, mm	10 or 15	13	25	25	25
Thickness h, mm					
fiberglass composite	10		0.8–3.3		2–10
carbon composite	—		0.5–2.5		—
boron composite	—		0.5–2.5		—

length of the specimen no less than 200 fibers should be laid in the direction of the load.

Polymeric composites reinforced with rigid fibers (glass, boron) are very sensitive to stress concentration. This should be taken into account in selection of the fillet radius for specimens with variable cross section. As is shown by studies of glass fiber composite specimens with gage lengths of constant cross section [222], a small fillet radius ($R < 200$ mm) results in a sharp reduction in the measured short-term static strength, but does not affect the value of the modulus of elasticity. Similarly, the fillet radius affects the short-term static strength of glass cloth reinforced composite specimens with variable gage length cross section (see also Section 2.2.2).

The dimensions of specimens with constant cross section specified in standards are presented in Table 2.3.2. (For notation see Fig. 2.2.4).

The ASTM Standard requirements take into account the anisotropy of the mechanical properties of fibrous polymeric composites. A large distance between the tabs and the gage length of the specimen assures sufficient volume of material to absorb the end effect. Large total specimen length essentially reduces the effect of bending, i.e., the effect of incorrect installation of the specimen in the testing machine grips. The length and width of a gage length of a specimen assure reliable and convenient fastening of the strain measuring instruments. It is desirable to raise the lower limit of specimen thickness and to use specimens of thickness $h > 1$ mm.

2.4. LOADING AT AN ANGLE TO THE REINFORCEMENT DIRECTION

2.4.1. Purpose and Details of Tests

Tests of fibrous polymeric composites with different structures at an angle to the reinforcement direction are carried out, in the main, in order to estimate the anisotropy of the mechanical characteristics of the material. For this purpose, it would have been reasonable to carry out an experimental determination of all characteristics of a lamina of the material tested with subsequent recalculation of the characteristics according to the theory of laminated materials. However, at present there is no reliable and conveniently realizable theory which would take into account the numerous effects of interaction of the laminae. Therefore, in practice, to obtain qualitative estimates it is necessary to test the multilayer material. In selection of specimen dimensions and experimental method, and also in the processing of test results, all the peculiarities of deformation and state of stress in the specimen under tensile or compressive loading at an angle to the reinforcement direction must be taken into account; this subject has been treated in Sections 2.1.4 and 2.3.1.

In standard testing machines and devices, free deformation of specimens during tension at an angle to the principal axes of elastic symmetry of the material is impossible even in the case of a hinged support. To eliminate this restriction of deformation of the specimen section, special attachments must be developed in which the specimen can rotate freely. In compression, free deformation of the ends of a specimen is out of question. Therefore, in practice the effects of the restriction of specimen deformation and of peculiarities of stress are taken into account through selection of the shape and dimensions of the specimen, mainly, its width.

2.4.2. Selection of Specimen Width

The presence of edge effect zones can have a significant effect on the average modulus of elasticity of the material $E_x^{\exp}$ measured experimentally on specimens of various width. In the case of tension of a strip of width b, reinforced at an angle $\pm\theta$ to the direction of action of the load x, the modulus of elasticity $E_x^{\exp}$ is determined by the formula [204]

$$E_x^{\exp} = \frac{E_x}{1 - \eta_{16}\eta_{61}\left(1 - \dfrac{\tanh \kappa}{\kappa}\right)} \qquad (2.4.1)$$

where $\kappa = b\sqrt{G^0/h'h^\circ G_{xy}}$; E_x, G_{xy}, η_{16}, η_{61} are the moduli of elasticity and shear and coefficients of mutual influence of the reinforcement layers on the xy axes; G^0 and h^0 are the shear modulus and thickness of the polymeric interlayer, respectively.

As can be seen from the formula (2.4.1) that the modulus of elasticity E_x^{exp} increases from E_x to $E_x/(1 - \eta_{16}\eta_{61})$, with increase of the parameter κ from 0 to ∞. Since κ is proportional to width b, an increase in specimen width must be accompanied by an increase in the averaged modulus of elasticity E_x^{exp}. This fact has repeatedly been confirmed experimentally. The limiting width of the specimen b_{lim}, above which no increase in E_x^{exp} is observed, depends on the parameter κ. It can be seen from formula (2.4.1) that the greater the stiffness of the reinforcement G_{xy} and the more compliant the binder, i.e., the smaller G^0, the greater is b_{lim}. The tensile strength $\sigma_x^{u(\text{exp})}$ at an angle to the reinforcement direction depends on the specimen width in a similar manner. In this case, there also is a limiting width above which $\sigma_x^{u(\text{exp})}$ practically does not increase.

The effective modulus of elasticity of a material with principal axes of elastic symmetry rotated by angle θ relative to the direction of the load, may be determined by the formula [164]

$$E_x = \frac{1}{s_{11}} = E_x^0(1 - p) \tag{2.4.2}$$

where $E_x^0 = \sigma_x(x,0)/\varepsilon_x(x,0)$ is the modulus of elasticity, determined from the experimentally measured stresses $\sigma_x(x,0)$ and strains $\varepsilon_x(x,0)$ on the axes of the specimen; $p = 6s_{16}^2/\{s_{11}[6s_{66} + s_{11}(l^2/b^2)]\}$; and s_{11}, s_{16}, s_{66} are coefficients of compliance.

It was assumed in the derivation of formula (2.4.2) that the points of the end sections on the specimen axis are fixed: $u(0,0) = 0$, $u(l,0) = 0$; $v(0,0) = 0$, $v(l,0) = 0$. The compliance coefficients s_{11}, s_{16}, and s_{66} are calculated at a fixed angle of rotation of the coordinate axes θ.

The calculations show that the error in determination of the modulus of elasticity, without consideration of the actual deformation scheme (i.e., according to the formula $E_x^0 = \sigma_x(x,0)/\varepsilon_x(x,0)$), characterized by the parameter p, can prove to be negligible for fiberglass composites, but considerable for boron and carbon composites, especially with a small l/b ratio [164].

The reason for reduction in strength with decrease in specimen width is a comparatively strong decrease in the active bearing area, because of partial separation of the polymer matrix along the edges of the specimen (for edge effect, see Section 2.1.4). To eliminate this effect, so-called continuous fiber specimens [62, 243] have been produced. A deficiency of these specimens is the presence of an irregular stress field in the gage section of the specimen. In research practice, these specimens have not been widely used.

The method of experimental determination of the edge effect zone width has been presented in [15]. Strengths $(\sigma_{x1}^u)'$ and $(\sigma_{x2}^u)'$ of two series of specimens of width b_1 and b_2, when $b_2 > b_1$, are determined. Specimens should be selected of such width that average values of strengths $(\sigma_{x1}^u)'$ and $(\sigma_{x2}^u)'$ would differ to a statistically significant degree. The edge effect zone width a is determined by the formula

$$a = \frac{1 - \dfrac{(\sigma_{x1}^u)'}{(\sigma_{x2}^u)'}}{2\left[\dfrac{1}{b_1} - \dfrac{1}{b_2} - \dfrac{(\sigma_{x1}^u)'}{(\sigma_{x2}^u)'}\right]}. \qquad (2.4.3)$$

The strength of the material is determined by the relationship

$$\sigma_x^u = (\sigma_x^u)'\left(1 - \frac{2a}{b}\right) \qquad (2.4.4)$$

where $(\sigma_x^u)'$ is the experimentally determined strength.

To obtain comparable data, specimens of width no smaller than the limiting width b_{lim} must be used. However, at present there are no recommendations for selection of the specimen width of fibrous polymeric composites with various reinforcement materials and layups.

Where wide specimens of materials with high κ values are used, difficulties arise in production of a uniform state of stress over the entire width. Therefore, for studying of materials with reinforcement layup at an angle to the direction of the load, testing of tubes, which can be treated as specimens of infinite width, is recommended. The specifics of testing tubular specimens, as well as possible causes of divergence of the test results for flat and tubular specimens, are examined in Section 3.5.

2.4.3. Processing of Test Results by Plotting Circular Diagrams

Results obtained in testing specimens cut at several angles are usually used for plotting $E_\theta = E(\theta)$, $\nu_\theta = \nu(\theta)$, $\sigma_\theta^u = \sigma^u(\theta)$, etc. When polar coordinates are used, these plots are quite complicated; this hampers comparison of test data with the corresponding graphs plotted according to the formulas of transformation coefficients a_{ij} and B_{ij} under rotation of the coordinate axes.

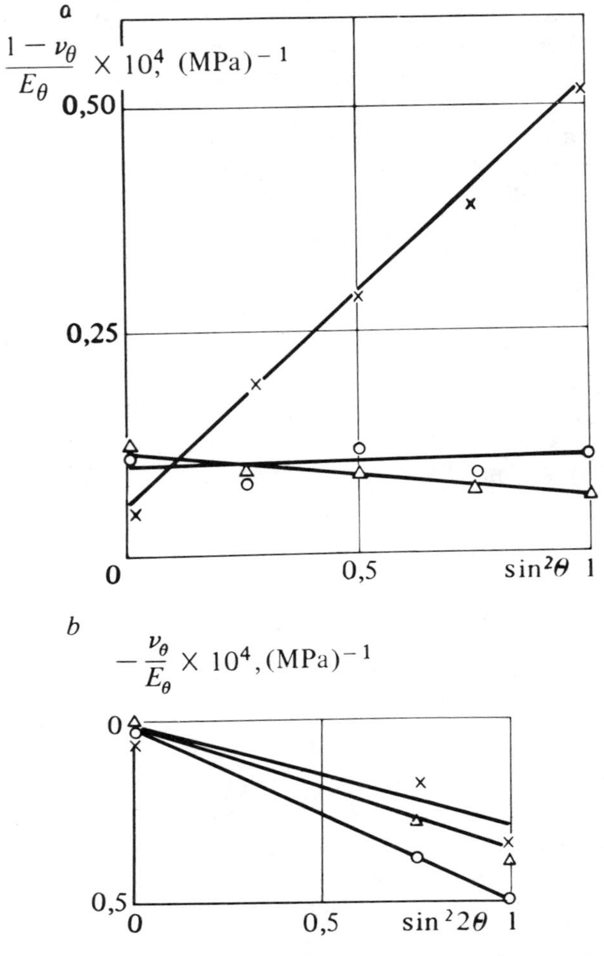

Fig. 2.4.1. Example of processing results of tensile tests at an angle with using coordinates. $\sin^2\theta, (1-\nu_\theta)/E_\theta(a)$ and $\sin^2\theta, -\nu_\theta/E_\theta(b)$. Layup: crosses—0°; open circles—0/90°; triangles—1 : 2.

A more clear representation of the scatter of the experimental results can be obtained by the use of the coordinate systems $\sin^2 2\theta$; $-\nu_\theta/E_\theta$ and $\sin^2\theta$; $(1-\nu_\theta)/E_\theta$, shown in Fig. 2.4.1.

The quantity $-\nu_\theta/E_\theta$ is a linear function of the amplitude $\sin^2 2\theta$, and the quantity $(1-\nu_\theta)/E_\theta$ is a linear function of the amplitude $\sin^2\theta$. Therefore, the scatter of test data can be decided from the density of grouping of experimental points relative to a (least squares) straight line. Analogously, one can act in analysis of the results of determination of the coefficients a_{13} and a_{23}.

The (least squares) coefficients a and b of the linear equation $y = a + bx$ drawn through a set of n experimental points (x_i, y_i), are determined by the relationship

$$a = \frac{s_{xx}s_y - s_x s_{xy}}{ns_{xx} - s_x^2}; \qquad b = \frac{ns_{xy} - s_x s_y}{ns_{xx} - s_x^2} \tag{2.4.5}$$

where

$$s_x = \sum_{i=1}^{n} x_i, \qquad s_y = \sum_{i=1}^{n} y_i; \qquad s_{xx} = \sum_{i=1}^{n} x_i^2, \qquad s_{xy} = \sum_{i=1}^{n} x_i y_i$$

Processing the test results in uniaxial tension or compression of specimens cut at an angle θ to the axis of orthotropy x using this method yields the following expressions for determination of the principal elastic constants:

$$E_0 = \frac{1}{k_1' - k_1}, \qquad E_{90} = \frac{1}{k_1' + k_2' - k_1},$$
$$G_0 = \frac{1}{k_2' + 2k_1' - 4k_1 - 4k_2}, \qquad \nu_0 = \frac{k_1}{k_1 - k_1'}. \tag{2.4.6}$$

where

$$k_1 = \frac{s_{11}s_2 - s_1 s_{12}}{ns_{11} - s_1^2}, \qquad k_2 = \frac{ns_{12} - s_1 s_2}{ns_{11} - s_1^2},$$

$$k_1' = \frac{s_{11}' s_2' - s_1' s_{12}'}{ns_{11}' - (s_1')^2}, \qquad k_2' = \frac{ns_{12}' - s_1' s_2'}{ns_{11}' - (s_1')^2},$$

$$s_1 = \sum_{i=1}^{n} \sin^2 2\theta_i, \qquad s_2 = -\sum_{i=1}^{n} \frac{\nu_{\theta i}}{E_{\theta i}},$$

$$s_{11} = \sum_{i=1}^{n} \sin^4 2\theta_i, \qquad s_{12} = -\sum_{i=1}^{n} \frac{\nu_{\theta i}}{E_{\theta i}} \sin^2 2\theta_i, \tag{2.4.7}$$

$$s_1' = \sum_{i=1}^{n} \sin^2 \theta_i, \qquad s_{11}' = \sum_{i=1}^{n} \sin^4 \theta_i,$$

$$s_2' = \sum_{i=1}^{n} \frac{1 - \nu_{\theta i}}{E_{\theta i}}, \qquad s_{12}' = \sum_{i=1}^{n} \frac{1 - \nu_{\theta i}}{E_{\theta i}} \sin^2 \theta_i.$$

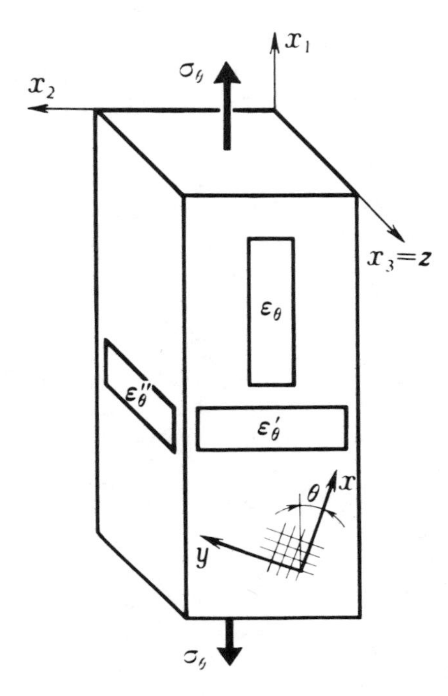

Fig. 2.4.2. Schematic arrangement of strain gages for measurement of longitudinal (ε_θ) and transverse (ε_θ' and ε_θ'') strains in a specimen cut at an angle θ to the reinforcement direction.

Figure 2.4.1 gives examples of processing the test results of a bidirectional (1 : 2) boron composite (reinforcement volume fraction $V_f = 0.57$) by the method presented above. The results obtained for boron composites with a 0° layup ($V_f = 0.37$) and a 0/90° ($V_f = 0.42$) layup are also shown in the figure.

2.4.4. Determination of Elastic Constants from Tests of Specimens of Two Series

Frequently, in order to determine the elastic constants of an orthotropic material, which is necessary for the solution of the plane problem, we are reduced to performing two series of test in uniaxial loading of specimens (Fig. 2.4.2) cut at angles θ_1 and θ_2 ($\theta_1 + \theta_2 \neq 90°$) to the x axis. Once the longitudinal ($\varepsilon_{\theta 1}$ and $\varepsilon_{\theta 2}$) and transverse ($\varepsilon_{\theta 1}'$ and $\varepsilon_{\theta 2}'$) strains and the corresponding stresses $\sigma_{\theta 1}$ and $\sigma_{\theta 2}$ have been measured, the elastic constants can be determined:

$$\frac{1}{E_x}=\frac{L_{\theta 1}p_2-L_{\theta 2}p_1}{p_2-p_1}+4(T_{\theta 1}-T_{\theta 2})p_1p_2(q_2-q_1),$$

$$\frac{1}{E_y}=\frac{L_{\theta 1}(p_2-1)-L_{\theta 2}(p_1-1)}{p_2-p_1}+4(T_{\theta 1}-T_{\theta 2})(p_1+p_2-1-p_1p_2)(q_2-q_1),$$

$$\frac{1}{G_{xy}}=\frac{L_{\theta 1}(2p_2-1)-L_{\theta 2}(2p_1-1)}{p_2-p_1}+\frac{T_{\theta 1}(2p_2-1)-T_{\theta 2}(2p_1-1)}{1-p_1-p_2},$$

$$\frac{v_0}{E_0}=\frac{T_{\theta 1}p_2-T_{\theta 2}p_1}{p_1-p_2},$$

where (2.4.8)

$$L_{\theta 1}=\frac{\varepsilon_{\theta 1}}{\sigma_{\theta 1}},\qquad L_{\theta 2}=\frac{\varepsilon_{\theta 2}}{\sigma_{\theta 2}},$$

$$T_{\theta 1}=\frac{\varepsilon'_{\theta 1}}{\sigma_{\theta 1}},\qquad T_{\theta 2}=\frac{\varepsilon'_{\theta 2}}{\sigma_{\theta 2}},$$

$$p_1=\sin^2\theta_1,\qquad p_2=\sin^2\theta_2,$$

$$q_1=\sin^2 2\theta_1,\qquad q_2=\sin^2 2\theta_2.$$

If the specimen dimensions permit fitting of a strain gage to it for measurement of deformation ε''_θ in the direction of the z axis (as shown in Fig. 2.4.2), the principal elastic constants $v_{zx}=-a_{13}E_x$ and $v_{xy}=-a_{23}E_y$ can be determined in these tests. The following should be determined for this purpose:

$$z_{\theta 1}=\frac{\varepsilon''_{\theta 1}}{\sigma_{\theta 1}},\qquad z_{\theta 2}=\frac{\varepsilon''_{\theta 2}}{\sigma_{\theta 2}} \tag{2.4.9}$$

and, further,

$$a_{13}=\frac{z_{\theta 1}p_2-z_{\theta 1}p_1}{p_2-p_1},\qquad a_{23}=\frac{z_{\theta 1}(p_2-1)-z_{\theta 2}(p_1-1)}{p_2-p_1}.$$

Frequently, the values $\theta_1=0°$, $\theta_2=45°$ are chosen. In this case,

$$E_x = \frac{1}{L_0} = \frac{\sigma_0}{\varepsilon_0}, \qquad \nu_{yx} = -\frac{T_0}{L_0} = -\frac{\varepsilon'_\theta}{\varepsilon_0}, \tag{2.4.10}$$

$$G_{xy} = \frac{1}{2(L_{45} - T_{45})} = \frac{E_{45}}{2(1 + \nu_{45})},$$

$$E_y = \frac{1}{2L_{45} + 2T_{45} - L_0 - 2T_0} = 2E_x(1 - \nu_{45}) - E_{45}(1 - 2\nu_{yx}),$$

$$\nu_{zx} = -\frac{z_0}{L_0} = -\frac{\varepsilon''_0}{\varepsilon_0},$$

$$a_{23} = 2z_{45} - z_0.$$

2.5. RESISTANCE TO INTERLAMINAR TENSION

Resistance to interlaminar tension of fibrous polymeric composites is experimentally studied on flat and ring specimens. There are several types of flat specimens for studying resistance to interlaminar tension. The simplest specimen consists of three parts: two metallic end pieces, the material being tested glued between them [Fig. 2.5.1(a)], where the reinforcement plane of the material is perpendicular to the direction of the tensile forces p^t. Resistance to interlaminar tension is determined by the formula

$$\sigma_z^{tu} = \frac{p^{tu}}{F} \tag{2.5.1}$$

where p^{tu} is the breaking load of the specimen due to interlaminar tension and F is the cross-sectional area of the specimen measured perpendicular to the direction of the load.

The principal deficiency of this test technique is the concentration of the interlaminar stresses in the material near the free edges of the specimen. It is possible to partially reduce stress concentration by cutting off the edges of metallic end pieces adjacent to the specimen and by filling the space with a polymeric binder [Fig. 2.5.1(b)]. This type of test is applicable only to materials in which the resistance to interlaminar tension is lower than the strength of adhesive joints between the specimen and the metallic end pieces.

A specimen with a circular recess (Fig. 2.5.2), i.e., with a preset gage section, is more reasonable than that just described. Specimens of this type can also be glued to two metallic end pieces; however, in that case,

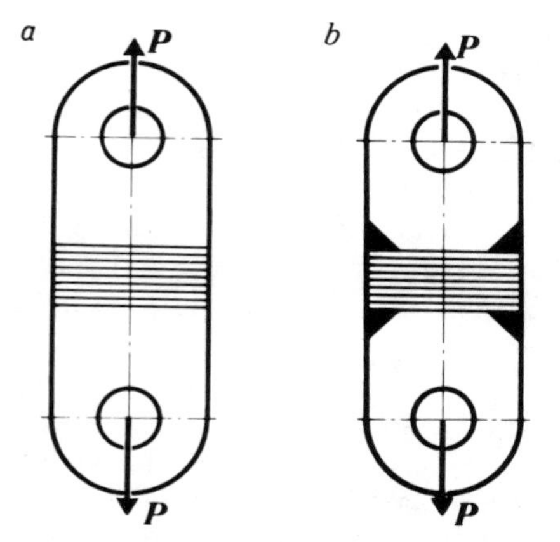

Fig. 2.5.1. Flat specimens for determination of the interlaminar tensile strength: (a) force transmission over the entire specimen length; (b) weakened boundary conditions.

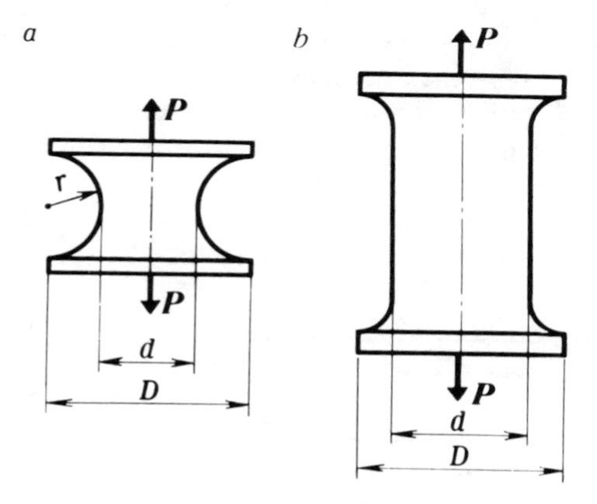

Fig. 2.5.2. Specimens for determination of interlaminar tensile strength: (a) with a circular groove; (b) with a cylindrical middle part.

when the thickness of the material being tested allows, it is more reasonable to prepare a specimen and the end pieces for fastening as a single unit.

The gage length of a specimen can be marked by a relatively narrow circular groove [Fig. 2.5.2(a)] or it can have a cylindrical segment of constant cross section [Fig. 2.5.2(b)]. In both cases, stress concentration takes place, which depends on the ratios r/d and D/d (for notation, see Fig. 2.5.2). For bars with a circular recess made of anisotropic materials, coefficients of stress concentration cannot be determined analytically; for bars with a circular recess made of isotropic materials, coefficients of stress concentration can be found, for example, in [177]. The actual resistance to transverse tension is determined as the product of resistance σ_z^{tu} calculated by the formula (2.5.1), and coefficient of concentration k.

A more theoretically complicated method of experimental determination of resistance to interlaminar tension in the testing of flat specimens has been suggested in [90, 165]. A strip specimen was tested (in [90] strips of 127 × 25 × h with metallic tabs were used), in which the reinforcement was laid in such a manner that under the axial tensile forces the specimen delaminated due to action of tensile interlaminar normal stresses σ_z^t. The complexity of realization of this test technique resides in the selection of the necessary reinforcement layup, i.e., in producing the necessary state of stress in the specimen. In [165] the fiber layup was $[\pm\theta/90°]_s$, in [90] the layup was $[+\theta/-\theta_2/+\theta/90°]_s$ with the angle θ equal to 25–26°. However, at the present time one problem is still unsolved: the specimens fail as a result of joint action of the interlaminar normal stresses σ_z^t and interlaminar tangential stresses τ_{yz} on the boundary between packets of reinforcement with a layup 90° and $\pm\theta$, i.e., in interlayers where interlaminar tangential stresses τ_{yz} are maximum, but not in the interlayer with coordinate $z = 0$, where stresses σ_z^t are maximum. The cause of these unsatisfactory results is a very coarse modeling of the distribution of interlaminar normal stresses σ_z through the specimen width.

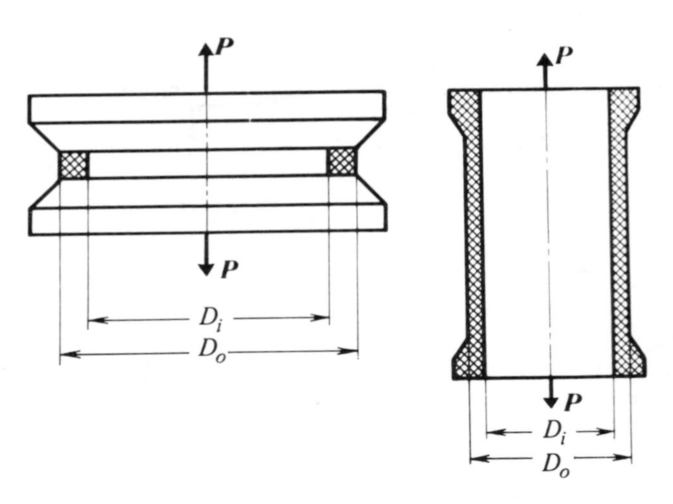

Fig. 2.5.3. Loading scheme of rings for determination of interlaminar tensile strength.

Fig. 2.5.4. Loading scheme of a cylindrical specimen for determination of interlaminar tensile strength.

Resistance to interlaminar tension is determined on ring specimens in the same manner as on flat ones. A ring with carefully machined side surfaces is glued to two metallic disks, which are extended as shown in Fig. 2.5.3. The method is applicable only to rings, wound of filaments or roving.

The method of tension of a cylindrical specimen with a preset gage length of a reduced cross section (Fig. 2.5.4) is more reliable, but less economical. In both cases, the method has the same disadvantages as in testing of flat specimens—stress concentration near the free edges of the specimen or near the reduced cross section. Resistance to interlaminar tension is determined by the formula (2.5.1), by substitution of $F = (\pi/4)(D_0^2 - D_i^2)$ (for notation, see Fig. 2.5.3 and 2.5.4). The method of determination of resistance to interlaminar tension in bending of rings is treated in Section 5.5.3.

2.6. TENSION OF RING SPECIMENS

2.6.1. Tension by Split Disks

In-plane tension of ring specimens is accomplished under internal pressure, which is produced by means of a rigid split disk [the so-called NOL ring method, Fig. 2.6.1(a, b)], by a flexible ring [Fig. 2.6.1(c)], or hydraulically [Fig. 2.6.1(d)]. Besides the loading schemes, the values to be determined in the experiment, force factors and variable geometric dimensions of specimens are also presented in Fig. 2.6.1.

The method of tension by means of a split disk has been standardized in the USA (ASTM D 2290-76) and it is widely used the world over. It was developed initially (in 1955) in the US Naval Ordnance Laboratory (NOL) for evaluation of the effect of methods of chemical treatment of glass fiber roving on the strength of glass fiber composites. The following ring dimension were established by the ASTM standard:

INSIDE DIAMETER, mm	WIDTH, mm	THICKNESS, mm
146.05 ± 0.05	$6.35 + 0.13$	1.52 ± 0.05
146.05 ± 0.05	$6.35 + 0.13$	1.52 ± 0.25
146.05 ± 0.05	$6.35 + 0.13$	3.18 ± 0.05

The cross-sectional dimensions of specimens must be examined somewhat critically. For example, studies of unidirectional epoxy fiberglass composites show that the specimen width plays a greater role in determination of the strength than in tests of flat specimens. Thus, with a winding angle of 86° (i.e., almost in the plane of the ring), the specimen width should be at least 15–20 mm, while at 55° it should be at least 80–120 mm. With smaller specimen widths, the strength turns out to be understated [73]. The measured

			a, b	c	d
Loading scheme			a b	c	d
Determinable characteristics			$E_{\theta}^{t}, \sigma_{\theta}^{tu}, \tau_{\theta r}^{u}$,	$E_{\theta}^{t}, \sigma_{\theta}^{tu}, \tau_{\theta r}^{u}$,	$E_{\theta}^{t}, \sigma_{\theta}^{tu}, \tau_{\theta r}$
Measurable values			$P, \Delta u, \epsilon_{\theta}^{t}$	p, ϵ_{θ}^{t}	p, ϵ_{θ}^{t}
Geometrical sized			$R/h, b, h$	$R/h, b, h$	$R/h, b, h$
Limitations	Structural	Layup	0°; 90°; 0/90°		
		Orientation	0°, 90°		
	Physical		For ϵ_{θ} : linear range of the curve $P \sim \epsilon_{\theta}$ or $p \sim \epsilon_{\theta}$		
	Geometrical		For σ_{θ}^{tu} : $0.08 \leqslant h/R \leqslant 0.18$ (for GFRP) $L \geqslant \frac{2}{3} R$	For σ_{θ}^{tu} : $0.08 \leqslant h/R \leqslant 0.18$ (for GFRP)	

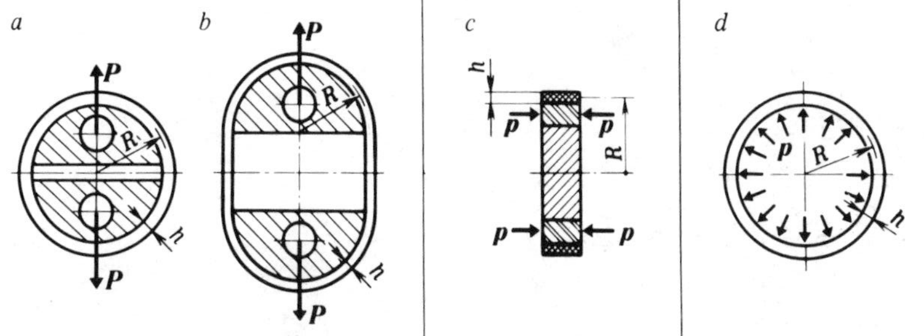

Fig. 2.6.1. Loading scheme of ring specimens under tension by means of: (a and b) rigid split disks; (c) a compliant ring; (d) hydraulic test.

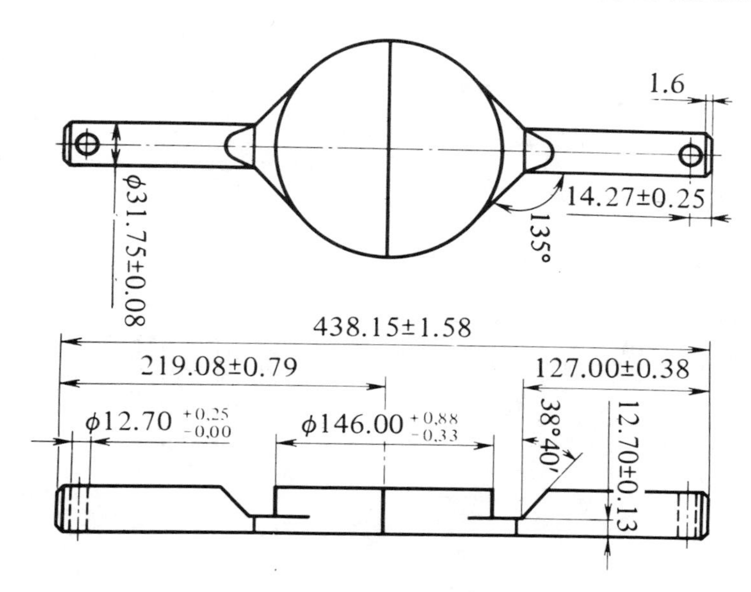

Fig. 2.6.2. Device for testing rings by means of rigid split disks according to ASTM D 2290-76 [11].

strength of rings is affected not only by the cross-sectional dimensions of the ring, i.e., the scale and edge effects, but also by the thickness-to-radius ratio h/R.

At the present time, the NOL method is used for determination of the modulus of elasticity, circumferential tensile strength, and shear strength. The fixture used for this purpose according to ASTM D 2290-76 is shown in Fig. 2.6.2; simpler variants are also possible (Fig. 2.6.3). In selection of the fixture, however, it must be taken into account that the effect of incorrect installation of the fixture decreases with increasesing distance between the fastening hinges in the testing machine. Because of rather limited size of the device its geometrical dimensions must be determined correctly.

In testing, a ring made by winding on a multi-section mandrel or cut from a cylinder is mounted on the split disk, which is installed in the testing machine by means of a fixture. This same fixture is used to test rings manufactured by winding on a split mandrel (split disk). Before a ring is mounted, its contact surfaces and the fixture are lubricated or friction-reducing inserts are used (for example, fluoroplastic bands). Experience has shown that the effect of friction in the determination of ring strength is negligibly small in the presence of, for example, graphite lubricant. The rate of movement of the testing machine grips according to ASTM D 2290-76 is 2.5 mm/min.

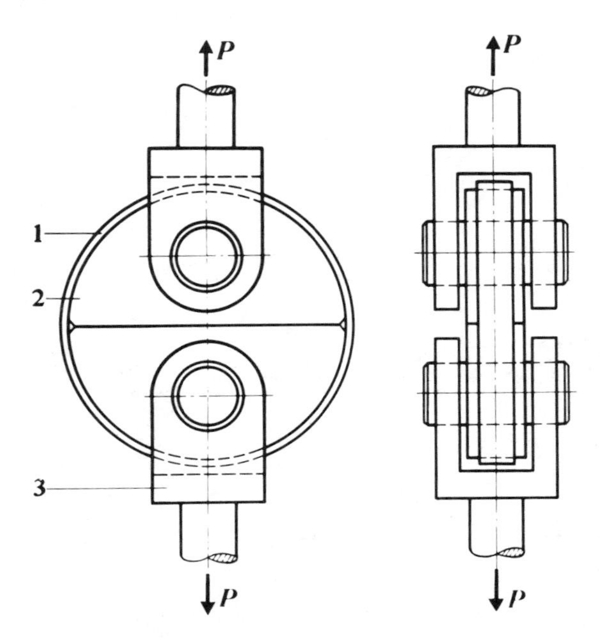

Fig. 2.6.3. Simplified device for testing rings using rigid split disks: (1) specimen; (2) split disk; (3) pull rod.

Split disks may also be hydraulically driven (see Fig. 2.6.4), in which case no testing machine is needed.

There are several ways of measuring deformations to determine the modulus of elasticity E_θ^t. The simplest method consists in the measurement of the increase in the gap between the half-disks. In this case, the modulus of elasticity E_θ^t is calculated by the formula

$$E_\theta^t = \frac{\Delta P}{2bh} \frac{\pi D}{2\Delta u} \tag{2.6.1}$$

where b is the ring width; h is the thickness; D is the average diameter; and Δu is the change in the distance between the half-disks with a load increase ΔP.

The initial distance between half-disks must be minimal (it is desirable to have all-round contact of the split disk), since with appearance of a gap between half-disks the corresponding section of the specimen not only stretches, but also deflects; the formula (2.6.1) does not allow for this peculiarity of deformation of the specimen.

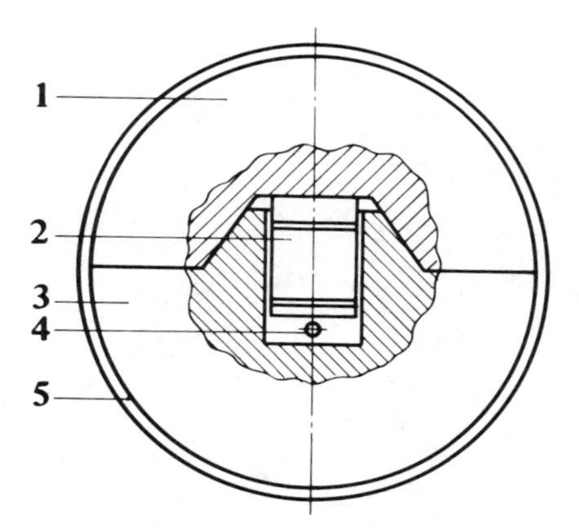

Fig. 2.6.4. Hydraulic drive of the device for testing rings by rigid split-disks [60]: (1 and 3) half-disks; (2) piston; (4) opening for liquid feed; (5) specimen.

The circumferential strain ε_θ^t can be measured by means of resistance strain gages glued to the outer surface of the ring.* In this case, attention must be focused on selection of the location of the resistance strain gages. Due to incomplete contact between the specimen and split disks, the effects of friction, and local deformations of the specimen near the splits in the disks, distribution of the circumferential strains of the ring is uneven. A characteristic example of uneven circumferential strain distribution is shown in Fig. 2.6.5, which shows that the degree of uneveness changes with the load and is especially evident in regions near the edges of the half-disks. It follows from this figure and from other research that it is most reasonable to glue resistance strain gages to sections loaded at an angle $\theta = 30$–$45°$ to the midplane of the split disk. Bonding of resistance strain gages is reasonable in all four quadrants of a specimen with subsequent averaging of their recordings. In this case, the modulus of elasticity is obtained by the formula

$$E_\theta^t = \frac{P}{2bh} \frac{1}{\varepsilon_\theta^t} \tag{2.6.2}$$

where P is the load and ε_θ^t is the mean strain measured by resistance strain gages.

The circumferential modulus of elasticity E_θ^t can also be determined in

* Resistance strain gages are the most widely used sensors of deformation. However, it has been shown in [263] that resistance strain gages are subject to greater error than, wire sensors wound together with the reinforcement. This may be caused by nonuniform deformability of the specimen and unreliable bonding of resistance strain gages.

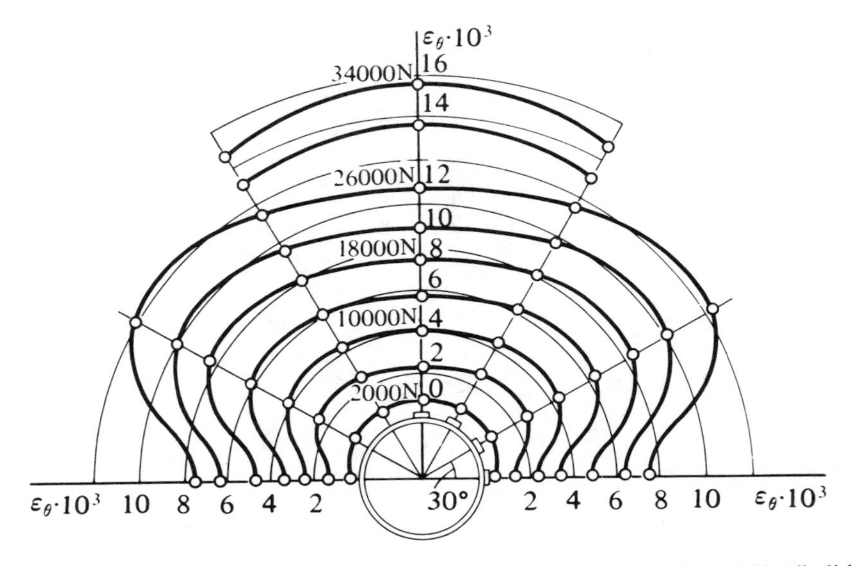

Fig. 2.6.5. Strain distribution ε_θ over the circumference of a ring in tests with a rigid split disk [115].

tests of an elongated ring specimen [Fig. 2.6.1(b)]. In this case, the strain ε_θ^t is measured by means of resistance strain gages which are glued in the middle of the two straight sections of the specimen. Since the distance between half-circumferences is rather large (it must be at least $\frac{2}{3}R$), the distribution of strains in straight sections of the specimen is uniform. The modulus of elasticity E_θ^t is calculated by the formula (2.6.2).

In determination of the strength σ_θ^{tu} the specimen is loaded with a given speed to failure and the breaking load P^{tu} is fixed. The strength is determined by the formula:

$$\sigma_\theta^{tu} = \frac{P^{tu}}{2bh}. \tag{2.6.3}$$

However, the strength calculated by (2.6.3) is only an apparent characteristic of the material. In tension of the specimen a gap is formed around the midplane of the split disk, and the specimen bends in this section. As a consequence of the change in radius of curvature of the ring, a concentration of stresses occurs. In testing of NOL rings [117], it has been stated that concentration of radial tensile stresses σ_r^t exerts a negligible effect, whereas interlaminar shear stresses $\tau_{\theta r}$ at specimen failure, when $\sigma_\theta^t \to \sigma_\theta^{tu}$, can exceed

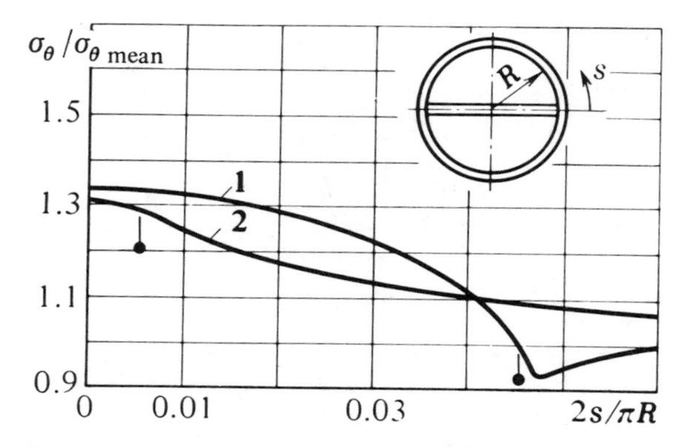

Fig. 2.6.6. Distribution of relative circumferential stresses $\sigma_\theta/\sigma_{\theta\,\mathrm{mean}}$ on the inner ring surface [117]: (1) glass fiber; (2) carbon composite. The point with a vertical line designates the coordinate of the edge of the half-disk.

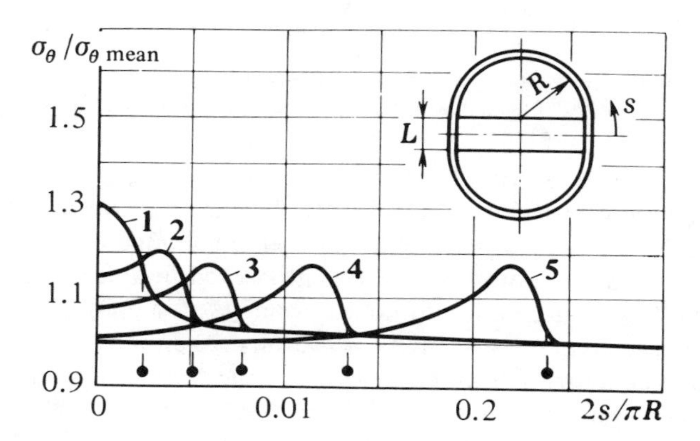

Fig. 2.6.7. Distribution of relative circumferential stresses $\sigma_\theta/\sigma_{\theta\,\mathrm{mean}}$ of an elongated ring specimen of organic composite (Kevlar 49), depending on the distance L between half-circumferences [117]. The points with vertical lines designate the coordinates of an edge of the half-disk.

the ultimate value for the material. As a result of bending, the inner material layers are overloaded and the outer layers underloaded. Curves of variation in relative circumferential stresses $\sigma_\theta^t/\sigma_{\theta\,\mathrm{mean}}^t$ on the inner (contact) surface of the specimen, calculated in [117] for epoxy glass fiber and carbon composites, are shown in Fig. 2.6.6. It follows from this figure that distribution of circumferential stresses depends very little on the material of specimens and for NOL rings ($R/h = 48.5$)$\sigma_\theta^t/\sigma_{\theta\,\mathrm{mean}}^t \leqslant 1.35$.

It can be seen from Fig. 2.6.7, in which the curves are plotted for epoxy

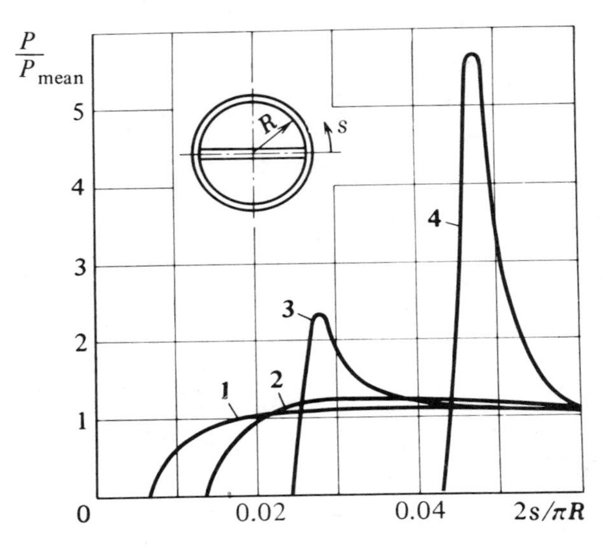

Fig. 2.6.8. The relative contact pressure p/p_{mean} between the ring and the split disk [117]: (1, 2) carbon composites of two types; (3) organic composite; (4) glass fiber composite.

organic composite (Kevlar 49), that the ratio $\sigma_\theta^t/\sigma_{\theta\,mean}^t$ decreases somewhat with an increase in the gap between the half-disks. Consideration of this condition has resulted in development of an elongated specimen [Fig. 2.6.1(b)]. It follows from Fig. 2.6.7 that it is reasonable to select a length of the middle straight section of no less than $\frac{2}{3}R$; for NOL rings, the length is 50 mm. However, impersections in such specimens, which are difficult to eliminate, frequently reduce to zero the advantage of a uniform state of stress in the specimen.

Concentration of stresses in NOL rings attenuates at a distance of about $0.1R$ from the midplane of the split disk. The contact pressure between the disk and the specimen changes in this section. It follows from Fig. 2.6.8 that the variation in contact pressure essentially depends on the material of the specimen. In Fig. 2.6.8 are presented the theoretical curves of variation in relative contact pressure p/p_{mean} near the midplane of the split disk (p is the internal pressure which would cause a circumferential stress σ_θ^t in the specimen equivalent to that formed in tension by the split disk). Materials for which the unsupported section of a specimen clings to the split disk (glass fiber, organic composites), i.e., low-modulus composites, are characterized by a sharp increase in relative contact pressure at a distance from the midplane of the split disk (Fig. 2.6.8). This can contribute to specimen failure [117]. An unsupported carbon composite (i.e., high-modulus composite) specimen section moves away from the half-disks and as a consequence contact pressure at the midplane of split disk does not practically change (Fig. 2.6.8).

In practice, it has been established [115, 117] that tests of NOL rings, i.e., thin rings, by means of split disks give strength values σ_θ^{tu} 5–20% lower than do hydrostatic tests; for epoxy composites, for example, the following strength values have been obtained [117]:

REINFORCEMENT	TENSILE STRENGTH σ_θ^{tu}, MPa, USING	
	SPLIT DISKS	HYDROSTATIC METHODS
S glass	1972	2330
Kevlar 49	1834	1737
Thornel 75 S	910	937
Thornel 400	1261	1461

The phenomenae, described above gain force [117] with the increase in the relative thickness h/R, the anisotropy and elongation at rupture of the material. Obviously, the failure mode will also change. For thin rings ($h < 2$ mm) separation into layers and helical unwinding are characteristic; in such cases, evaluation of the material strength is impossible.

It has been established upon more general study of the state of stress in a ring [172–174] that the stresses σ_θ and $\tau_{\theta r}$ do not practically depend on the angle θ, except in the region around the midplane of split disk, i.e., at angles near to $\theta = 0$. The length of this region depends on the relative ring thickness, characterized by the value $\bar{r}_1 = R_i/R_0$ and the elastic characteristics of the material. Outside this region, the tangential stresses are zero. Close to section $\theta = 0$ the tangential stresses change very sharply both with coordinate θ and with radius R. With the increase in the relative thickness h/R and decrease in anisotropy $E_\theta/G_{\theta r}$ the stress concentration in this region increases. A change in anisotropy E_θ/E_r has a lesser effect on the maximum values of the normal and tangential stresses. The appearance of a gap between the ring and the split disk (mutual withdrawal of the contact surfaces) decreases stress concentration [174] somewhat.

The distribution of the circumferential stresses σ_θ in the $\theta = 0$ section differs essentially from constant, which is characteristic of sections, remote from the section $\theta = 0$. Concentration of the circumferential stresses can be estimated by means of the coefficient of concentration [172]:

$$k = \sigma_{\theta\,\max}/\sigma_0$$

where

$$\sigma_0 = \frac{1}{1 - \bar{r}_1} \int_{\bar{r}_1}^{1} \sigma_\theta \, dr. \tag{2.6.4}$$

A gap formed between the half-disks during loading has little effect on the state of stress in the ring [172].

Numerous efforts have been made toward improvement of this method; they have principally been aimed at achieving uniformity of load. Thus, to eliminate bending strain on the inside surface of the ring, notches are made at the midplane [201, p. 281; 225]. This ensures failure of the ring in a specific section, and the effect of bending is decreased. However, uncontrollable stress concentration is introduced, and the possibility of failure of the ring due to interlaminar shear increases. Efforts have also been made to replace the split disk with several rigid sectors. Fig. 2.6.9 shows a device consisting of four sectors (3), connected by means of pull rods (2) and pins (5) to balance beams (1), which in turn are connected by means of pins (6) to forks (7), which are inserted into the testing machine grips. With the use of this device, a sharp change in the circumferential strains is observed close to the sections with coordinates $\theta = 0$ and $\theta = \pi/2$, but this change is less than that exhibited under tension by split disk. In this case, it is recommended that the resistance strain gages for measurement of ε_θ^t be placed in sections with coordinates $\theta = \pi/4$.

The advantages of the four-sector device are discovered in strength tests, since more accurate results can be obtained because of a decrease in stress concentration. However, the complexity of fabrication of the multisector devices and their operation is not always compensated by this advantage [176].

The circumferential stresses σ_θ^t in the case of use of multisector devices are determined by the formula (for notation see Fig. 2.6.10; n is the number of sectors in the semi-circumference) [155]:

$$\sigma_\theta^t = \left\{1 - \frac{2\pi}{n}\left(\frac{3G_{\theta r}}{E_\theta}\right)^{1/2}\left[\frac{\cosh(\lambda_2\theta)}{\sinh\left(\frac{\pi\lambda_2}{n}\right)} - \frac{2G_{\theta r}}{(E_\theta/E_r)^{1/2}} \frac{\cosh(\lambda_1\theta)}{\sinh\left(\frac{\pi\lambda_1}{n}\right)}\right]\frac{z}{h}\right\}\frac{P}{2bh} \tag{2.6.5}$$

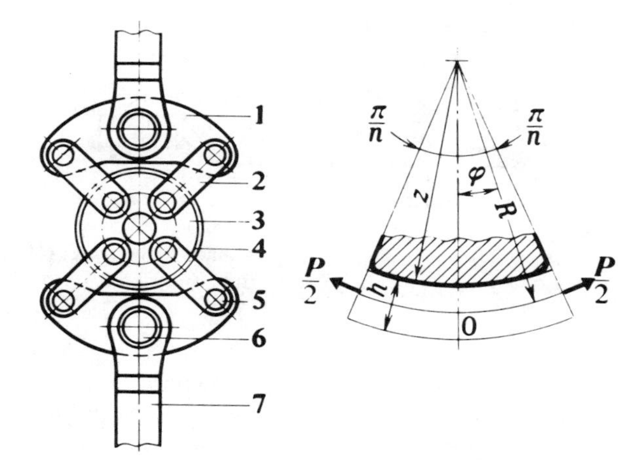

Fig. 2.6.9. A four-sector device for tensile loading of a ring [176]: (1) balance beam; (2) pull rod; (3) rigid sector; (4) ring specimen; (5 and 6) pins; (7) fork.

Fig. 2.6.10. Design of n-sector device for tensile tests of rings [155].

where

$$\lambda_1 = \left(\frac{3E_r}{G_{\theta r}}\right)^{1/2} \frac{R}{h}, \qquad \lambda_2 = \alpha\lambda_1, \qquad \alpha = \frac{2G_{\theta r}}{(E_r E_\theta)^{1/2}}.$$

The actual stress at failure of the ring is determined, with allowance for the concentration coefficient, by [155]

$$\sigma_{\theta\,\max} = \kappa\sigma_{z=0} = \kappa\frac{P}{2bh} \tag{2.6.6}$$

where

$$\kappa = 1 + \frac{\pi}{n}\left(\frac{3G_{\theta r}}{E_\theta}\right)^{1/2}\left[1 - \frac{2G_{\theta r}}{(E_r E_\theta)^{1/2}}\right].$$

2.6.2. Loading by Means of a Compliant Ring

Loading schemes where the working medium which causes a uniform pressure on the inner surface of the ring is rubber or liquid have found wide application. These methods are intended for determination of the modulus

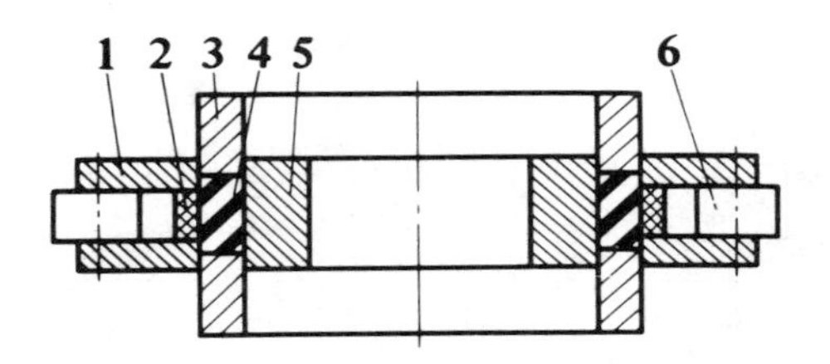

Fig. 2.6.11. Device for loading a ring by means of an inner rubber ring: (1) outer disk; (2) specimen; (3) pressure ring; (4) rubber ring; (5) inner ring; (6) spacer.

of elasticity and strength of the material in the circumferential direction. The phenomena connected with bending of a ring on the midplane between the rigid half-disks is eliminated by the use of uniform internal pressure. Therefore, the error associated with these methods depends less on the relative thickness of a ring h/R and the anisotropy of the material.

In loading by a compliant ring, the pressure on the specimen is created by means of compression of the rubber ring* in a closed space (Fig. 2.6.11); to eliminate friction, the contact surfaces are lubricated or special inserts are used. The compressive forces applied to the end faces of the rubber ring and the circumferential or radial strains are measured in the experiment. In order to avoid errors** in conversion of the compressive force, the device is calibrated by means of a steel ring (with resistance strain gages) substituted for the test item, in order to determine the contact pressure. Strains ε_θ^t are determined by means of resistance strain gages and the radial displacements are measured by means of a deformation convertor. The modulus of el .sticity is calculated by the formula

$$E_\theta^t = \frac{pD_i}{2h\varepsilon_\theta^t} \tag{2.6.7}$$

where p is the specific pressure created by the rubber on the inner surface of the ring, D_i is the inside diameter of the ring, and ε_θ^t is the strain measured on the outer surface of the ring or by means of a displacement convertor.

The strength is determined by the formula

*Working medium in the form of a disk is also used; in this case, the required testing machine power is considerably higher.

**For technical rubbers, Poisson's ratio is less than 0.5; this introduces an error in the analytical determination of pressure p. It should be taken into account also that the mechanical characteristics of rubbers change with time.

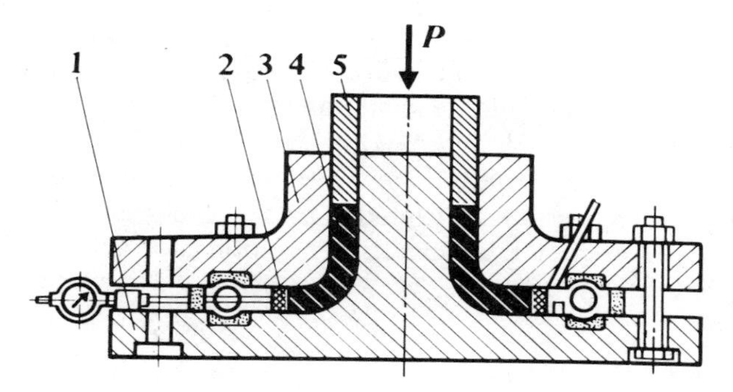

Fig. 2.6.12. Device for tensile loading of ring specimens by means of a profiled ring [132]: (1) base; (2) specimen; (3) cover; (4) profiled ring; (5) sleeve.

$$\sigma_\theta^{tu} = \frac{p^u D_i}{2h} \tag{2.6.8}$$

where p^u is the specific pressure at failure of the ring.

In this device calibration the specific pressure is determined by the formula

$$p = \frac{1}{2}\varepsilon_\theta^t E_{st}\left(\frac{R_o^2}{R_i^2} - 1\right) \tag{2.6.9}$$

where R_o and R_i are the outside and inside radii, respectively, of the steel ring; ε_θ^t is the strain determined by resistance strain gages on the outer surface of the steel ring; and E_{st} is the modulus of elasticity of steel.

Loading by means of a rubber ring, which permits the attainment of high specific pressures (about 200 MPa and higher), requires very careful preparation of the inner surface of the specimen and a very well designed device: the rubber is capable of entering the tiniest irregularity in the surface of the specimen. This leads to the development of axial forces which can cause delamination of the ring.

A modified loading process by means of a compliant ring is accomplished in the device described in Reference [132]. In this device (Fig. 2.6.12), loading is transmitted to the specimen through a shaped ring made of highly elastic polyurethane. The method of calibration and technical facilities of the device are described in [132].

2.6.3. Hydrostatic Loading

Hydrostatic tests, in which the internal pressure is created by a fluid, are free of the deficiencies inherent in the previously described methods. A flexible insert (of rubber or similar flexible material) is placed inside the test ring. The pressure p is exerted by the hydraulic system. The pressure and the circumferential strain (or radial displacement of the outer surface) are measured. The modulus of elasticity E_θ^t and strength σ_θ^{tu} of the material are determined by formulas (2.6.7) and (2.6.8). In this case, the greatest accuracy in processing the test results is ensured. The basic deficiency of the method is the necessity of a special, comparatively complicated and expensive equipment to produce the pressure.

Chapter 3
Compression Testing
Testing of Tubular Specimens
Bearing Tests

3.1. COMPRESSION OF FLAT SPECIMENS

3.1.1. Basic Relationships

Compression testing of fibrous polymeric composites, especially in the direction of the reinforcement, is a widespread form of testing. The apparent simplicity of loading, measurement of the load and deformation, and the simplicity of the analytical apparatus are attractive. Compression is a very widespread form of deformation in structures. Therefore, correct procedure is no less important in a compression test than in a tensile test.

At the present time, there are only a few standards (ASTM D 3410-75, GOST 25.602-80) pertaining to compression testing of fibrous polymeric composites which hold good for unidirectional and orthogonally reinforced fiber glass, boron, and carbon composites. The use of respective standards intended for rigid plastics (GOST 4651-78, ASTM D 695-69, DIN 53457, etc.) is unreasonable, since these standards do not take into account the specific features of fibrous polymeric composites.

In general, all the basic relations concerning tensile testing of fibrous polymeric composites (see Section 2.1.3) in the directions of the principal axes of elastic symmetry of the material are retained in compression, with due consideration taken of the direction of deformation. At the same time, tensile and compression tests of fibrous polymeric composites differ qualitatively more than respective tests of isotropic materials. Along with specific features, described in Sections 2.1.4 and 2.3.1, the behavior of an anisotropic body under tension or compression at various angles to the principal axes of elastic symmetry of the material, interlaminar stresses and the effect of structural symmetry under compression) there are peculiarities dictated by the mode of loading and material structure. These peculiarities must be taken into account in selecting the specimen shape and dimensions and the method of loading, and in evaluation of material strength, and of the failure mode. The effect of these peculiarities increases with increase in the anisotropy of the material. This is treated in more detail in the subsequent sections of this chapter.

3.1.2. Peculiarities of Deformation

In the general case, a compression stress-strain curve of fibrous polymeric composites is nonlinear. However, in composites reinforced with rigid fibers (glass fiber, boron, and carbon composites), in loading in the directions of reinforcement the stress-strain curve is practically a straight line (with the usual measurement errors) from the start of loading up to failure of the material. Sometimes, characteristic breaks are also observed in the compression stress-strain curve just as intensile tests (see Fig. 3.1.2). If such a break appears shortly before failure of the material, it is explained by a loss of stability of part of the reinforcement. At lower loads a break in the compression stress-strain curve develops as a consequence of inaccuracy in fabrication and installation of the specimen in the testing machine, and also as a consequence of restriction of its deformation. In loading perpendicular to the plane of reinforcement layup, a physical nonlinearity is found in the stress-strain curve; its degree is determined by the properties of the polymer matrix. Nonlinearity of the curve is also observed under compression of unidirectional composites in the reinforcement direction as a consequence of poor adhesion of reinforcement and matrix (Fig. 3.1.1), when failure of the material starts with disturbance of interface bond long before reaching the ultimate load.

Studies of the deformed state of prismatic specimens under compression [187, 222] show that the longitudinal and transverse strains at the supported surfaces considerably exceed the strains in the middle section of the specimen. Stresses corresponding to these strains can exceed the resistance of fibrous polymeric composites in the transverse direction. This will cause premature failure of the specimen. High-strength unidirectional composites are particularly sensitive to the effect of such stress concentration owing to their high strength anisotropy. Because of the nonuniformity of the strain field, the numerical values of the modulus of elasticity under compression obtained during tests of short specimens can differ appreciably from the values of the modulus of elasticity under tension. This is not a consequence of material properties, but rather is caused by experimental errors and incorrect selection

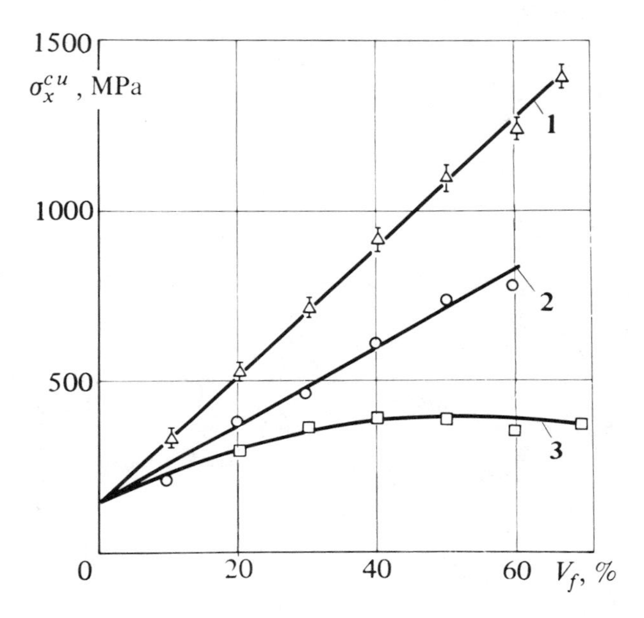

Fig. 3.1.1. Effect of carbon fiber type and surface treatment on compression diagram [87]: (1) type 2, treated; (2) type 1, treated; (3) type 1, untreated.

of the specimen dimensions (Fig. 3.1.2). This error is particularly significant in those cases where the modulus of elasticity is calculated, not from measurements in the gage length of the specimen with a uniform deformation field over its entire length, but from measurements of the approach of the supporting surfaces of the testing machine to which the compression force is applied.

In compression of fibrous polymeric composites with asymmetrical reinforcement layup it is impossible to exclude all restriction of deformation, since at least one of the support surfaces in the testing machine remains stationary relative to the longitudinal axis of the specimen. Therefore, testing of fibrous polymeric composites with an asymmetrical reinforcement layup under compression is unreasonable.

3.1.3. Modes of Failure

Failure modes of fibrous polymeric composites under compression depend on the reinforcement scheme of the material, the mechanical characteristics of its components, the relative specimen dimensions (which determine the total loss of stability), the direction of load relative to the reinforcement,

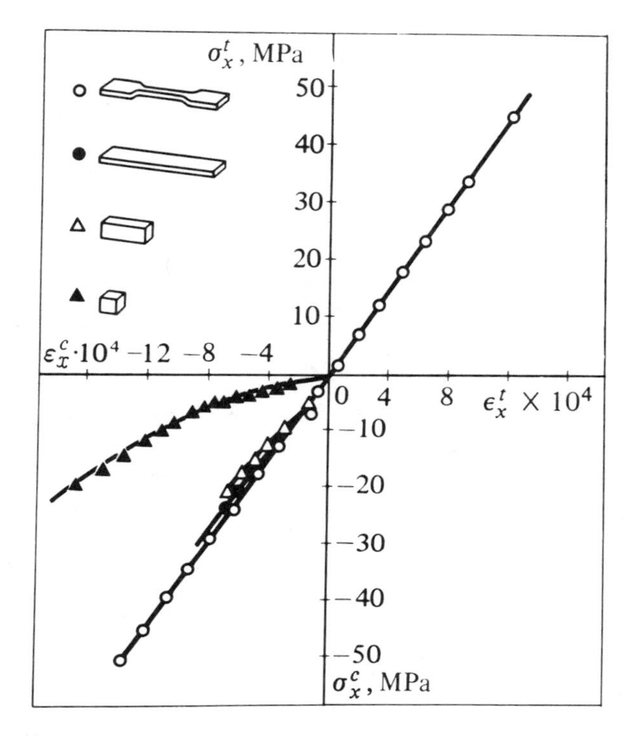

Fig. 3.1.2. Tension (t) and compression (c) diagrams of specimens of various shapes made of glass cloth/polyester composite [233]. *Open circles*: Specimen according to Federal Specification L-P-406 with reduction in cross section of middle part. *Filled circles*: Specimen according to Federal Specification L-P-406. *Open triangles*: Bonded specimen with balsa filler (ASTM D 695-54). *Filled triangles*: Bonded cube of thin sheets according to DIN 53454.

and variations introduced in production of the bulk material and fabrication of the specimen.

In compression of unidirectional composites in the direction of reinforcement, depending on the stiffness of the matrix, three basic failure modes of specimens can be observed [80]. For composites with a matrix of low-modulus binders (modulus of elasticity $E_m = 15$–25 MPa), local buckling of fibers is the critical mode of failure. In this case, production variations (initial waviness of reinforcing fibers, nonuniform reinforcement spacing) markedly affects the strength of the material under compression. In testing of composites with a matrix of medium stiffness ($E_m = 200$–700 MPa), transverse rupture of the material is due to difference between the Poisson's ratios of the material components and to nonuniform distribution of transverse strains over the specimen length. In this case, eventual failure of the

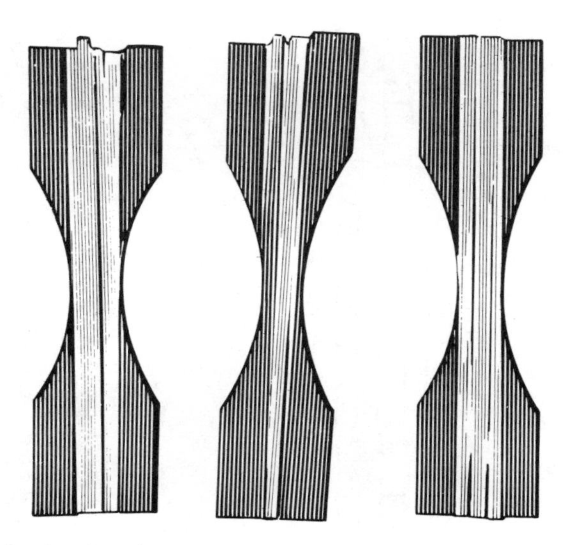

Fig. 3.1.3. Delamination of specimens of unidirectional glass fiber composite under compression.

specimen can also result from local buckling of the fibers, but the critical load will be determined by the characteristics of the already partly distorted material. Finally, in testing unidirectional composites with a rigid matrix ($E_m > 2000$ MPa), the material will fail in compression (shearing of the fibers at an angle of 45°) without any local buckling of the reinforcement.

If the load is directed perpendicular to the direction of the reinforcement, the material usually will fail because of loss of stability at stresses which correspond to the shear strength of the matrix. In this case, the specimen breaks into prismatic pieces.

The modes of failure described above are the principal ones. They can be accompanied by a series of other phenomena—by inelastic and nonlinear behavior of the fibers and especially of the matrix, interlaminar stresses, surface ply separation, overall loss of stability, destruction of the end faces, splitting across the layer (Fig. 3.1.3). Diverse combinations of all these phenomena can make it very difficult to establish the failure mode.

Let us treat surface ply separation in more detail. Surface ply separation is caused by local loss of stability, accompanied by breakage of the binder. A model of surface ply separation is shown in Fig. 3.1.4. An expression of the lowest value of the critical stress σ_{cr} has the form:

$$\sigma_{cr} = \frac{2\pi^2}{3} E_x \; {\frac{\kappa_0^2}{2}}^{1/5} \tag{3.1.1}$$

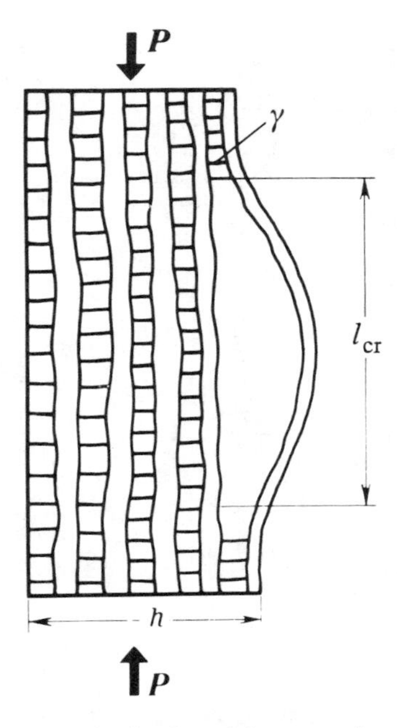

Fig. 3.1.4. The model of delamination of the composite under compression.

where $\kappa_0 = 36\gamma/(\pi^4 E_x l_{cr})$; γ is the specific surface energy at failure according to Griffith; l_{cr} is the length of the peeled section upon local loss of stability.

The thickness of the peeled section can be preset by the reinforcement scheme; the advantageous thickness of peeling h^* for unidirectional materials is equal to

$$h^* = \frac{1}{2}\left(\frac{\gamma l_{cr}^4}{\pi^4 E_x}\right)^{1/5}. \tag{3.1.2}$$

It should be noted that even in the case of incomplete delamination, the buckling of a side surface of the specimen, though not perceived visually, can result in considerable errors in measurement of deformations.

In contrast to unidirectional composites, materials reinforced at an angle to the longitudinal specimen axis fail due to shear without any destruction of the end faces. In this case, the entire shear load is taken by the matrix. The final failure of the specimen in this case can occur from shearing at an angle 45° to the specimen axis, but in contrast to compression along the axis of reinforcement the failure is a consequence of shear failure of the matrix between reinforcement layers.

Because of the diversity of factors determining the ultimate failure of the specimen, it is more conventional to determine the compression strength of fibrous polymeric composites as the breaking load divided by the cross-sectional area of the specimen, than is the case in the testing of isotropic materials. Without indication of the failure mode, test results for fibrous polymeric composites under compression are not mutually comparable.

Friction over the supporting surfaces of the specimen shows up strongly in the test results of compression tests. While for metals elimination of friction over the supporting surfaces promotes preservation of specimen shape during the test and an increase in measured compression strength, a decrease in friction on the end surfaces of fibrous polymeric composites specimen results in a decrease in the measured strength.

3.1.4. Strain Rate

Test data are known only for composites reinforced with glass cloth. At low strain rates, the short-term static strength increases quite rapidly with increase in strain rate and the greatest scatter of readings is observed in this range. At strain rates of $\dot{\varepsilon}^c = 0.0008$ sec^{-1} and above, experimentally obtained strength values are more stable, increasing negligibly with increase in strain rate. Low strength values at low strain rates can be explained by creep in the material, which shows up especially strongly when the load corresponds to approximately 0.7 σ^{cu}. If the strain rate $\dot{\varepsilon}^c > 0.008$ sec^{-1}, the effect of creep fails to appear. These conclusions on the role of the polymer binder during compression of fibrous polymeric composites have been confirmed by studies over a quite broad range of strain rates and environmental temperatures [222]. Under compression the modulus of elasticity of a material and elongation at rupture are less sensitive to change in strain rate.

ASTM Standard D 3410-75 recommends a strain rate of $\dot{\varepsilon}^c = 0.042$ sec^{-1}; if the testing machine does not permit a constant strain rate $\dot{\varepsilon}^c$ to be maintained, the recommended speed of movement of a movable grip of the testing machine is equal to $V_l = 1.3$ mm/min.

3.2. SPECIMEN SHAPE AND DIMENSIONS

3.2.1. Introductory Remarks

Compression test results of fibrous polymeric composites depend to a still greater extent on specimen shape and dimensions than do tensile test results. In various standards (GOST 4651-68, ASTM D 695-69, DIN 53457) for compression tests of rigid plastics, prismatic and cylindrical specimens of

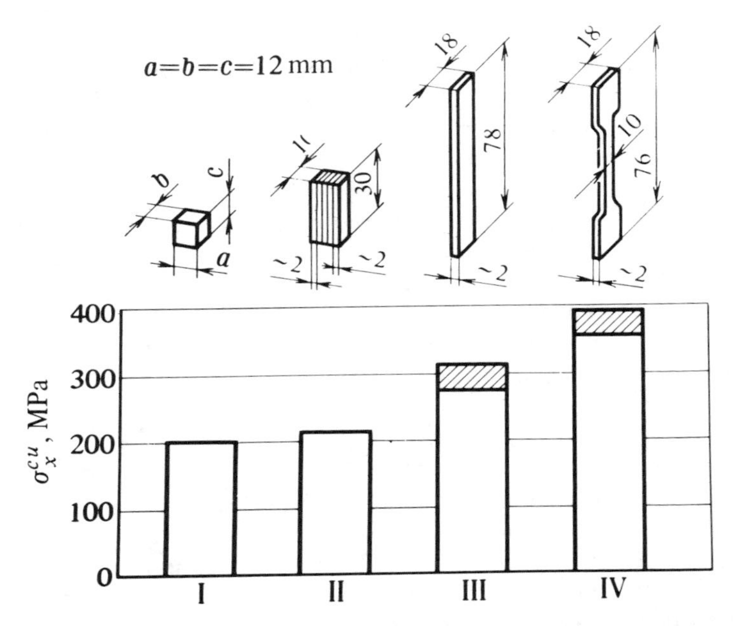

Fig. 3.2.1. Compression strength of specimens of various shapes [233]: (I) according to DIN 53454; (II) according to ASTM D 695-54; (III and IV) according to Federal Specification L-P-406. The scatter is cross-hatched.

small sizes are specified (the largest dimension of the base is 10–15 mm, height 15–55 mm). The small dimensions of standardized specimens do not permit qualitative measurement of deformations during testing, nor do they ensure sufficient uniformity of deformation over the gage length of the specimen, nor eliminate the effect of restriction of deformation and squeezing of the supported surfaces of the specimen. If deformations of the supported surfaces of a specimen are restricted, a break is observed in the stress-strain curve (Fig. 3.1.2), and the measured strength and modulus of elasticity turn out to be understated. Fig. 3.2.1 shows how specimen shape and dimensions affect the measured compression strength.

It is rather difficult to accurately fabricate small specimens. All deviations from the asigned geometric shape result in eccentricity of the applied load and reduction in the breaking forces. Moreover, the mechanically treated surface of small specimens is comparatively large (the ratio of the outer surface to the volume of geometrically similar prisms and cylinders is inversely proportional to their linear dimensions). This also decreases the breaking load. The scale effect also appears in compression tests (see Section 3.2.2). In view of all these deficiencies, the use of standardized specimens

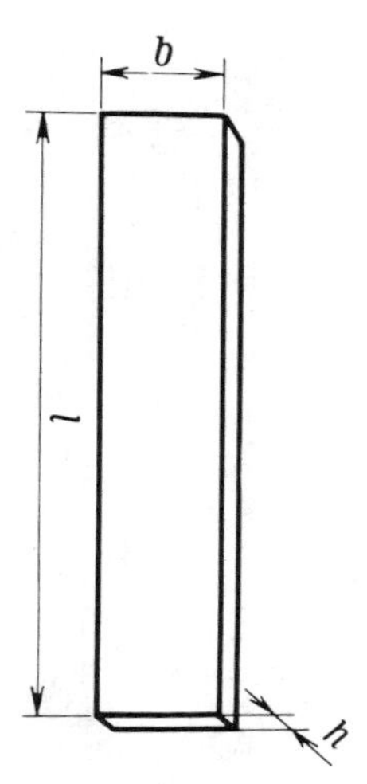

Fig. 3.2.2. Bar or strip specimen.

for rigid plastics in testing fibrous polymeric composites should be considered inappropriate.

In studies of elastic and strength properties of fibrous polymeric composites under compression specimens of various types are used. Stiffness is studied using relatively long specimens in the form of strips or specimens with variable cross section; in this case, special fixtures that preventing loss of stability in the specimen assure a sufficient zone of uniform stress for reliable measurement of deformations. Strength measurement is accomplished using strip-type specimens or specimens with variable cross section, i.e., with assigned sections of failure. Specimens with a circular cross section and sandwich beams are somewhat more practical.

3.2.2. Bars and Strips

Bars and strips are very simple specimen shapes. (Fig. 3.2.2). Their thickness h usually equals the thickness of the sheet from which the specimen is cut. The condition of transition to a continuum must be satisfied, i.e., the numbers of reinforcing layers must be greater than the required minimum (see Section 1.2.4). The specimens are cut in the direction of the principal axes of elastic symmetry of the material. Material structure (reinforcement layup) must be symmetric relative to the longitudinal axis of the specimen across both the width and the thickness.

During strength testing of bars and strips, failure must be assured over the

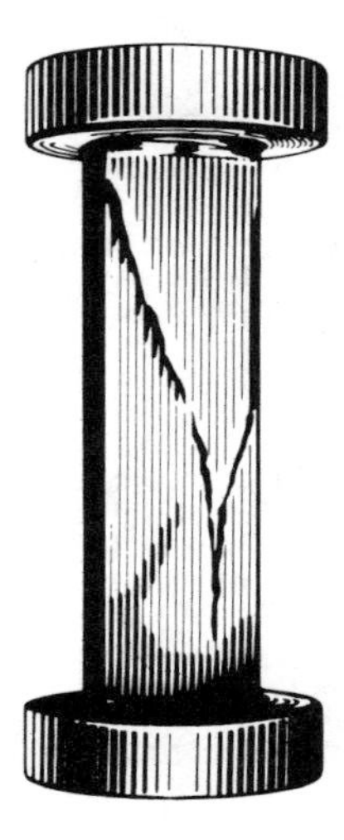

Fig. 3.2.3. Specimen with metal holders.

given gage length. This is accomplished by protecting the end sections of the specimen with tabs and by selecting a method of loading and a device design that will safeguard the specimen against longitudinal delamination. Tab material and dimensions are selected according to the same criteria as in the case of tension (see Section 2.2.2). Metal holders are also used to protect the end sections of the specimen during loading; they are attached with cold-setting resin or with Wood's metal and tightly envelop the ends of the specimen (Fig. 3.2.3).

The dimensions of strips and bars are selected with allowance for concentration of normal stress σ_y and tangential stresses τ_{xy}. The gage length of the specimen is chosen so as to eliminate loss of stability (see Section 3.2.6). Investigations show [272] that during loading of a strip of orthotropic material over the end faces (for loading schemes, see Section 3.3), the concentration of stresses σ_y and τ_{xy} becomes noncritical at a dimensional ratio of $l/b > 5$; during loading over the side faces of a specimen the concentration becomes noncritical at $l/b > 4$. The relative critical gage length of the specimen l_{cr}/h, eliminating the loss of stability, changes over a wide range and must be calculated for each concrete material tested (see Section 3.2.6).

The experiment showed that during determination of compression strength, carefully fabricated strips with tabs are equivalent to specimens with variable cross section i.e., to specimens with a preset section of failure. In Reference [272], the following results have been obtained during testing of unidirectional composites:

	σ_x^{cu}, MPa		
	STRIPS WITHOUT TABS	STRIPS WITH TABS	SPECIMENS WITH VARIABLE CROSS SECTION
Carbon composite, 0°	620	790	830
Boron composite, 0°	620	1440	1550

ASTM Standard D 3410-75 recommends strip-type specimens of the following dimensions (in mm) for the determination of strength and elastic characteristics of unidirectional and orthogonally reinforced composites under compression in the directions of the principal axes of elastic symmetry (for notation, see Fig. 3.2.2):

	l	b	h
Boron composite	139.7	6.35	1.5–2.0
Carbon composite	139.7	6.35	1.5–3.0
Glass fiber composite	139.7	6.35	3.2–4.0

The gage length of these specimens is 12.7 mm, but with allowance for transition sections of tabs it becomes 25.4 mm. The load is applied to the side faces of the specimen.

To determine the elastic constants the specimens, strips of nonstandard dimensions are also used. For example, in [222, p. 87] specimens of dimensions $60 \times 30 \times h$ (mm) are recommended; these do not watch the requirement given above.

3.2.3. Specimens with Variable Cross Section

For fibrous polymeric composites specimens with variable cross section are not standardized; there are only practical recommendations on selection of their shape and dimensions. Shape and dimensions of specimens with variable cross section depend on the purpose of test: whether strength or elastic constants are being determined. Profiling of the gage length of a specimen is accomplished in the same way as under tension (see Section 2.2.2).

Specimens for testing glass cloth composites are shown in Fig. 3.2.4 [in Fig. 3.2.4(a) for materials more than 8 mm thick; in Fig. 3.2.4(b) for materials of thickness $5 \leqslant h \leqslant 8$ mm]. These specimens have a sufficiently large fillet radius ($R = 160$ mm) to eliminate stress concentration, and the variable

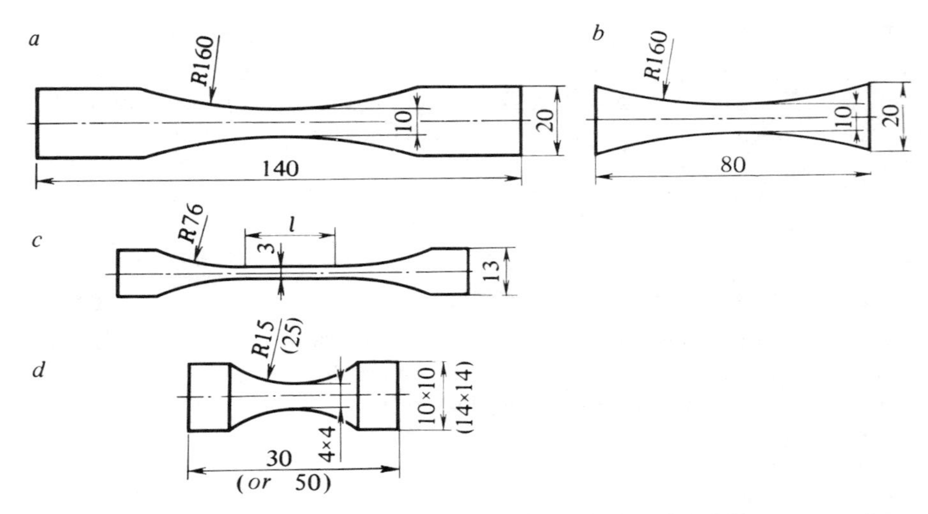

Fig. 3.2.4. Specimens with variable cross sections for compression tests of fibrous composites.

width eliminates the appearance of large deformations on the support surfaces of the specimen. Specimens of this shape are used for glass cloth composites in loading parallel to the principal axes of elastic symmetry of the material. If the load is directed perpendicular to the reinforcement plane, a specimen thickness of 10 mm is recommended [76].

In tests of high-modulus dibrous polymeric composites, specimens with a sharply distinguished gage length [Fig. 3.2.4(c)] are sometimes used. The use of such specimens is dictated by the necessity to ensure that the failure of high-strength composites take place in the gage length of the specimen without additional protection of the ends with tabs. However, as shown by experiment, splitting across the width of its gage length is frequently observed during tests of unidirectional composites in the reinforcing direction without protection of specimen ends.

A specimen for compression tests of unidirectional glass fiber composites in the reinforcement direction is shown in Fig. 3.2.4(d). The shape and dimensions of this specimen are selected from the following considerations [208]: Splitting of the heads of the specimen across the width of the gage length is eliminated (the inequality $\sigma_x^{cu} F^c < \tau_{xz}^u F^{spl}$ should be satisfied, where σ_x^{cu} is the short-term static compression strength of the material of the specimen; F^c is the calculated cross-sectional area of the gage length of the specimen; τ_{xz}^u is the shear strength of the material parallel to the fibers; F^{spl} is the splitting area of the end sections). Bearing of the support surfaces is eliminated (for glass fiber specimens, bearing of the support surfaces will be eliminated at $F^c/F = 0.15$–0.25, where F is the area of the support surfaces of the specimen). And loss of stability is eliminated because of the relatively small specimen height. The effect of the fillet radius on stress concentration and, consequently, on the measured strength of a specimen of unidirectional fiberglass composite is insignificant.

An advantage of specimens with variable cross section is a clearly defined gage length. The principal shortcomings are the nonuniformity of the state of stress and more laborious fabrication. Tabs are used for protection of the end parts of specimens with variable cross section; on selection of their material and dimensions, see Section 2.3.3.

3.2.4. Specimens with a Circular Cross Section

Specimens with circular cross-sectional area are used for determination of the elastic constants and compression strength of some unidirectional composites, mainly those which can be easily processed mechanically (for example, carbon composites). Tests show that the scatter of measured values of specimens with a circular cross section is smaller, and the absolute strength values are greater than those of specimens with a rectangular cross section [131]. Special fittings are used to prevent crushing and delamination of end faces and to ensure failure of a specimen in a reference section; the cross section of the gage length is made smaller than the cross section of the ends of the specimen (Fig. 3.2.5). Selection of the optimum fitting shape has been described in Reference [68]. The fittings are fabricated of mild steel, and they have a transition zone in the form of a truncated cone. If the cone abutts smoothly on the surface of the specimen [Fig. 3.2.5(a)], the specimen fails through the section at the point of transition as a consequence of stress concentration. The end section of the specimen must precisely correspond to the shape and dimensions of the opening drilled into the fitting, without the slightest gap. Epoxy glue is used to fasten the end fittings.

The specimen and fitting shape developed ensures sufficiently reliable and reproducible experimental results. The dimensions (in mm) of specimens and fittings for epoxy carbon composites given below were presented in Reference [68]. For notation, see Fig. 3.2.5(b).

	d	d_1	D	D_1	a	b	l_1	l	L
Specimen A	11.0	8.4	19	13	15	25	13	76	106
Specimen B	9.4	7.2	19	11	15	25	13	58	88
Specimen C	6.4	4.6	13	8	15	25	13	38	68

Since fabrication of specimens with circular cross section is rather laborious, they are not widely used, regardless of their advantages.

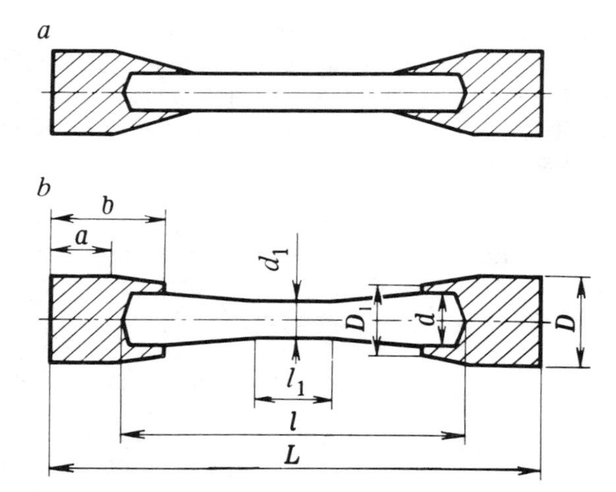

Fig. 3.2.5. Specimens with circular cross section and metal end fittings [68]: (a) Constant cross section; transition to the end fittings is gradual. (b) Reduced cross section in the gage length; transition section of the end fitting is in the form of a truncated cone.

3.2.5. Sandwich Beams

Sandwich beams are used for determination of elastic constants and strength in compression. As in the case of tensile tests, sandwich beams are loaded in four-point bending. The main advantage of sandwich beams is the absence of all phenomena connected with loading of the specimens (stress concentration, buckling of the specimen, etc.). For more detail on sandwich beams, see Section 5.1.

All specimens of the shapes and dimensions described above are suitable for determination of the compression strength of a material. For determination of the elastic constants, only those specimens are usable which have a sufficient gage length of constant cross section, in which a uniform state of stress is assured.

3.2.6. Dimensions of Specimen Gage Length

Only specimens with variable and circular cross sections have a clearly defined gage length. The gage length of other types of specimens with constant cross section constitutes a part of the total length of the specimen. In both cases, the length of a specimen with an assigned gage length is selected with allowance for the end effect and specimen fastening conditions. There are fewer numerical data on the dimensions of the gage length of a

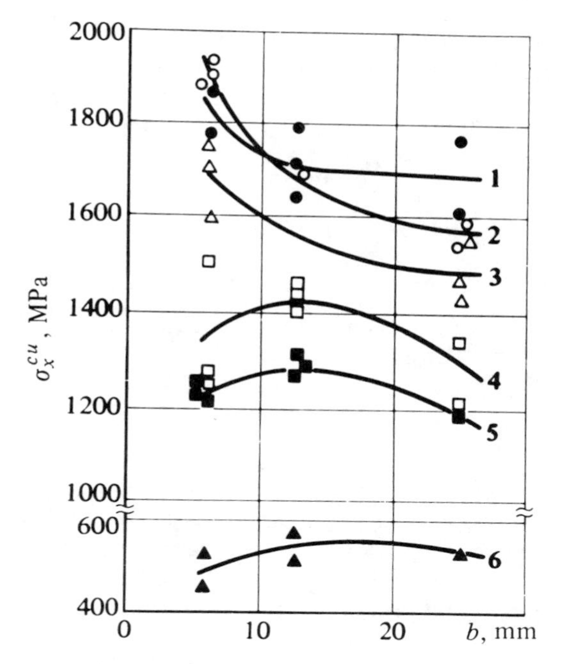

Fig. 3.2.6. Effect of specimen width and length and number of layers of reinforcement on compression strength σ_x^{cu} of unidirectional boron epoxy composites ($V_f = 50\%$) [131]; the number of layers in a specimen and the gage length are as follows: (1) 12 layers, 12.7 mm; (2) 18 layers, 12.7 mm; (3) 18 layers, 25.4 mm; (4) 6 layers, 12.7 mm; (5) 12 layers, 25.4 mm; (6) 6 layers, 25.4 mm.

specimen for compression testing than for tensile testing. The cross-sectional dimensions of specimens are approximately the same in both types of loading. There are no data at all on the length of the end effect zone.

Results obtained in tests of unidirectional boron composites are presented in Fig. 3.2.6. The strength of the 12- and 18-ply short specimens is approximately the same, and it decrease with increase in specimen width, i.e., with the cross-sectional area. The strength of 6-ply specimens is less than the strength of 12- and 18-ply specimens, especially with long specimens. The strength of long specimens is less than that of short specimens: in 18-ply composites by approximately 10%; in 12-ply by 30%; and in 6-ply by a factor of almost three. Obviously buckling of the specimens as a consequence of eccentricity in loading will have a greater effect the test results in the latter case.

Compression strength values obtained in testing specimens of unidirectional carbon and boron composities with a constant cross section of various dimensions are given below [272]:

	σ_x^{cu}, MPa						
	WIDTH,[a] mm			THICKNESS,[b] mm			
MATERIAL	5	10	20	1.0	2.5	4.0	7.2
Carbon composite, 0°	690	720	600	960	920	910	780
Boron composite, 0°	1110	1030	930	—	—	—	—

[a] Length 70 mm, thickness 2 mm.
[b] Length 70 mm, width 10 mm.

It follows from these data that the increase in the width of a strip of a constant length, i.e., decrease in the l/b ratio below a definite limit, results in decrease of the measured strength. An analogous phenomenon is observed with the increase in thickness of a strip of both constant length and width.

The specimen gage length is established from the following requirements: The state of stress in the gage length should be uniform; the loss of stability must be eliminated: and proper location of strain gages must be assured.

A uniform state of stress in the specimen gage length is assured by gradual transmission of the compressive forces, which is obtained through selection of the shape and size of specimen end sections or by means of tabs. As can be seen from Fig. 3.3.3, the required length of specimen end sections exceeds the specimen gage length by several times.

To eliminate loss of stability, the free length (between tabs, testing machine grips, or other supporting devices) of the specimen gage Section must be less than the critical length l_{cr}, which can be calculated by the formula

$$l_{cr} = 0.907h \sqrt{E_x^c \left(\frac{1}{\sigma_{cr}^c} - \frac{1.2}{G \times z} \right)} \qquad (3.2.1)$$

where E_x^c and G_{xz} are the moduli of elasticity and shear of the material being tested; σ_{cr}^c is the critical compression stress (for unidirectional materials for which the stress-strain curve is rectilinear right up to failure, $\sigma_{cr}^c = \sigma_x^{cu}$).

Equation (3.2.1) was obtained for the case of the specimen with both ends free. The length l_{cr} calculated by this formula is somewhat overestimated, since the tabs and grips of devices stiffen the specimen fastening.

It is seen from Fig. 3.2.7, that the greater the strength of a material tested, the more the measured strength depends on the accuracy of selection of the gage length of a specimen. Therefore correct calculation of the gage length, with due allowance for the loss of stability, is necessary.

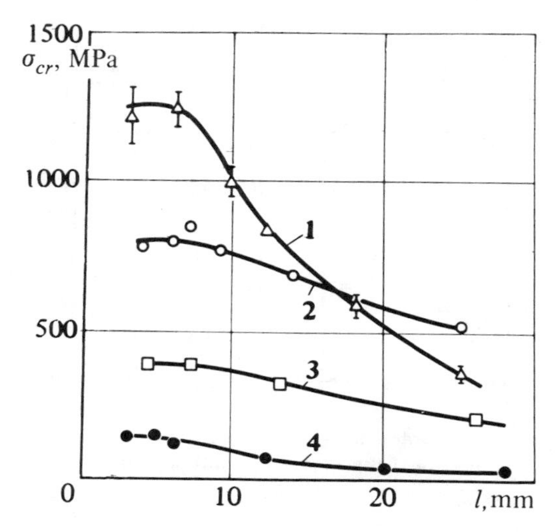

Fig. 3.2.7. The effect of the gage length on the critical compression load (unidirectional carbon composite, $V_f = 60\%$) [87]: (1) type 2, treated; (2) type 1, treated; (3) type 1, untreated; (4) pure binder.

The condition for location of strain gages also complies with the three mentioned requirements. In some cases, this may cause technical difficulties. For example, ASTM D 3410-75 specifies a specimen gage length equal to 12.7 mm for determination of both strength and elastic constants. From this consideration, the standard specifies that the base of resistance strain gauges must not exceed 1.6–3 mm. If this requirement is technically unrealizable, specimens of other dimensions must be used for determination of elastic constants and tested in devices which prevent their loss of stability (see Section 3.3.4).

3.3. LOADING

3.3.1. End Loading

In compression tests of fibrous polymeric composites special attention should be paid to the method of loading. Three methods of specimen loading in compression tests can be distinguished:

1. By axial (normal) forces on the end faces of the specimens [Fig. 3.3.1(a)]
2. By normal and tangential forces on the side faces of the specimen [Fig. 3.3.1(b)]
3. By joint forces on the end and side faces of the specimen [Fig. 3.3.1(c)]

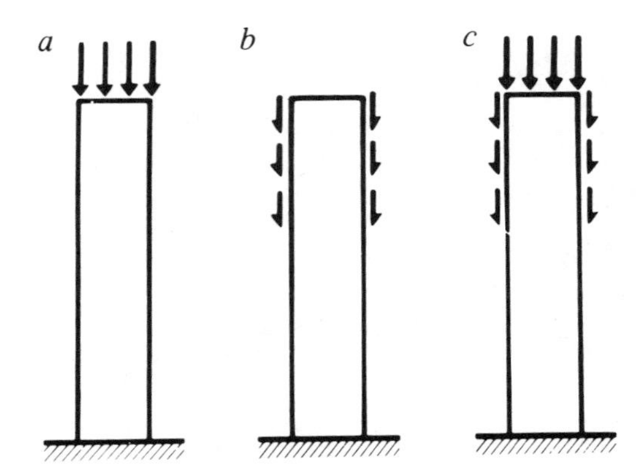

Fig. 3.3.1. Schemes of compression loading of specimens: (a) with normal forces on the end faces; (b) with tangential forces on the side faces; (c) width simultaneous normal and tangential forces.

In end loading of a specimen, the compression load should be applied through flat parallel polished support surfaces. The tangent of the angle of deviation of the direction of the load from the longitudinal axis of the specimen must not exceed 0.001. To satisfy this rigid requirement and eliminate random errors in installation of the specimen, numerous devices have been developed for compression testing. Precision of the loading and installation of the specimen in these devices is ensured by high accuracy in fabrication of the guide and support surfaces. Specimens can stand free or be supported by their sides in the devices; their ends can also be fixed in grooves, providing precise loading [188, p. 64],

Investigation [171] show that, even with the most careful working of the end surfaces of a specimen, it is impossible to ensure complete contact between the end surface of the specimen and the surface of the hinge-mounted platens of the testing machine. In most cases, the specimen is initially compressed along one of the edges of its end faces and it also fails close to it before achieving complete contact with the support surfaces, especially with materials with a high-modulus matrix. The effect of poor alignment and incomplete contact of the end surfaces of the specimen with the testing machine plantens can be seen in Fig. 3.3.2, in which compression stress-strain curves of 6-, 12-, and 18-ply boron composites are presented. Thin specimens prove to be considerably more sensitive to the loading conditions; for example, 6-ply specimens huckle under about one-third the load achieved with 18-ply specimens.

In machining of boron composites, especially for unidirectional panels,

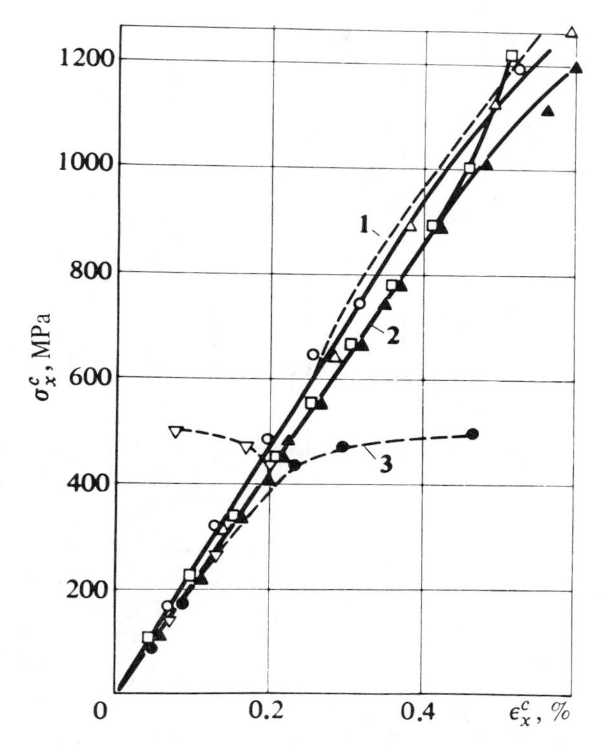

Fig. 3.3.2. Compression stress-strain curves of boron epoxy composites (strain gages glued to opposite faces of the specimens) [131]; number of layers in specimens: (1) 18; (2) 12; (3) 6.

it is impossible to obtain a smooth specimen end face: the ends of the boron fibers project above the surface of the matrix. This not only disrupts contact between the end faces of the specimen and the testing machine plantens, but because of stiffness of the boron fibers it also creates stresses in the specimen material which are impossible to evaluate. Therefore, it is recommended that testing machine platens for boron composite tests be fabricated of mild steel, although this will result in rapid wear of their surfaces.

In connection with the deficiencies noted, a tendency is observed to abandon end loading and substitute simultaneous end loading and loading over side faces. During end loading, attempts have been made at elimination of premature failure of the specimen material by introducing an intermediate element—an end cap on the specimen. Metal end caps, carefully fitted to the specimen [68], have proved to be the most effective. Attempts have also been made to cover specimens entirely with a hardened binder [82].

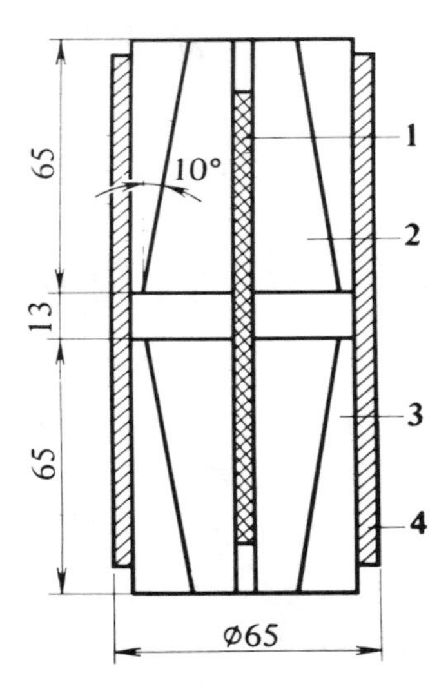

Fig. 3.3.3. Fixture for loading specimens on side surfaces [171]: (1) specimen; (2) split collet-type grips; (3) tapered sleeve; (4) cylindrical shell.

3.3.2. Loading over Side Faces

This type of loading is specified in ASTM Standard D 3410-75. In compression loading of specimens with tangential forces special fixtures are necessary. The fixture (Fig. 3.3.3) consists of self-tightening split collet-type grips (2) and tapered sleeves (3), which are placed inside a cylindrical shell (4). The tabbed parts of the specimen (1) are inserted into the grip cavities with the grips in a partly opened position. After being manually closed, the grips are fitted into the sleeves. A 12.7 mm thick spacer is placed between the top and bottom grips. The entire assembly is placed into a cylindrical shell (4) which fits snugly around the tapered sleeves. The fixture, with the specimen, is then installed between the self-centering plantens of the testing machine. The tapered sleeves are preloaded with a small force (the stress on the gage length of a specimen is approximately 40 MPa), after which the spacer is removed. With further axial loading of the tapered sleeves, the grips tightly squeeze the tabbed side faces of the specimen and thus, by means of tangential forces, they transmit the axial compression force to the gage length of the specimen.

The data of Table 3.3.1, which presents the strengths of carbon composites obtained by various methods, give evidence of the quality of this fixture. The main shortcoming of fixtures with collet-type grips is that, to ensure

Table 3.3.1. Compression Strength of Unidirectional Graphite Composites [171].

MATERIAL	FIBER CONTENT V_f, %	E_x^c, GPa	σ_x^{cu}, MPa UNSUPPORTED	TABBED BAR	NOL RING
Celanese/Epi-Rez 508	60	5.52	398	787	780
Celanese/ERLA-4617	60	5.52	—	930	—
Morganite II/Epi-Rez 508	50	2.07	772	982	1007
Morganite II/ERLA-4617	50	2.07	—	1090	—

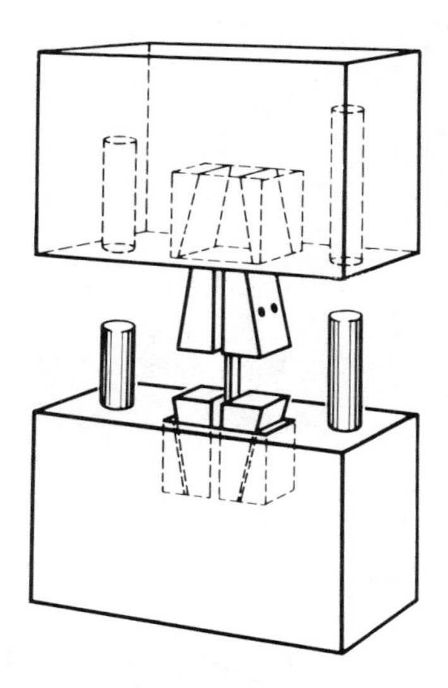

Fig. 3.3.4. IITRI fixture with flat wedge grips [96].

their free movement, high-quality machining of their surfaces is required. With this in mind, a fixture is recommended in reference [96] in which collet-type grips are replaced by flat wedge grips, which by means of bolts are fixed to the specimen (Fig. 3.3.4). Test results for composites of four different types are presented in Table 3.3.2. Three types of carbon composites were tested—high-strength Modmor II/Narmco 5206 and Thornel T 300/Narmco 5208 and high-modulus Hercules 3002 M, as well as boron composite AVCO 5505. The specimens had three different layups: unidirectional 0° (12-ply), unidirectional 90° (20-ply) and $[0/\pm 45/0/90]_s$ (18-ply). Several loading methods have been compared: a sandwich beam, a Celanese-type fixture (approximately corresponding to the fixture of ASTM D 3410-75)

Table 3.3.2. Average Compression Test Results of Several Composites, Obtained in Various Fixtures [96].

MATERIAL AND FIXTURE	TEMPERATURE, °C	STRENGTH, MPa, FOR A LAYUP:			MODULUS OF ELASTICITY, MPa, FOR A LAYUP:		
		0°	90°	$[0/\pm45/90]_s$	0°	90°	$[0/\pm45/90]_s$
Modmor II/Narmco 5206							
sandwich beam	21	972	195	683	135	11.8	75.8
flat wedges	21	1007	170	655	141	10.1	65.5
Celanese	21	—	228	—	—	9.9	—
sandwich beam	177	848	119	579	130	8.3	71.7
flat wedges	177	889	116	586	137	5.1	62.7
Hercules 3002 M							
sandwich beam	21	690	148	414	170	8.6	92.4
flat wedges	21	669	226	407	155	8.3	81.4
sandwich beam	177	572	129	379	170	9.4	86.2
flat wedges	177	634	188	400	172	7.5	67.6
Thornel T 300/Narmco 5208							
sandwich beam	21	1703	246	765	159	12.1	84.8
flat wedges	21	1503	250	786	159	11.3	86.2
Celanese	21		148			13.0	
sandwich beam	177	1475	197	662	148	12.1	82.7
flat wedges	177	1420	210	662	155	11.0	97.2
AVCO 5505							
sandwich beam	21	2496	203	1627	217	24.5	125.5
flat wedges	21	1351	245	1227	192	26.8	111.0
sandwich beam	177	1965	163	1262	203	27.9	110.3
flat wedges	177	869	137	1041	203	12.3	96.5

and the IITRI fixture with flat wedges suggested in [96]. It follows from the table that in testing of high-strength materials, especially boron composites, the last fixture yields somewhat lower results than does a sandwich beam specimen. This is evidence of incomplete force transmission to the specimen.

3.3.3. Combined Loading

Experiments show that the most complete and also the most time-consuming type of loading is combined loading over the end and side faces of the specimen simultaneously. Results of testing three different types of fibrous polymeric composites—relative strengths (strength under combined loading is equal to unity) and coefficients of variation—are presented in Table 3.3.3.

It can be seen from Table 3.3.3 that combined loading ensures not only

Table 3.3.3. Strength of reinforced materials under combined loading [272].

MATERIAL	RELATIVE STRENGTH UNDER LOADING:			VARIATION COEFFICIENT, %, UNDER LOADING:		
	AT THE END FACES	AT THE SIDE SURFACES	COMBINED	AT THE END FACES	AT THE SIDE SURFACES	COMBINED
Carbon composite, 0°	0.59	0.92	1.00	9.0	8.2	7.9
Boron composite, 0°	0.66	0.86	1.00	9.0	9.1	6.3
Three-dimensional carbonized carbon composite	0.54	0.89	1.00	16.8	7.5	4.0

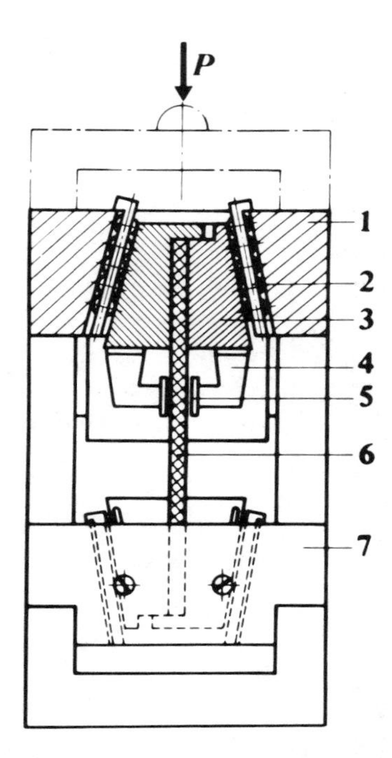

Fig. 3.3.5. Fixture for combined compression loading of fibrous composites: (1) upper grip; (2) roller bearing; (3) wedge inserts; (4) supports protecting against loss of stability of specimens; (5) fluoroplastic tabs; (6) specimen; (7) lower grip.

the highest, but also the most stable results with the smallest scatter. A schematic diagram of the fixture used in [272] is shown in Fig. 3.3.5. The angle of inclination of the wedge-type inserts is selected according load distribution over end and side faces of the specimen. The load over the side faces of the specimen must not exceed the transverse compression strength of the material. It is known empirically that the load at the specimen end faces must not exceed 40–50% of the breaking load (in the absence of side load). In operating fixtures, the angle of inclination is 14–17°. The end parts of the specimen are protected by tabs.

3.3.4. Prevention of Buckling

This specimens ($h < 3$ mm) are tested in a special fixture which prevents buckling of the specimen. Several such fixtures have been developed—by the ASTM, by the DEL, etc. [62, 149, 206], but the principle of action is the same: the lateral surfaces of the specimen are in contact with prismatic bulges, which prevent buckling of the specimen but does not restrict deformation in its plane. To avoid brooming of the ends of the specimens, the end parts of the specimens may be reinforced by metal tabs glued to their lateral surfaces, or by metal caps.

3.4. COMPRESSION OF RINGS

Compression of rings in their planes is accomplished under external pressure. The methods of generating external pressure used in practice are shown in Fig. 3.4.1. The determinable experimental values, force factors, and variable geometric specimen dimensions are also shown in the figure.

In testing of a ring with a rigid split disk [Fig. 3.4.1(a)] equations (2.5.2) and (2.5.3) are used for processing the results. However, the peculiarities of this type of test should be taken into account. In compression loading, in parts of the ring, along with compressive stresses, there are also bending stresses whose magnitude depends on the mechanical characteristics of the material and the relative specimen dimensions, as well as on the structure of the fixture. In standard fixtures there is usually some gap between half-frames. Experiment shows that, as in the case of tension of rings by means of a split disk, the maximum strains in the ring are observed at the gaps of the fixture. At some distance from the gap the strains level out (Fig. 3.4.2). Some variation is ascribed to the practically unavoidable eccentricity of the applied compression load. More uniform strain curves are obtained if the specimen is supported in the fixture and also from inside. But the best results are obtained in testing rings in fixtures with half-frames and girder locks. It is possible, using locks, to prevent any increase in horizontal diameter of the ring under load and the measured strength values increase [208].

With regard to the determination of the modulus of elasticity E_θ^c, all the remarks made about tensile tests using split disks (Section 2.5.1) hold true. Because of the indicated disadvantages, testing of rings by means of split disks may be recommended only for qualitative evaluation of elastic and strength properties of different materials.

Loading with external pressure by means of a compliant (rubber) ring [Fig. 3.4.1(b)] and by hydraulic methods [Fig. 3.4.1(c)] is accomplished as in tensile testing of rings. Adjustment of a sealing insert under hydraulic pressure is provided around the outer surface, but resistance strain gages are mounted on the inner specimen surface. In compression loading by compliant ring the latter forms an elastic base for the specimen and to a certain extent increases the critical stress at which the ring specimen loses stability. This effect can be increased through proper selection of rubber

			a	b	c	d
Loading scheme			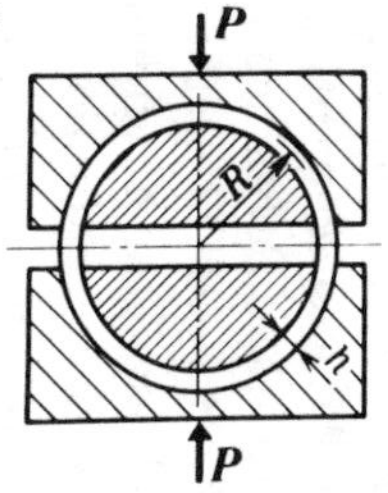	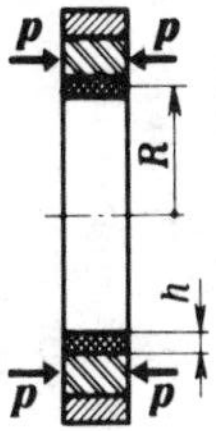	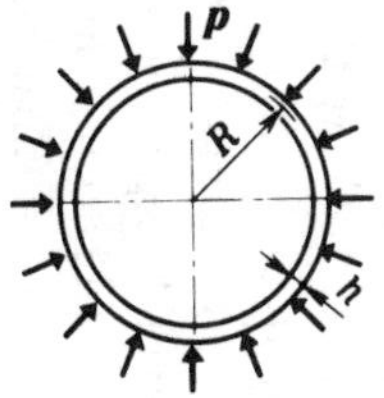	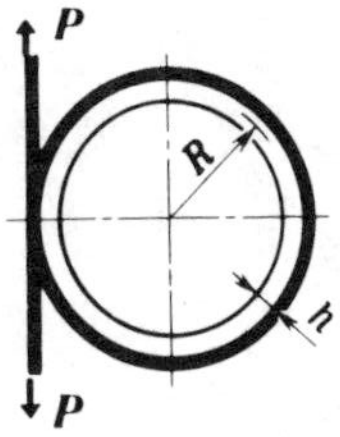
Determinable characteristics			E_θ^c, σ_θ^{cu}	E_θ^c, σ_θ^{cu}	E_θ^c, E_r, σ_θ^{cu}	E_θ^c, σ_θ^{cu}
Measurable values			P, Δu, ϵ_θ^c	p, ϵ_θ^c	p, ϵ_θ^c	P, ϵ_θ^c,
Geometrical sized			R/h, b,h	R/h, b,h	R/h, b,h	R/h, b,h
Limitations	Structural	Layup	0°; 90°; 0/90°			
		Orientation	0°, 90°			
	Physical		For E_θ:linear range of the curve of $P \sim \epsilon_\theta$ or $p \sim \epsilon_\theta$			
	Geometrical		For σ_θ^{cu} : $0.08 \leqslant h/R \leqslant 0.18$ (for GFRP)			

Fig. 3.4.1. Schemes of loading ring specimens with external pressure by means of a rigid split disks (a), a compliant ring (b), hydraulically or with levers (c), and a steel starp (d).

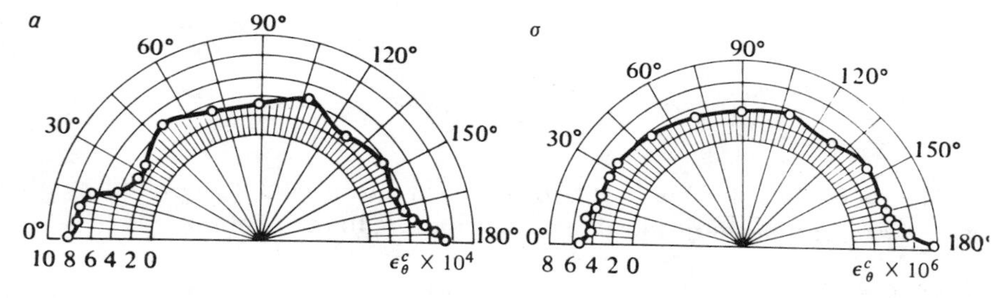

Fig. 3.4.2. Distribution of strain ε_θ^c on the inside surface of the ring at various load levels [208]; the stress is equal to: (a) 66 MPa; (b) 230 MPa.

hardness. In testing under external pressure, requirements of careful processing of the outer and end surfaces of the specimen increase.

The external pressure may also be applied by means of mechanical fixtures [190, 201, 208]. The load on the specimen is created by means of 72 identical loading levers, which are connected by a loading plunger to the punch of the testing machine. The advantage of multi-lever fixture is its simple construction (although it is time-consuming to fabricate) and the correctness of the experimental results obtained, even though the nonuniformity of the load may reach 10%. This nonuniformity increases over time and necessitates periodic adjustment and calibration of the fixture. In contrast to hydraulic tests and loading by means of a compliant ring, the direction of the load remains constant in the multi-lever fixture—toward the center of the fixture—and this direction does not follow the possible changes of the specimen shape in its plane. However, to judge by the experimental results, these changes are not critical. The capacity of multi-lever fixtures is structurally limited (for example, the capacity of the fixture described in [208] for testing of rings with inner diameter 150 mm does not exceed 20 metric tons). This in turn puts restrictions on selection of the maximum specimen thickness.

The compression strength of a ring specimen is determined by the formula

$$\sigma_\theta^{cu} = \frac{k(P^u - \Delta P)}{2\pi bh} \tag{3.4.1}$$

where P^u is the breaking load measured by a dynamometer; ΔP is the resistance of the multi-lever fixture running idle (determined in calibration of the fixture); b and h are the width and thickness of the ring respectively; and k is the coefficient of calibration of the fixture.

At $h/R > 0.05$ the strength of a ring specimen must be obtained from the formula

$$\sigma_\theta^{cu} = \frac{k(P^u - \Delta P)}{2\pi bh}\left(1 + \frac{h}{R}\right). \qquad (3.4.2)$$

The coefficient of calibration is determined during compression testing of a steel calibration ring and calculated according to the relation

$$k = \frac{2\pi bh}{1 + \dfrac{h}{R}} \frac{\varepsilon\theta E_{\text{steel}}}{P - \Delta P} \qquad (3.4.3)$$

where E_{steel} is the modulus of elasticity of steel and ε_θ is the circumferential strain of the calibration ring under the load P.

In the determination of the strength of a ring under compression in the multi-lever fixture, the loading rate is assumed to be equal to $\dot{\varepsilon}_\theta = 0.01\ \text{min}^{-1}$ [208].

Uniform compression over the outer surface of the ring may also be created by means of a steel strap [Fig. 3.4.1(d)] wrapped around the ring specimen. A slot at the intersection of the ends of the strap allows the completion of the loading surface. The contact surfaces of the specimen and the strap are lubricated and a shim is placed at the intersection of the strap ends to prevent bending of the ring at this location. This method is used for determination of compressive strength, calculated from the formula

$$\sigma_\theta^{cu} = \frac{P^u}{bh} \qquad (3.4.5)$$

where P^u is the ultimate tensile load applied to the ends of the strap.

In compression testing of rings under external pressure, selection of the relative thickness h/R of the specimen is the main problem. In studying compressive characteristics it is necessary to take into account the fact that the relationship between breaking pressure and relative thickness has three well-defined regions. In the first region (thin-walled rings) the load-carrying capacity is exhausted due to loss of stability. The critical pressure at which loss of stability takes place may be evaluated from the formula [231, p. 168; 237, p. 157]

$$p_{\text{cr}} = \varphi p_{\text{cr}}^* \qquad (3.4.6)$$

where $p_{\text{rr}}^* = 3E_\theta I/R^3$ is the critical pressure per unit of length of the ring axis, determined without allowance for shear; $\varphi = 1/(1 + 0.4\kappa^2)$; R is the

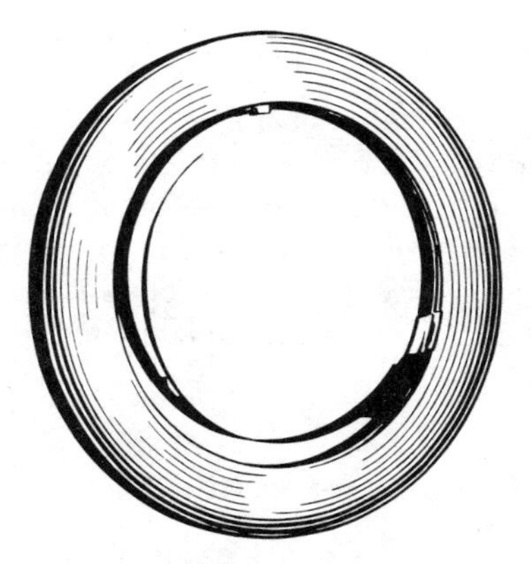

Fig. 3.4.3. Delamination on the inner ring surface.

mean radius of the ring; κ is a parameter of anisotropy and geometry of the specimen ($\kappa = (h/R)\sqrt{E_\theta/G_{\theta z}}$).

In the second region, compression failure of the ring occurs. In the third region (thick-walled rings), in analyzing strength it is necessary to consider not only σ_θ, but also σ_r. The boundaries of the three regions depend on the anisotropy of the material being tested.

In loading a ring with external pressure it is often impossible to correctly evaluate the compression strength due to peeling of the inner layer (Fig. 3.4.3). Peeling and subsequent loss of stability of the inner layer occur when the difference between the energy of the inner layer when it maintains a circular configuration and the energy of the same layer after peeling and loss of stability is higher than the inner layer bond energy. At layer thickness h_0, which is determined by the layup, the critical stress is equal to [227]

$$\sigma_{cr} = 0.916E_\theta\left[\left(\frac{h_0}{R_i}\right)^2 + \kappa\left(\frac{h_0}{R_i}\right)^{-1}\right]^{1/2} \tag{3.4.7}$$

where $\kappa = 4.77\gamma/(E_\theta R_i)$; γ is the specific fracture energy according to Griffith; and R_i is the inner radius of a ring.

Delamination as a result of peeling can propagate in two ways: with helical winding, it may initiate unwinding; in a circular layup, subsequent layer-by-layer peeling can occur.

The method of external pressure is used also for determination of the modulus of elasticity in the radial direction, E_r. In this case a short cylinder tightly mounted on a rigid mandrel is loaded with external pressure p. During testing the pressure and circumferential strain ε_θ are measured. The modulus of elasticity E_r is determined from the formula [28]:

$$E_r = \frac{ph}{\varepsilon_\theta R}(1 - \nu_{zr}\nu_{rz}) \tag{3.4.8}$$

where p is the external pressure; h is the thickness of the cylinder wall; ε_θ is the circumferential strain measured on the outer cylinder surface; R is the mean radius of a cylinder; and ν_{zr}, ν_{rz} are Poisson's ratios.

The use of equation (3.4.8) requires preliminary determination of the Poisson's ratios ν_{zr} and ν_{rz} of the material being tested.

3.5. TESTS OF TUBULAR SPECIMENS

3.5.1. Axial Tension and Compression

During tests of tubular specimens mechanical characteristics of filament wound materials with various fiber layups are evaluated. Tubular specimens can be loaded in axial tension and compression, in shear, with internal or external pressure. Also, these loads can be applied both separately and in various combinations, for example, axial tension with internal pressure or torsion. Methods of loading tubular specimens under tension and compression using internal and external pressure are considered in Section 3.5.2. Torsion of thin-walled tubes is treated in Section 4.2. Analysis of complex loading of tubular specimens is beyond the scope of this book.

During testing of tubular specimens in axial tension or compression it is possible to determine strengths σ_z^{tu} and σ_z^{cu}, the moduli of elasticity E_z^t and E_z^c, and Poisson's ratio $\nu_{\theta z}$. Tubular specimens are often considered as flat specimens of infinite width which are devoid of edge effect (see Section 2.1.4), and are used for evaluation of the mechanical characteristics of sheet composites with various reinforcement layups. Such an approach is permissible only by convention. First, flat and tubular specimens are fabricated by different methods, and elimination of the effects of this difference is hardly possible. It is more difficult to ensure constant thickness of the article, as well as content and direction of reinforcement, in the fabrication of tubular specimens, and variations in their characteristics may markedly affect the test results. Secondary, in an anisotropic shell the axes of the stress tensor and the strain tensor coincide only in particular cases, but under uniaxial tension or compression the state of stress in a tubular specimen differs from the state of stress in a flat specimen. Because of these differences, tubular specimens can not be considered equivalent to flat specimens. This is confirmed by the results of numerous experiments. For example, in testing 7-ply (90, 45, 135, 0, 135, 45, 90°) and 13-ply (90, 0, 0, 45, 135, 45, 90, 135, 45, 135, 0, 0, 90°) carbon composites on flat specimens with variable cross section ($l \times b = 57.2 \times 12.7$ mm) and on tubular specimens ($L = 203.2$ mm, $l = 101.6$ mm, $D_i = 57.2$ mm), the following results were obtained [97]:

MATERIAL AND LOADING SCHEME	$\sigma^{t(c)u}$, MPa		$E_z^{t(c)u} \times 10^{-3}$, MPa	
	FLAT SPECIMEN	TUBE	FLAT SPECIMEN	TUBE
Carbon composite, 7-ply				
tension	138	217	30.3	49.0
compression	131	177	34.5	44.8
Carbon composite, 13-ply				
tension	281	295	55.8	57.2
compression	165	208	54.5	57.9

The differences among the results presented in the table are due not only to the fabrication methods of the specimens. The materials tested also differ in thickness, and 7-ply and 13-ply carbon composites differ in stacking sequence as well.

In axial loading of a cylinder of finite length from a homogeneous, cylindrically anisotropic material, the stresses in the case where the anisotropy axis of the material coincides with a geometrical axis z are calculated from the following formulas [126]:

$$\sigma_z = \frac{P}{T}\left\{1 - \frac{m}{a_{33}}\left[a_{13} + a_{23} - \frac{1 - c^{k+1}}{1 - c^{2k}}(a_{13} + ka_{23})\rho^{k-1} - \frac{1 - c^{k-1}}{1 - c^{2k}}(a_{13} - ka_{23})c^{k+1}\rho^{-(k+1)}\right]\right\}, \tag{3.5.1}$$

$$\sigma_r = \frac{P_m}{T}\left[1 - \frac{1 - c^{k-1}}{1 - c^{2k}}\rho^{k-1} - \frac{1 - c^{k-1}}{1 - c^{2k}}c^{k+1}\rho^{-(k+1)}\right], \tag{3.5.2}$$

$$\sigma_\theta = \frac{P_m}{T}\left[1 - \frac{1 - c^{k+1}}{1 - c^{2k}}k\rho^{k-1} + \frac{1 - c^{k-1}}{1 - c^{2k}}kc^{k+1}\rho^{-(k+1)}\right], \tag{3.5.3}$$

$$\tau_{\theta r} = \tau_{\theta z} = \tau_{rz} = 0. \tag{3.5.4}$$

In the equations (3.5.1)–(3.5.4),

$$T = \pi R_o^2(1 - c^2) - \frac{2\pi m R_o^2}{a_{33}}\left[\frac{1 - c^2}{2}(a_{13} + a_{23}) - \frac{(1 - c^{k+1})^2}{1 - c^{2k}} \cdot \frac{(a_{13} + ka_{23})}{k + 1} - \frac{(1 - c^{k-1})^2 c^2}{1 - c^{2k}} \cdot \frac{a_{13} - ka_{23}}{k - 1}\right],$$

$$c = \frac{R_i}{R_o}, \qquad k = \sqrt{\frac{\beta_{11}}{\beta_{22}}} = \sqrt{\frac{a_{33}a_{11} - a_{13}^2}{a_{22}a_{33} - a_{23}^2}},$$

$$m = \frac{a_{23} - a_{13}}{\beta_{11} - \beta_{22}}, \qquad \rho = \frac{r}{R_o} \qquad \text{(where } r \text{ is the current radius),}$$

$$\beta_{11} = a_{11} - \frac{a_{13}^2}{a_{33}}, \qquad \beta_{22} = a_{22} - \frac{a_{23}^2}{a_{33}},$$

$$a_{11} = \frac{1}{E_r}, \qquad a_{22} = \frac{1}{E_\theta}, \qquad a_{33} = \frac{1}{E_z},$$

$$a_{12} = -\frac{\nu_{r\theta}}{E_\theta} = -\frac{\nu_{\theta r}}{E_r}, \qquad a_{13} = -\frac{\nu_{rz}}{E_z} = -\frac{\nu_{zr}}{E_r},$$

$$a_{23} = -\frac{\nu_{\theta z}}{E_z} = -\frac{\nu_{z\theta}}{E_\theta}.$$

If $a_{13} \neq a_{23}$, the stress σ_z is distributed over the cross section nonuniformly and is accompanied by stresses σ_r and σ_θ in longitudinal sections; in the determination of the strength $\sigma_z^{t(c)u}$ the stresses σ_r and σ_θ must be considered stabilizing. The only case where tension-compression of an anisotropic tube is not accompanied by the appearance of radial and circumferential stresses is the case where the cross section of the tube coincides with the plane of transverse isotropy of the material (for example, in testing tubes from a unidirectional material where reinforcement is oriented strictly along the tube axis).

A tube made by transverse winding is monotropic. In this case, the relationships (3.5.1)–(3.5.4) are somewhat simplified:

$$E_r = E_z, \qquad \nu_{rz} = \nu_{zr}, \qquad \nu_{\theta r} = \nu_{\theta z}, \qquad \nu_{r\theta} = \nu_{z\theta},$$

$$a_{11} = a_{33} = \frac{1}{E_z}, \qquad a_{22} = \frac{1}{E_\theta}, \qquad a_{12} = a_{23} = -\frac{\nu_{\theta z}}{E_z} = -\frac{\nu_{z\theta}}{E_\theta},$$

$$a_{13} = -\frac{\nu_{rz}}{E_z}, \qquad \beta_{11} = \frac{1}{E_z}(1 - \nu_{rz}^2), \qquad \beta_{22} = \frac{1}{E_\theta}(1 - \nu_{z\theta}\nu_{\theta z}),$$

$$k = \sqrt{\frac{E_\theta}{E_z}\frac{1 - \nu_{rz}^2}{1 - \nu_{\theta z}\nu_{z\theta}}}.$$

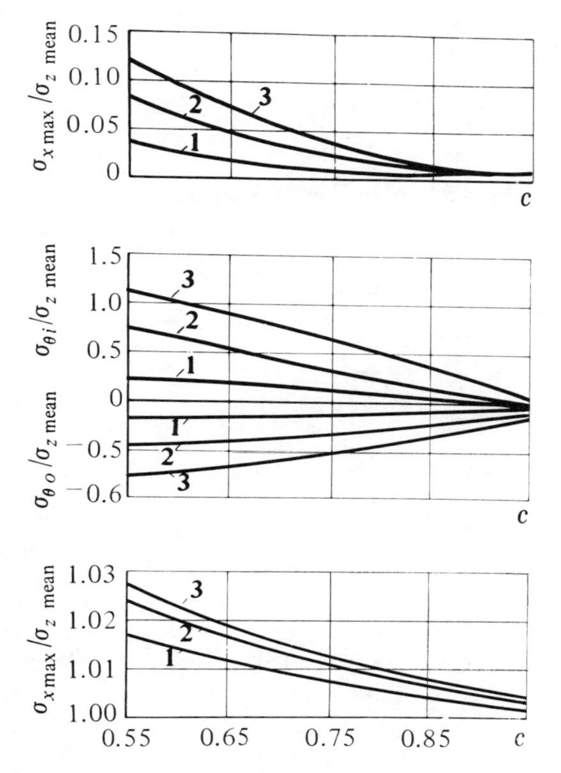

Fig. 3.5.1. Relative values of maximum stresses in tubular specimens of various relative thicknesses [30].

For given characteristics of the material being tested, the only possibility of controlling stabilizing stresses σ_r and σ_θ is by variation of relative thickness c (or R/h):* with increasing c, i.e., with transition to thinner-walled specimens, the effect of stresses σ_r and σ_θ decreases.

With an increase in the relative thickness c, distribution of stresses σ_z over the specimen thickness becomes more uniform. In Fig. 3.5.1, the relations of stresses $\sigma_{r\,\mathrm{max}}/\sigma_{z\,\mathrm{mean}}$, $\sigma_{\theta\,\mathrm{max}}/\sigma_{z\,\mathrm{mean}}$, and $\sigma_{z\,\mathrm{max}}/\sigma_{z\,\mathrm{mean}}$ (where $\sigma_{z\,\mathrm{mean}} = P/F$)

*The relationship between c and R/h is as follows:

$$c = \frac{2\dfrac{R}{h} - 1}{2\dfrac{R}{h} + 1} \quad \text{and} \quad \frac{R}{h} = \frac{1 + c}{2(1 - c)}.$$

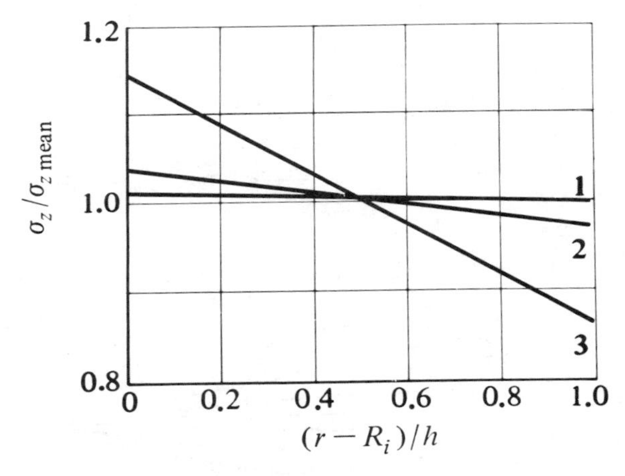

Fig. 3.5.2. Relative axial stresses in the gage length of the helically wound specimen (helical angle 30°) loaded in uniaxial tension [166]. Materials: (1) fiberglass ($E_L = 41.4$ GPa, $E_T = 13.8$ GPa, $G_{LT} = 5.12$ GPa); (2, 3) carbon composite ($E_L = 137.8$ GPa, $E_T = 6.9$ GPa, $G_{LT} = 4.14$ GPa). $R/h = 10$ (curves 1, 3) and $R/h = 40$ (curve 2). Directions: L—along, T—across the layer.

are shown for three materials with the following characteristics [30]:

	$E_\theta \times 10^{-4}$, MPa	$E_z \times 10^{-4}$, MPa	ν_{rz}	$\nu_{\theta z}$	k
Glass fiber composite	6.83	2.21	0.35	0.11	1.68
Boron composite	13.90	1.86	0.35	0.034	2.58
Carbon composite	9.00	0.69	0.35	0.033	3.41

It is clear from Fig. 3.5.1 that stresses σ_r may be considered negligible at $c > 0.80$ ($R/h > 4.5$), but the effect of stresses σ_θ must be evaluated in each concrete case by comparing it not only with stresses $\sigma_{z\,\mathrm{mean}}$, but also with the circumferential strength σ_θ^u of the material being tested.

Studies of the state of stress in a cylinder made by helical winding [166], show that the concept of a thin-walled tube is a function of the anisotropy of the material. For example, it follows from Fig. 3.5.2 that a fiberglass composite specimen can be considered to be thin-walled, i.e., to have a uniform state of stress in its gage length, at a ratio $R/h = 10$, but that a carbon composite specimen only at $R/h = 40$.

The modulus of elasticity $E_{z\,\mathrm{mean}}$ is experimentally determined from the formula

$$E_{z\,\text{mean}} = \frac{\sigma_{z\,\text{mean}}}{\varepsilon_{zo}} = \frac{P}{F} \cdot \frac{1}{\varepsilon_{zo}} \tag{3.5.5}$$

where $\sigma_{z\,\text{mean}} = P/F$ is the mean stress in the axial direction and ε_{zo} is the axial strain of the outer surface of a specimen.

Actually, by taking into account the anisotropy of a specimen, we obtain:

$$E_z = \frac{\sigma_{zo} - \nu_{\theta z} \alpha_{\theta o}}{\varepsilon_{zo}}.$$

For the previously mentioned materials at $c > 0.55$ ($R/h > 1.7$) the ratio $E_z/E_{z\,\text{mean}} < 1.03$, i.e., the error is negligible.

Much greater error is possible in the determination of Poisson's ratio $\nu_{\theta z}$. Experimentally, Poisson's ratio is determined from the formula

$$\nu_{\theta z\,\text{mean}} = -\frac{\varepsilon_{\theta o}}{\varepsilon_{zo}} \tag{3.5.6}$$

where $\varepsilon_{\theta o}$ and ε_{zo} are the strains of the outer surface of a specimen in the circumferential and axial directions, respectively.

In fact, allowing for the anisotropy of the specimen, we have

$$\nu_{\theta z} = \frac{\varepsilon_{\theta o} \nu_{z\theta} \sigma_{\theta o}}{\varepsilon_{zo}(\sigma_{\theta o} - \nu_{z\theta}\sigma_{zo}) + \varepsilon_{\theta o}\sigma_{\theta o}\nu_{z\theta}}.$$

It can be seen from Fig. 3.5.3, where the error in the determination of Poisson's ratio $\nu_{\theta z}$ is estimated by relationship $\nu_{\theta z\,\text{mean}}/\nu_{\theta z}$, that it is impossible to determine Poisson's ratio $\nu_{\theta z}$ for essentially anisotropic materials (curves 2 and 3) in this way.

It is possible to reduce the error in determination of Poisson's ratio $\nu_{\theta z}$ by measuring the strains on the outer ($\varepsilon_{\theta o}$) and inner ($\varepsilon_{\theta i}$) surfaces of the specimen. In this case, Poisson's ratio $\nu_{\theta z}$ is experimentally determined by the formula

$$\nu_{\theta z}^* = -\frac{\varepsilon_{\theta o} + \varepsilon_{\theta i}}{2} \cdot \frac{1}{\varepsilon_{zo}}. \tag{3.5.7}$$

In so doing, the condition $\varepsilon_{zo} = \varepsilon_{zi}$ should be satisfied. As can be seen from Fig. 3.5.3, where the relations $\nu_{\theta z}^*/\nu_{\theta z}$ for the materials discussed above are shown by a dashed line, the error in the determination of Poisson's

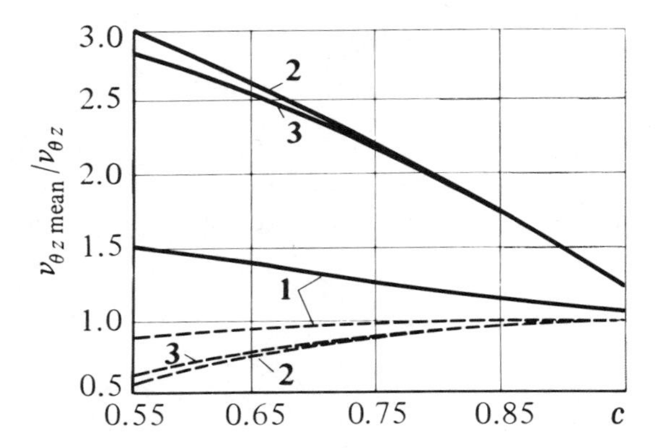

Fig. 3.5.3. The error in determination of Poisson's ratio on fiberglass (1), boron (2), and carbon (3) composite tubes [30].

ratio $\nu_{\theta z}$ in these cases is consierably smaller. In practice, however, measurement of the strains on the inner surface of a tubular specimen is not always possible.

From the point of view of experimental technique, each loading mode—tensile or compressive—has its own advantages and disadvantages. In tension, the main difficulty lies in fastening the specimens in the testing machine. During tensile testing, tubular specimens are fixed in the testing machine by means of special grips which must ensure reliable specimen fastening, eliminate failure of specimen end parts as a result of transverse compression and splitting, and at the same time introduce the smallest possible distortions into the specimen's state of stress. These requirements are controversial. Reliable fastening and protection of end parts of the specimen require the use of massive and rigid grips; however, solely because of the high rigidity of the grips considerable decrease in the measured strength of the specimens can be observed. For example, in testing carbon composite flat and tubular specimens, the following results have been obtained (on the left the values σ_z^{tu} are given for a 0° layup, on the right for 0/90°) [257]:

	σ_z^{tu}, MPa
Flat specimens	1056/587
Tubular specimens	648/455

When rigid grips are used, circumferential and radial deformations of the end parts of a specimen are restrained, so that appreciable bending stresses are initiated there which are even transmitted to the grips and result in premature failure of the specimen. It has been noted in [257] that the state of stress in the end parts of the specimen can be improved by the use of an adhesive joint, where grips are glued to the specimen.

If a tubular specimen has thicker end parts, the grip design must ensure application of the tensile or compressive force in the mid-plane of the specimen wall. Otherwise (for example, in fastening the specimen by means of screw threads on the outer surface of the thicker end parts) bending stresses develop in the specimen, with noticeable effect even at a significant distance from the fastened section.

The length of a tubular specimen in tensile tests is determined with an allowance for the edge effect. Investigations [257] show that the concentration of stresses at the points of attachment of the specimen fastening depends only weakly on the reinforcement layup (the layups 0, 145, 90° have been investigated); with a correct grip design it is quickly attenuated. Fig. 3.5.4 shows the distribution of the relative stresses $\sigma_z/\sigma_{z\,\mathrm{mean}}$, $\sigma_\theta/\sigma_{z\,\mathrm{mean}}$, and $\tau_{\theta z}/\sigma_{z\,\mathrm{mean}}$ in the edge effect zone, calculated for a tubular carbon composite specimen. It follows from the figure that the edge effect is attenuated within a region of length $2R$, and the full length L (between the grips) of the specimen can be calculated from the empirical formula

$$L = l + 4R \tag{3.5.8}$$

where l is the gage length and R is the mean radius of the specimen.

In axial compression tests of tubular specimens, there are no problems connected with attachment of the specimen, except the necessity of preventing bending moments. At the same time, all the peculiarities of testing of flat specimens are preserved: the concentration of deformations around the end surfaces, total or local loss of stability, and bearing and splitting of the end faces. Consequently, in compression tests the relative thickness h/R of the specimen must be increased (over the respective value in tensile tests), so that additional difficulties in ensuring of a uniform state of stress are created. To prevent bearing and splitting, the specimen end surfaces are frequently reinforced with overwinding of a unidirectional or woven material. However, this process is ineffective and laborious, since fracture is observed on the boundary between specimen and overwinding. In practice, profiled supports (Fig. 3.5.5) have proved usable. The tubular specimen is placed into the gap between the outer (1) and inner (2) cups, the walls of which are of varying thickness. The inner cup has an air outlet, the outer one an outlet for release. The gap between specimen walls and support is filled with a cured epoxy resin or Wood's metal.

Measurement of deformations on tubular specimens is carried out analo-

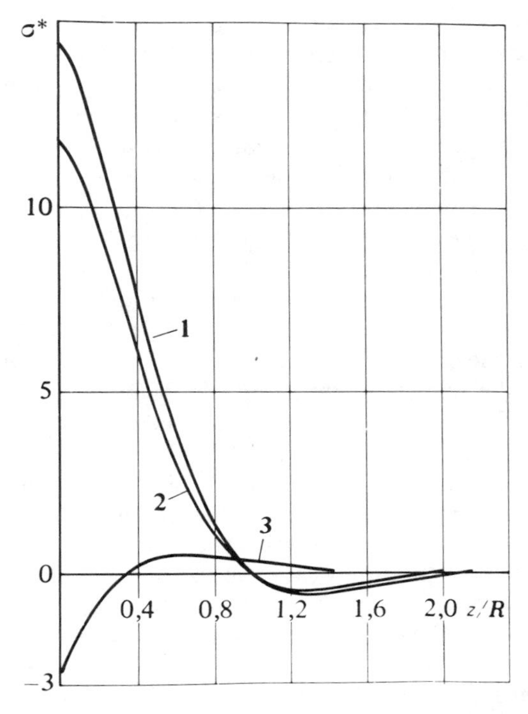

Fig. 3.5.4. Maximum relative stress in the end part of the helically wound specimen (helical angle 30°) loaded in uniaxial tension [166]. $R/h = 10$; material—carbon composite ($E_L = 137.8$ GPa, $E_T = 6.9$ GPa, $G_{LT} = 4.14$ GPa). On the ordinate axis are plotted:

$$1 - \frac{\sigma_z}{\sigma_{z\,\text{mean}}} \times 10^2; \qquad 2 - \frac{\tau_{\theta z}}{\sigma_{z\,\text{mean}}} \times 10^4;$$

$$3 - \frac{\sigma_\theta}{\sigma_{z\,\text{mean}}} \times 10^2, \quad \text{where} \quad \sigma_{z\,\text{mean}} = P/F.$$

gously to measurement of deformations in tension tests of flat specimens. The tensiometers are placed, as a rule, across the diameter of the specimen section; the number of pairs of tensiometers depends on the test purpose and specimen dimensions.

3.5.2. Tests under Internal and External Pressure

Testing of tubes under internal or external pressure serves for quality control of filament wound articles at winding angles other than zero, and

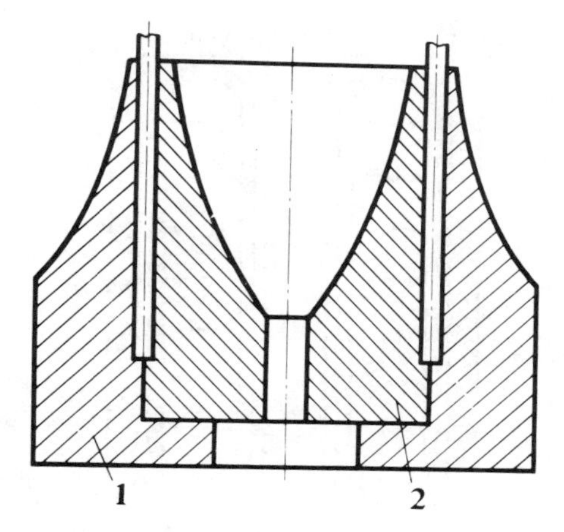

Fig. 3.5.5. Profiled support of tubular specimens in compression tests [113]: (1) outer sleeve; (2) inner sleeve.

for investigation of mechanical properties of materials, mainly the circumferential modulus of elasticity E_θ and strength σ_θ^u. In the internal pressure loading of a tube plugged with a cap (Fig. 3.5.6) and not bearing the axial load, a biaxial state of stress is created in each element of its wall: both circumferential (σ_θ) and radial (σ_r) stresses act in it. If the tube is thin-walled, the radial stresses can be neglected.

Uniformity of the state of stress in a tube wall depends on the anisotropy of the material E_z/E_θ and on the ratio R/h. It can be seen from Fig. 3.5.7, for example, that to ensure an equal distribution of circumferential stresses over the thickness of the wall, the relative thickness of carbon composite specimen must be reduced fourfold, compared to a glass fiber composite specimen, i.e., the concept of a thin-walled tube is dependent on the anisotropy of the material.

In order to ensure tightness of the specimen right up to failure (disruption of the integrity of the material in this case connected with the appearance of a break in the tension diagram of the material), elastic bladders are placed inside the tubes. The inserted plugs are sometimes used as plungers in a hydraulic system.

The pressure inside the tubes is created hydraulically. The working liquid of such a system is usually water or oil; sometimes an inert gas (for example, nitrogen) is used. In testing, a tube can be put into a water bath or some other liquid, which permits a constant ambient temperature to be maintained, and where gas is used, controls the tightness of specimens. The rate of increase in the pressure of a hydraulic system should be constant.

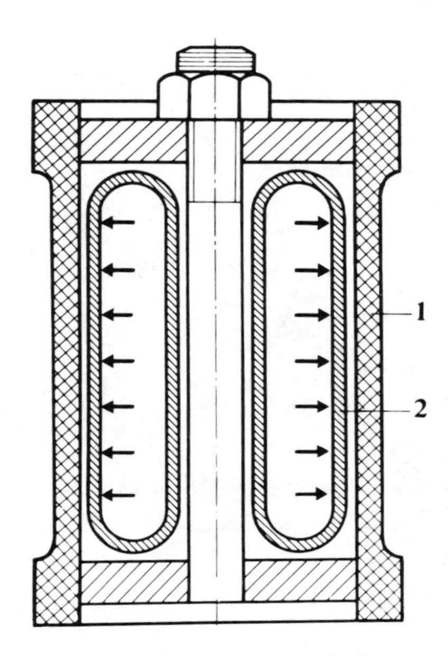

Fig. 3.5.6. Loading scheme of tubes with internal pressure [39]: (1) specimen; (2) elastic bladder of hydrosystem.

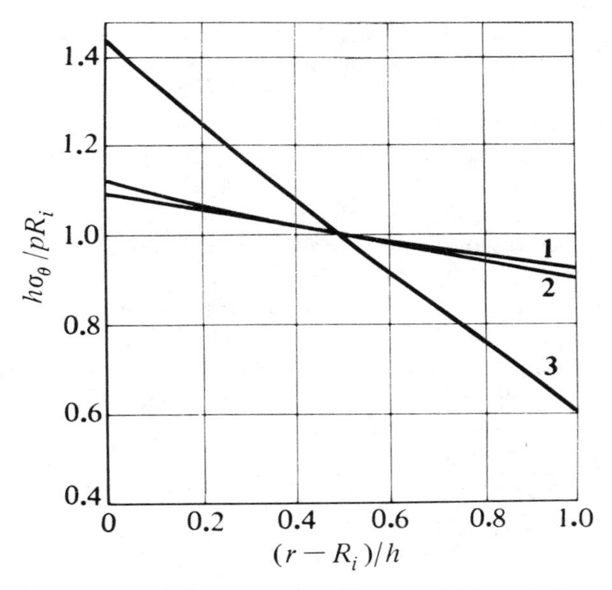

Fig. 3.5.7. Relative circumferential stresses in the gage length of a helically wound specimen (helical angle 60°) loaded with internal pressure [166]. Material: (1) glass fiber composite ($E_L = 41.4$ GPa, $E_T = 13.8$ GPa, $G_{LT} = 5.12$ GPa); (2, 3) carbon composite ($E_L = 137.8$ GPa, $E_T = 6.9$ GPa, $G_{LT} = 4.14$ GPa). $R/h = 10$ (curves 1 and 3) and $R/h = 40$ (curve 2).

Circumferential stresses in thin-walled tubes and the modulus of elasticity are calculated by the formula

$$\sigma_\theta = \frac{pD}{2h} \tag{3.5.9}$$

$$E_\theta = \frac{pD}{2h\varepsilon_\theta} \tag{3.5.10}$$

where p is the pressure in a hydraulic system; D is the mean diameter of the specimen; h is the wall thickness of the specimen; and ε_θ is the circumferential strain under internal pressure p measured by resistance strain gages.

The circumferential and radial stresses in thick-walled tubes under internal pressure are determined by Lamé's formulas:

$$\sigma_\theta = \frac{pR_i^2}{R_o^2 - R_i^2}\left(1 + \frac{R_o^2}{r^2}\right), \tag{3.5.11}$$

$$\sigma_r = \frac{pR_i^2}{R_o^2 - R_i^2}\left(1 - \frac{R_o^2}{r^2}\right) \tag{3.5.12}$$

where R_i, R_o, and r are inside, outside, and current radii of the specimen, respectively.

The circumferential and radial stresses are maximum at $r = R_i$:

$$\sigma_{\theta\,\max} = \frac{R_o^2 + R_i^2}{R_o^2 - R_i^2}p \tag{3.5.13}$$

$$\sigma_{r\,\max} = -p. \tag{3.5.14}$$

Equations (3.5.13) and (3.5.14) are used to determinate the maximum stresses at specimen failure, when $p = p^u$.

Tests of tubes under external pressure are less widespread than tests under internal pressure; however, this method has been standardized [ASTM D 2586-68 (Rev. 1974)], and it is intended for qualitative evaluation of the effects of reinforcement and polymer binder content, winding schemes, and cure cycle on the strength and for quality control of the item. The standard is intended for items with a reinforcement content no less than 50% by weight.

Testing of tubes using external pressure is accomplished in a pressure chamber provided with the equipment necessary for hydrostatic tests. The hydraulic system must ensure an even increase in pressure not exceeding 28 MPa/min.

According to ASTM D 2586-68 (Rev. 1974) the length of the specimen should be 152 mm, the inner diameter 67 mm, and the wall thickness 10 mm ($h/R \approx 0.3$, i.e., a thick-walled tube). These specimen dimensions have been checked on epoxy and *E*-glass materials. For other materials, it is recommended to recalculate the wall thickness in accordance with the material characteristics. The specimen ends are capped with sealing plugs. It should be taken into account in production of the sealing plugs that the part which enters the specimen is not cylindrical but forms a half-roll; its dimensions are given in ASTM D 2586-68 (1974) standard. If the plugs are a cylindrical, then in loading the stress concentration is initiated in the specimen by restricting deformation, and the measured strength is somewhat reduced [144].

The maximum circumferential stresses (at $r = R_i$) at failure of the specimen are determined from Lamé's formula:

$$\sigma_{\theta\,\max} = -\frac{2R_o^2}{R_o^2 - R_i^2} p^u \tag{3.5.15}$$

where p^u is the pressure at rupture of the specimen.

3.6. BEARING

3.6.1. Standard Method

Bearing tests of fibrous polymeric composites are used to evaluate the behavior of the material in joints (bolted, riveted, etc.) and bearing surfaces of structures, and they only indirectly characterize the mechanical properties of the material. These tests determine the load at which bearing deformation does not damage the integrity of a part of structure. The standardized bearing test method is very simple: a metal pin is inserted into the calibrated opening in the specimen, to which a tensile or compression load is applied; displacement of the specimen, i.e., deformation of the opening in the material and the compression or tensile load applied are measured.

The initial sections of bearing curves for fibrous polymeric composites are similar to the initial sections of the stress-strain curves under tension or compression, a typical bearing curve is shown in Fig. 3.6.1. However, at present, there is not a single method, even for isotropic materials, for direct determination of bearing strength from tension or compression stress-strain curves. Bearing tests are carried out under tensile or compression

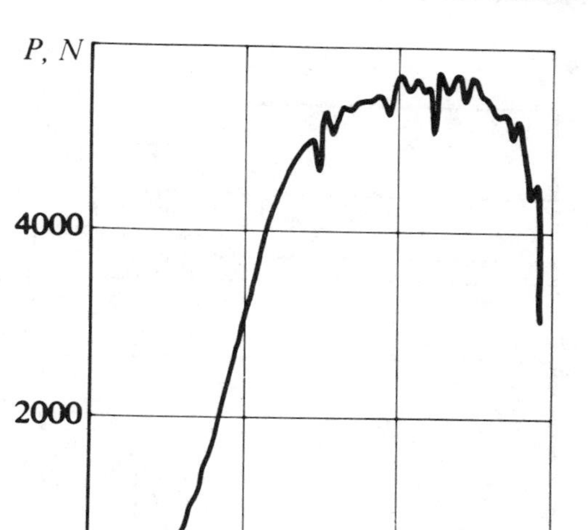

Fig. 3.6.1. Typical bearing curve [196].

loads directed along the longitudinal axis of the specimen. Bearing strength is greater under compression than under tension. Therefore, for a complete evaluation of material behavior bearing tests with both types of loading should be conducted. The direction of the load relative to the principal axes of the material is selected in accordance with the purpose of the test.

The specimens for bearing tests of fibrous polymeric composites have not been thoroughly standardized; in the U.S.A., ASTM D 953-69 standard for rigid plastics is used. As an illustration, in Fig. 3.6.2, a specimen in accordance with ASTM D 953-69 is shown. Thick specimens ($h = 6$ mm) give more accurate results, but for materials characterized by brittle failure it is sounder practice to use thin speciments, which are less inclined to premature failure. In testing fibrous polymeric composites, the width of specimens must be established experimentally, since the optimum value of $\lambda = d/b$, i.e., the value at which bearing failure is ensured, depends on the properties of the material [196].

Selection of the diameter of the opening d depends on sepecimen thickness h. For precise measurement of the deformation special fixtures and sensitive measuring instruments are necessary (for example, the value of a division of the indicator should be no more than 0.002 mm). The carbon composite specimens [Fig. 3.6.2(b)] used for tests in [97] have one peculiarity compared

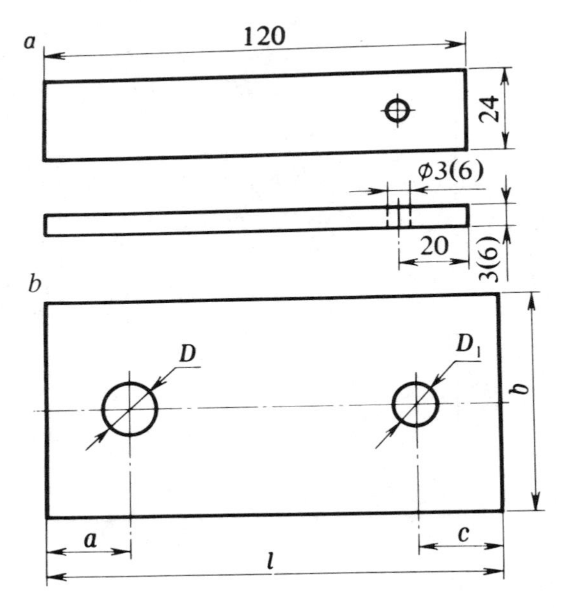

Fig. 3.6.2. Specimens for bearing tests of rigid plastics (a) and carbon composite (b) [11, 97].

with standard ASTM specimens: the diameter of one of the openings is somewhat larger than the other; in this manner a reference section is fixed.

In the ASTM standard, the following recommendations on conduct of bearing tests are presented. Bearing strain $\varepsilon_{\text{bear}}$ is measured by stages until the strain of the opening reaches $\varepsilon_{\text{bear}} = 4\%$, and then at maximum load. The bearing stress σ^u_{bear} and the tangential modulus of elasticity E_t are determined as follows:

$$\sigma^u_{\text{besr}} = \frac{p}{dh}, \tag{3.6.1}$$

$$E_t = \frac{\sigma_{\text{bear}}}{\varepsilon_{\text{bear}}}. \tag{3.6.2}$$

Equation (3.6.1) is conventional, since an even distribution of the normal stresses over the contact surfaces is assumed.

The recommended loading rate (for non-reinforced rigid plastics) is 1 mm/min; the loading rate is most often, determined individually, depending on the mechanical properties of the material being studied.

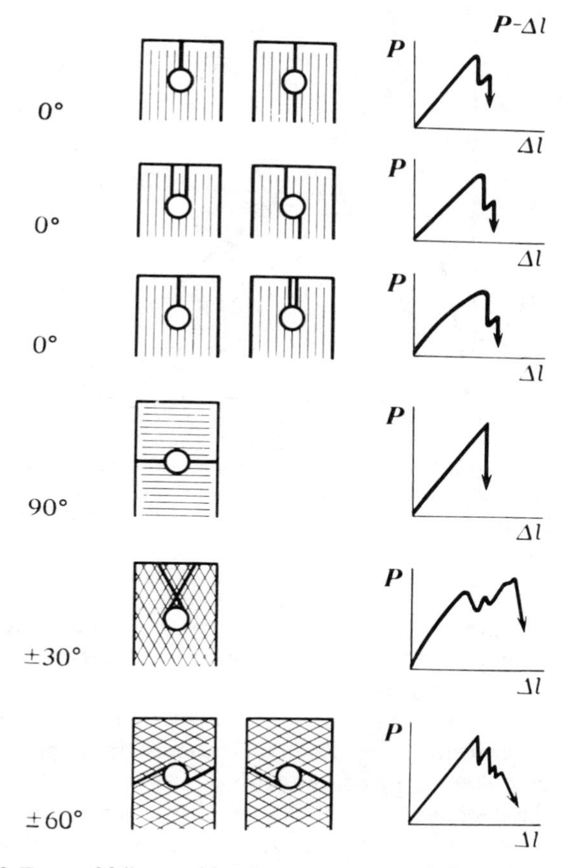

Fig. 3.6.3. Types of failure and bearing curves versus reinforcement layup [245].

3.6.2. Failure Diagram

Bearing strength is a very arbitrary characteristic of a material, since in transmission of the load to the sheet through the pin a very complex state of stress is set up around the pin. At a certain stress plastic deformation is observed near the pin and folds are formed, but the load absorbed by the specimen continues to increase, even with clear failure of the material (see Fig. 3.6.1). In the general case, failure of the specimen can occur as a consequence of bearing, breaking, shearing, or a combination of these types of failure. Furthermore, the development of cracks can also depend on the reinforcement layup (Fig. 3.6.3) and the dimensions of the specimen, i.e., the location of the opening.

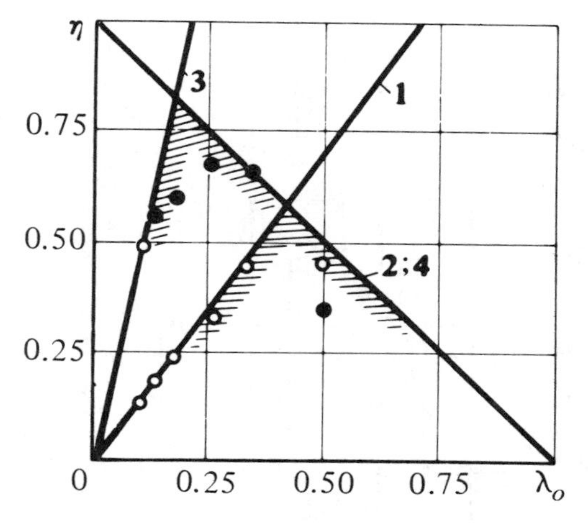

Fig. 3.6.4. Typical bearing test diagram [196]; material—SVAM; (1, 3) calculated straight lines of bearing η_{bear}; (2, 4) calculated straight lines of breakage η_{break}. Direction of the load relative to the fibers: $\alpha = 0°$ (curves 1, 2) and 45° (curves 3, 4). Experimental points: *open circle*: $\alpha = 0°$. *Filled circle*: $\alpha = 45°$.

The numerical characteristics of bearing also depend on the reinforcement layup. Thus reinforcement laid at an angle of 145° markedly increases the bearing strength. The edge effect, i.e., the effect of interlaminar stresses appearing around the edge of an opening, also depends on the reinforcement layup. Stress concentration, i.e., the state of stress in a specimen depends on the relative diameter of the opening d/b.

In [196, p. 92] it is suggested that the λ versus η failure diagram (Fig. 3.6.4) to used to evaluate the load-carrying capacity of the material, where

$$\lambda = d/b, \tag{3.6.3}$$

$$\eta = P^u/P, \tag{3.6.4}$$

with λ a parameter and η the strength utilization factor. Here d is the diameter of the opening in the specimen; h is the specimen thickness; b is the specimen width; P^u is the breaking load for specimen with an opening; $P = bh\sigma_x^{t(c)u}$ is the calculated breaking load for an intact specimen (without an opening) of cross-sectional dimensions b and h; $\sigma_x^{t(c)u}$ is the tensile (t) or compression (c) strength of the material.

If failure is caused by a break along the minimum section, the strength utilization factor is

$$\eta_{\text{break}} = \frac{1-\lambda}{k_{\text{break}}}. \tag{3.6.5}$$

The stress concentration factor at failure k_{break} is calculated from the formula

$$k_{\text{break}} = \frac{\sigma_x^{t(c)u}}{\sigma_{\text{mean}}} = \frac{(1-\lambda)bh}{P^u}\sigma_x^{t(c)u} \tag{3.6.6}$$

where σ_{mean} is the mean stress in the specimen.

Equation (3.6.5) describes a curve in λ, η coordinates (Fig. 3.6.4).

If the failure is caused by bearing, the strength utilization factor is

$$\eta_{\text{bear}} = \frac{\sigma_{\text{bear}}^u}{\sigma_x^{t(c)u}}\lambda. \tag{3.6.7}$$

If σ_{bear}^u is a characteristic of the material which is independent of specimen dimensions, (3.6.7) is a straight line through the origin in λ, η coordinates (Fig. 3.6.4).

Thus, the $\eta = \eta_{\text{break}}$ and $\eta = \eta_{\text{bear}}$ curves restrict the possible region of failure (cross hatched in Fig. 3.6.4). Bearing tests are conducted on specimens of various widths, i.e., with various $\lambda = d/b$ ratios. Only those specimens for which the experimental points fall on the straight line $\eta = \eta_{\text{bear}}$ can be assumed to fail in bearing. This line fitted to the experimental points will be used for determination of the angular coefficient $\sigma_{\text{bear}}^u/\sigma_x^{t(c)u}$ in the formula (3.6.7) and, consequently, for determination of σ_{bear}^u.

* * *

The basic information on tensile, compression and bearing test methods treated in Chapters 2 and 3 is presented in Summary Table I. The purpose of this and subsequent summary tables is to show pictorially the capabilities of each of the test methods presented, to cite the primary references, list the measured and determined, values and to promote efficient experimental planning.

Comparison of methods from economical point of view—the amount of material used for fabrication of specimens, the cost of specimens and testing—is not presented in a table. These values vary greatly depending upon the purpose of test, cost of test materials (production parts vs. laboratory samples), sophistication of the equipment in the laboratory and qualifications of technical personnel.

Summary Table I. Tensile, Compression, and

METHOD	STANDARDS	CHARACTERISTICS DETERMINED		EXPERIMENTALLY MEASURED	
		ELASTIC CONSTANTS	STRENGTH	IN ELASTIC CONSTANT DETERMINATION	IN ST[...] DETER[...]
FLAT AND TUBULAR SPECIMENS					
TENSION	ASTM D 3039-74 BS 2782, Part 10, Method 1003 DIN 53392 GOST 11262-76 GOST 25.601-80	E_x^t, E_y^t, E_z^t, ν_{xy}^t, ν_{xz}^t, ν_{yz}^t	σ_x^{tu}, $_y^{tu}$, σ_z^{tu}	P^t (load), ε_x^t, ε_y^t, ε_z^t (strains)	P^{tu} (load [...] failure o[...]
at angle θ to the principal axes of elastic symmetry	—	E_θ^t, ν_θ^t indirectly G_{xy}, G_{xz}, G_{yz}	σ_θ^{tu}	P^t (load), ε_θ^t (strain)	The same
COMPRESSION	ASTM D 3410-75 GOST 25.602-80	E_x^c, E_y^c, E_z^c, ν_{xy}^c, ν_{xz}^c, ν_{yz}^c	σ_x^{cu}, σ_y^{cu} σ_z^{cu}	P^c (load), ε_x^c, ε_y^c, ε_z^c (strains)	P^{cu} (load [...] sion failu[...] specimen[...]
at angle θ to the principal axes of elastic symmetry	—	E_θ^c, ν_θ^c indirectly G_{xy}, G_{xz}, G_{yz}	σ_θ^{cu}	P (load), ε_θ^c (strain)	The same
BEARING	—	$E_t^{t(c)}$	$\sigma_b^{tu(cu)}$	$P^{t(c)}$ (load) $\varepsilon_b^{t(c)}$ (strain)	$P^{tu(cu)}$ (loa[...] failure of
interlaminar tension	—	—	σ_z^{tu}	—	P^{tu} (load a[...] specimen laminar t[...]
RING SPECIMENS					
TENSION by rigid split disks (NOL method)	ASTM D 2290-76 ASTM D 2291-76	E_θ^t	σ_θ^{tu}	P (load), ε_θ^t (circumferential strain)	P^u (load at[...] ring)
with a compliant ring or hydrostatically	—	E_θ^t	σ_θ^{tu}	p (internal pressure), ε_θ^t (circumferential strain)	p^u (interna[...] failure of
COMPRESSION by rigid split disks (NOL method)	—	E_θ^c	σ_θ^{cu}	P (load), ε_θ^c (circumferential strain)	P^u (load at[...] ring)
hydrostatically	—	E_θ^c	σ_θ^{cu}	p (external pressure), ε_θ^c (circumferential strain)	p^u (externa[...] failure of
hydraulically on a rigid mandrel	—	E_r	—	The same	—
TESTING OF TUBES					
with internal pressure	—	E_θ	σ_θ^u	p (internal pressure), ε_θ (circumferential strain)	p^u (interna[...] specimen[...]
with external pressure	ASTM D 2586-68	—	σ_θ^u	—	p^u (externa[...] specimen[...]

t Methods for Fibrous Polymeric Composites.

RECOMMENDED SPECIMEN SHAPES		TESTING EQUIPMENT	DEFICIENCIES AND RESTRICTIONS OF METHOD	SECTION OF THIS BOOK
OR DETERMINATION ELASTIC CONSTANTS	FOR DETERMINATION OF STRENGTH			
tangular parallel- iped (strip), speci- n with constant- ss-section gage gth, sandwich am (see summary ble III)	Rectangular parallel-epiped (strip), specimen with constant- or variable-cross-section gage length, sandwich beam	Tensile testing machine with special grips	Difficulty in fastening specimens in tests of high-strength unidirectional materials	2.1–2.3
tangular parallelepiped (strip), tubular cimen, sandwich beam		The same	Test results depend on choice of specimen dimensions and loading method; results of testing flat and tubular specimens are not comparable	2.4 and 3.5.1
tangular parallelepiped (strip), specimen h constant- or variable-cross-section gage gth, bars of circular cross section, sandwich am		Tensile testing machine with fixture for centering load and prevention of stability loss	Possibility of change from one type of failure to another (compression, shearing, stability loss)	3.1–3.3
tangular parallelepiped (strip), tubular cimen, sandwich beam		Tensile testing machine with fixture for loading without constriction of deformation, load-centering, and prevention of stability loss	Results of testing flat specimens depend on choice of specimen dimensions are loading method; test results on flat and tubular specimens are not comparable	2.4–3.5.1
angular strip with a calibrated opening		Tensile machine with fixture for installation of specimen.	Method is suitable for qualitative comparison of tested materials	3.6
	Flat, tubular, or circular specimen	Tensile machine with special grips	Account of stress concentrations is necessary	2.5 and 5.5.3
-walled ring ($h/R < 0.1$)		Tensile machine, rigid split disks or similar devices	Nonuniform distribution of circumferential strains	2.6
The same		Sources of pressure: hydraulic system or compliant ring inside the specimen	Complex testing equipment, in particular, hydraulic; high accuracy required in production of specimens	2.6
The same		Tensile machine, rigid split disks	Nonuniform distribution of circumferential strains	3.4
The same		Sources of pressure: hydraulic system, compliant ring, level system, steel tape	Complex testing device; possibility of change from one type of failure to another (compression, stability loss); necessity of auxilliary experiments	3.4
, thin-walled	—	Source of pressure: hydraulic system, special testing device		3.4
walled wound tube ($h/R < 0.1$)		Device for hydraulic tests	Limited use for thick-walled tubes	3.5.2
	Thick-walled wound tube	The same	For qualitative comparison of different materials	3.5.2

Chapter 4
Shear Testing

4.1. METHODS OF STUDYING SHEAR RESISTANCE

Polymeric composites with a laminated and fibrous structure have a typical shortcoming—their low shear resistance, especially in planes where the properties of the material are determined by the matrix. Low shear resistance is not only low shear stiffness, but also low shear strength. For many structures even small tangential stresses can cause loss of load carrying capacity. Therefore, correct determination of shear characteristics is of considerable singificance.

One of the principal difficulties in the development of methods of shear testing of fibrous polymeric composites is to provide pure shear in the specimens and thereby assure sufficient precision of the methods used in processing the experimental results. As pointed out in [2], these difficulties in the development of test methods might already have been overcome, were it not for such considerations as facilities for the fabrication of specimens of a given shape, consumption of material during fabrication (this is of equal importance in the development of new materials), and simple specimen fastening and loading mechanisms.

The difficulty of establishing pure shear in a specimen increases with increasing anisotropy and inhomogeneity of the material. As these characteristics increase so does the end effect zone, and in testing of high-modulus and high-strength fibrous polymeric composites there may be cases where it is impossible to obtain adequate zones in the specimen for measurement with a uniform state of stress. Therefore, shear testing of high-modulus and high-strength fibrous polymeric composites requires special attention.

Shear test methods substantially depend on the structure of the material and the orientation of the load to the axes of symmetry of the material. Depending on the orientation of the external shear forces to the direction

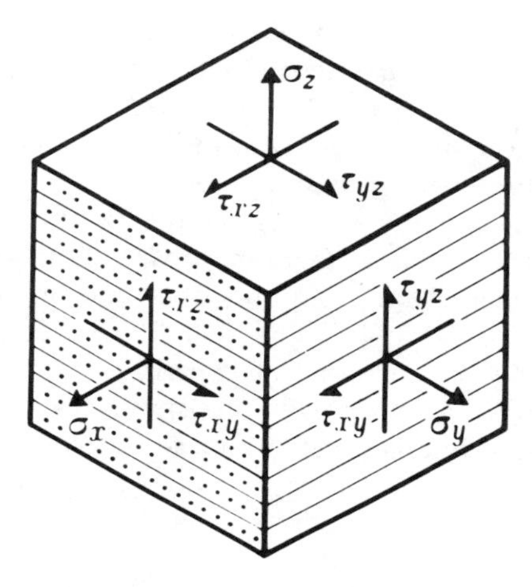

Fig. 4.1.1. Tangential stresses in the plane of the reinforcement: $\tau_{xy} = \tau_{yx}$; interlaminar stresses $\tau_{xz} = \tau_{zx}$ and $\tau_{yz} = \tau_{zy}$.

of fiber layup in fibrous polymeric composites two varieties of shear are distinguished. If the x and y axes coincide with the reinforcement plane, shear caused by stresses $\tau_{yx} = \tau_{xy}$ is called *in-plane shear*, and shear from stresses τ_{xz} or τ_{yz} is called *interlaminar shear* (Fig. 4.1.1). In the study of in-plane shear characteristics there are less stringent requirements on the temperature and time conditions than in the study of interlaminar shear, where the characteristics of the material are primarily determined by the temperature- and time-sensitive matrix. While in-plane shear can be studied on thin specimens, specimens of greater thickness are necessary for study of the interlaminar shear characteristics. This creates additional, mainly technological, difficulties for specification of fibrous polymeric composites. It should be noted that for multidirectionally reinforced composites the concept of interlaminar shear loses its meaning because of the existence of interlayer bonds. For these materials, special methods of estimating shear characteristics are sometimes needed.

At present there is a variety of methods of experimental determination of the elastic constants and shear strengths. However, there are still at the same time, no economical methods of testing of all types of polymeric composites for simultaneous determination of the elastic constants as well as shear strengths. Moreover, specimen shape and dimensions depend greatly on the purpose of the test: whether the elastic constants or the shear strength are being determined.

Because of the complexity of the entire set of questions connected with experimental determination of shear characteristics, there are practically no standardized methods of shear testing of fibrous polymeric composites. Standards ASTM D 2344-76 and ASTM D 2733-70 for interlaminar shear strength determination and ASTM D 3518-76 for determination of the in-plane shear modulus constitute an exception.

For study of the shear characteristics four types of specimens are used: tubular specimens, rods or bars, plates or strips, and rings and ring segments. Depending on specimen shape and dimensions and the purpose of the test, the specimens are loaded by torsion, bending, and uniaxial or biaxial tension-compression. The methods of experimental determination of the shear characteristics of fibrous polymeric composites are divided into three groups: in-plane shear, interlaminar shear and shearing-off tests. Respective stress states in a specimen can be formed by various means. Detailed descriptions are given in later sections of this chapter. Here we shall limit ourselves to a short general survey.

In-plane shear characteristics are studied using torsion of thin-walled tubes, panel shear testing, square plate twisting and tension of anisotropic strips.

Torsion of thin-walled tubular specimens is a comparatively general method, i.e., if permits estimation of both shear strength and stiffness. However, the use of the method is limited by the large consumption of material and the need for special equipment to fabricate and test the specimens. Because tubular specimens are produced mainly by winding, this method does not permit estimation of the shear characteristics of flat specimens. In order to study the characteristics of these products, one must resort to test methods for plates, rods, and bars. The main advantage of torsion of thin-walled tubes is the uniform state of stress produced, as a result of which this method is a control method for determination of in-plane shear characteristics. The method is treated in Section 4.2.

The panel shear test can be accomplished in several ways: plate shear test in a four-part frame, panel shear test, and rail shear test. Four-part frame and panel shear tests consume large quantities on the material. Furthermore, shear tests in a four-part frame have quite noticeable technical drawbacks. Therefore, rail shear tests have become very widespread. Plate shear tests permit determination of both shear modulus and shear strength; they are treated in detail in Section 4.3.

Square plate twist is a widely used method. This method is economical and experimentally well substantiated. However, the method is mainly employed for determination of elastic constants (combined with testing of a plate in bending it is possible to determine 18 elastic constants [118]). Square plate twist is treated in Section 4.4.

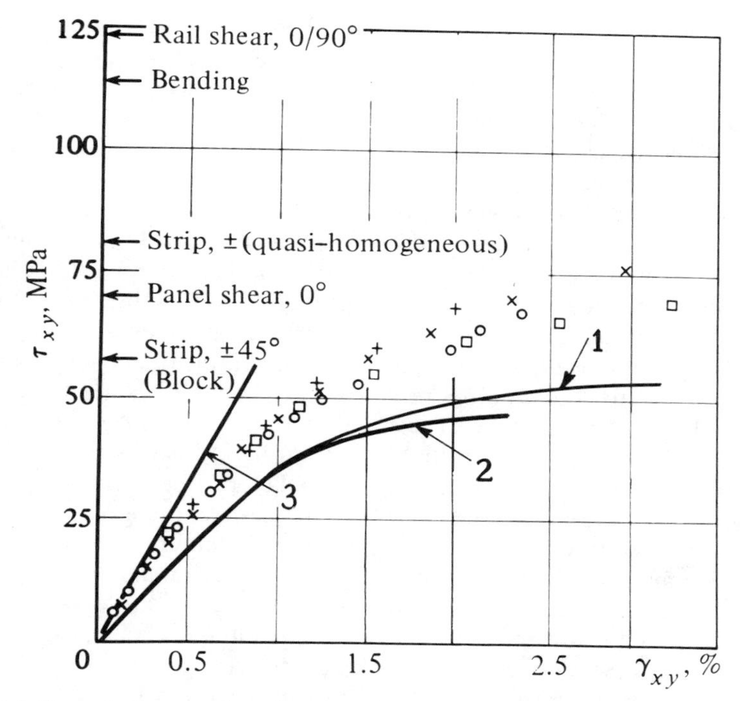

Fig. 4.1.2. Stress-strain curves for carbon-epoxy composite in shear. Test methods and material layup:

Panel shear, four-part frame: ○, 0°; □, 90°. Rail shear: +, (0/90°). Tension of a strip: *crosses*, (±45°). Torsion: tube of circular (1) and quadratic (2) cross section with layup [$0°/90°_2/0°/90°_2/0°/90°_2/0°$]; square plate twist (3).

Tension of an anisotropic strip is the simplest and most economical method for determination of the in-plane shear modulus, but it is somewhat imprecise. It is treated in Section 4.5.

In the study of shear characteristics, the relationships between the normal and tangential stresses and the respective strains under biaxial loading can be used. The common modes of loading are treated in Section 4.7.

Experimental comparisons [234] of these methods (panel shear test in a four-link rig, rail shear test, tension of an anisotropic strip, and square plate twist) show that in determination of the in-plane shear modulus, all methods yield quantitatively comparable results (see the strain curves in Fig. 4.1.2). In the determination of shear strength, the panel shear test and tension of an anistropic strip are quantitatively comparable, but the strengths obtained by the rail shear test and three-point bending tests distinctly stand out. The causes of these discrepancies are outlined in subsequent sections of the chapter.

The possible ways of studying the interlaminar shear characteristics of polymeric composites are few compared to the set of methods for study of the in-plane shear. This is a result of material structure and the difficulty of forming the necessary state of stress.

Elastic constants of a laminated and fibrous polymeric composite in interlaminar shear are determined primarily by the "work" of a polymeric interlayer, but the strength is determined using the adhesive force at the matrix-reinforcement interface and tangential stresses acting on that interface. Therefore, in the experimental determination of interlaminar shear strength, it is important to know the actual magnitude of tangential stresses which can lead to failure of the specimen. The maximum value of tangential stresses depends on the type of test and loading scheme, on the specimen shape and dimensions, and on all deviations from idealized structure caused by manufacturing (irregular reinforcement layup, fiber waviness, voids). Analytical evaluation of all these factors is impossible; therefore, experimentally determined interlaminar shear characteristics are mean values and are applicable mainly for a qualitative comparison of materials.

The oldest method of study of interlaminar shear characteristics as well as shearing and splitting is tension or compression of grooved specimens. This method is economical and simple to perform, but it is very sensitive to specimen quality and conduct of the experiment; it is treated in Section 4.6. Study of interlaminar shear characteristics in three-point bending tests is widespread, but highly unreliable; it is treated in detail in Chapter 5. A simple method of determination of the interlaminar shear strength is torsion of waisted specimens; it is treated in Section 4.9.5.

The methods of determination of the shearing-off resistance of polymeric composites cannot be treated as forming an independent kind of test, and their treatment in a special section is justified only from the point of view of the purpose and technology of an experiment. Owing to a complex stress-strain state, shearing-off tests are inapplicable for study of shear characteristics. This method is treated in Section 4.8.

All the test methods mentioned above (except the three-point bending test) can be called direct methods: in testing according to these methods, the elastic constants or strength are calculated directly from the experimentally measured values. These methods and the results obtained can be easily checked and analyzed. In practice, however, it is just those shear test methods where results are obtained by laborious recalculation that are in widespread use. Among these methods are bending tests and torsion of bars with a straight and circular axis.

Table 4.1.1. Shear Characteristics of a Unidirectional Organic Epoxy Composite (Kevlar 49, $V_f = 65\%$), obtained by various methods [38].

	SHEAR STRENGTH		SECANT SHEAR MODULUS AT = 0.5%	
METHODS	τ^u, MPa	v, %	G, MPa	v, %
Torsion of 90° wound thin-walled tubes	31.3	6.0	1744	2.1
Torsion of 0° wound rod of circular cross section:				
strain measured with resistance strain gages	31.3	7.2	1965	3.6
angle of twist measured	32.5	7.5	1758	1.0
Tension of ±45° off-axis laminates:				
midplane symmetry (7 layers)	29.4	2.0	1923	4.7
midplane symmetry (7 layers, repeated)	27.9	2.4	1889	0.8
nonsymmetrical (8 layer)	31.7	3.6	1875	3.9
Tension of 10° off-axis laminates:				
with coated edges	19.4	3.6	2082	5.7
with bare edges	19.1	9.4	1903	6.3
Three-point bending				
prismatic bars	36.8	5.1	—	—
ring segments	38.4	4.1	—	—
Tension of grooved bars	7.7–23.5	2.8–22.0	—	—

In bending of bars it is possible to determine the interlaminar shear modulus G_{xz}^b and strength τ_{xz}^{bu}. For determination of the interlaminar shear modulus G_{xz}^b bars with a rectangular cross section are tested in three-point bending and the experimental results are processed according to formulas that take transverse shears into account. This method yields satisfactory results, but it requires well planned experimental methodology. It has been established in [228] that in bending tests the data scatter is greater than in square plate twist and torsion of bars, and the shear modulus is usually lower than expected.

For study of the interlaminar shear strength τ_{xz}^{bu} in three-point bending, bars with a small relative span l/h, i.e., bars which are known to fail in interlaminar shear, have been tested. Owing to the indefinite character of the specimen state of stress, bending of bars with small values of l/h can be used only for approximate comparison of interlaminar shear resistance of various materials. In practice, equations for calculation of the interlaminar shear strength of prismatic bars sometimes hold good for ring segments and even intact rings. Studies have shown that failure mechanisms of prismatic bars, ring segments with concavity up or down, and intact rings of polymeric composites are qualitatively different. Bending test methods for prismatic bars, rings, and ring segments are treated in greater detail in Chapter 5.

Torsion tests of orthotropic bars with straight or circular axes make it possible to determine the shear moduli in two mutually perpendicular planes of the material; the ability to determine shear strength is rather limited. Torsion tests and processing of results are painstaking and sensitive to experimental error, but the results of properly conducted experiments are comparable with results obtained by other methods. Torsion test methods for bars with straight and circular axes are treated in Section 4.9 and 4.10. An understanding of various methods of determination of these characteristics can be obtained from the data presented in Table 4.1.1.

In calculations using equations of the theory of the elasticity of an anisotropic body, it must be kept in mind that in practice the loading scheme frequently does not correspond to that assumed in the analytical solution. For example, in an analytical solution of the torsion problem it is usually assumed that the torque is applied integrally to the end faces of the specimen. In practice, in the majority of cases, transmission of the torque is accomplished by tangential forces on the side faces and their distribution is not always known. These phenomena, which are difficult to evaluate analytically, show up in the dimensions of the end effect zone. Significant deviations from the calculation scheme can be observed also in panel shear tests and in square plate twist.

4.2. TORSION OF THIN-WALLED TUBES

Torsion tests of thin-walled tubes produced by filament or fabric winding permits determination of the shear characteristics in the plane of the reinforcement layup: the shear modulus $G_{\theta z}$ sufficiently reliably and the strength $\tau_{\theta z}^{u}$ less precisely, since different failure mechanisms in torsion are possible depending on the relative thickness of the tube.

In torsion of thin-walled tubes, geometrically accurate and correctly assembled in the testing machine, the tangential stresses around the circumference and along the length of the specimen are distributed uniformly. If specimen wall thickness h is sufficiently small compared to mean radius R, it is possible to neglect the change in shear deformation through the wall thickness. In torsion the concept of a thin-walled tube is a function of the material anisotropy E_z/E_θ. In Fig. 4.2.1, distribution of relative tangential stresses $\tau_{\theta z}/\tau_{\theta z\text{mean}}$ is shown, where $\tau_{\theta z\text{mean}}$ is calculated from (4.2.3) through the wall thickness in the gage length of the specimen made of two different

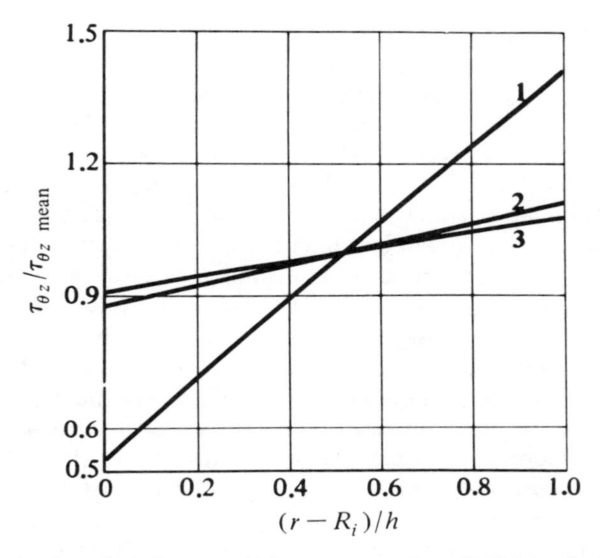

Fig. 4.2.1. Distribution of relative tangential stresses over the wall thickness in the gage length of a helically wound specimen (helical angle 60°) [166]. Material: (1, 2) carbon composite ($E_L = 137.9$ GPa, $E_T = 6.9$ GPa, $G_{LT} = 4.14$ GPa); (3) glass fiber composite ($E_L = 41.4$ GPa, $E_T = 13.8$ GPa, $G_{LT} = 5.12$ GPa). $R/h = 10$ (curves 1, 3) and $R/h = 40$ (curve 2).

materials. It can be seen from Fig. 4.2.1 that glass fiber composite specimens with $R/h = 10$ and carbon composite specimens with $R/h = 40$ are equivalent with regard to distribution of tangential stresses, i.e., in this case, the relative wall thickness of a carbon composite specimen must be four times smaller. At $R/h = 10$, the gradient of tangential stresses in the wall of a carbon composite specimen is essential and must be taken into account in processing of experimental results.

Since specimen wall thickness h is usually considerably smaller than the length of a specimen, the end effect zone is also small. For example, it has been shown [166] that the zone of perturbation of radial displacements is attenuated at a distance $z \leqslant 1.3R$. Therefore, because of the homogeneous state of stress, torsion of thin-walled tubes is the most precise method of studying in-plane shear characteristics. Homogeneity of the state of stress allows the use of a simple relation for determination of the in-plane shear modulus $G_{\theta z}$ (z is the specimen axis, θ is a circumference with mean radius R)

$$G_{\theta z} = \frac{M_T}{I_p} \cdot \frac{l}{\Delta\varphi} \tag{4.2.1}$$

or

$$G_{\theta z} = \frac{M_T}{I_p} \cdot \frac{1}{\varepsilon_{+45^\circ} - \varepsilon_{-45^\circ}} \tag{4.2.2}$$

where M_T is the applied torque; I_p is the polar moment of inertia $[I_p = \frac{1}{2}\pi(R_o^4 - R_i^4)$ or, at $R/h > 10$, $I_p \approx 2\pi R^3 h]$; R, R_o, R_i are the mean, outer and inner radii of the tubular specimen; $\Delta\varphi = \varphi_1 - \varphi_2$ is the angle of twist in radians over the gage length l; ε_{+45° and ε_{-45° are strains measured at angles $+45^\circ$ and -45° to the specimen axis, respectively.

The accuracy of processing the experimental results by (4.2.1) and (4.2.2) depends not only on the correctness of choice of the relative thickness h/R of the tubular specimen but also on the structure of the material and the mutual location of the principal axes of the material and the specimen. The hypothesis of plane cross Sections, on the basis of which relations (4.2.1) and (4.2.2) were obtained, holds true only for materials with the plane of elastic symmetry perpendicular to the longitudinal axis of the specimen. For other materials, the error introduced by warping of the cross section has to be taken into account. For comparatively thin-walled specimens (glass fiber tubes for torsion tests have, as a rule, $h/R < 1/10$), this error is small.

In the fabrication of specimens and selection of their thickness, it must be kept in mind that the number of layers must be sufficient for transition to a continuous medium. In the range of loads studied, it is necessary to exclude the loss of stability of the specimen. Minimum deviation of the shape of the cross section from circular and, particularly, elimination of differing wall thicknesses must be striven for. (In the calculation formulas, the specimen radii enter in the fourth order!) Scatter in the magnitude of wall thickness is especially undesirable when localized methods are used for determination of the angle of twist (for example, by means of Martens strain gages).

Equation (4.2.2) is more universal than (4.2.1), for it allows control of the state of stress in the specimen. For this reason, $+45^\circ/0^\circ/-45^\circ$ rosettes of strain gages are used. The state of pure shear in the specimen is achieved when the readings of strain gages bonded at angles $\pm 45^\circ$, are numerically the same $(|\varepsilon_{+45^\circ}| = |\varepsilon_{-45^\circ}|)$ and the axial deformation is zero $(\varepsilon_{0^\circ} = 0)$, i.e., specimen deformation in the axial direction is not restricted. To control uniformity of deformation around the specimen circumference, several (3–4) $+45^\circ/0^\circ/-45^\circ$ rosettes are located on one of its sections. To exclude the effect of manufacturing variations, it is reasonable to test a specimen at two loading regimes—in the direction of winding and in the opposite direction.

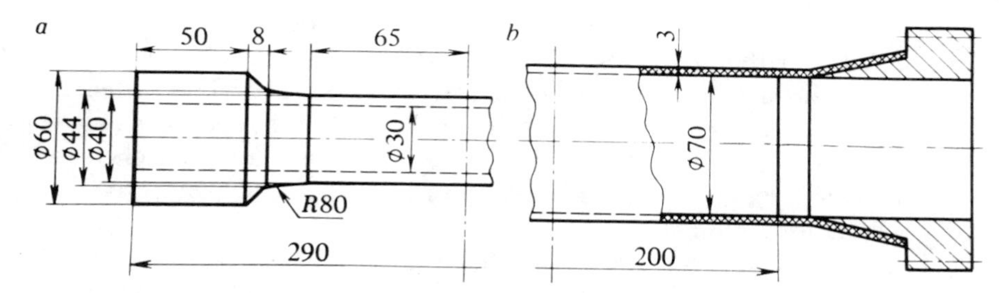

Fig. 4.2.2. Tubular specimens for torsion tests: (a) glass fabric would (thin-walled) [215]; (b) glass filament would (thin-walled) [193].

Several attempts are known for the use of tubular specimens for determination of shear strength $\tau_{\theta z}^{u}$. Thin-walled specimens, as a rule, lose stability long before their failure and so the formula for calculation of stresses,

$$\tau_{\theta z} = \frac{M_T}{W_p} \tag{4.2.3}$$

turns out to be unsuitable; here the moment resistance in torsion $W_p = (\pi/2R_0)(R_o^4 - R_i^4)$ or, at $R/h > 10$, $W_p \approx 2\pi R^2 h$.

In order to produce shear failure, it is necessary to pass to the region of relative thicknesses at which the specimens can not be considered thin-walled. Thus, in [215], specimens with $h/R \approx \frac{1}{4}$ were used; the data turned out to be comparable to the results of determination of the shear strength by the panel shear test. However, with increase in relative thickness the possibility of failure due to interlaminar shear increases. Therefore, for study of the shear properties of filament wound materials, flat specimens, produced by winding into polyhedral mandrels, were frequently used [231]. Dependence of shear modulus $G_{\theta z}$ and strength $\tau_{\theta z}^{u}$ on the angle between the reinforcement direction and the axis of the specimen and the results of experimental verification are presented in [269].

The shape and dimensions of tubular specimens for torsion testing have not been standardized. Two specimens are shown in Fig. 4.2.2, produced by filament and fabric winding and intended for determination of the elastic and strength characteristics. Loading, especially in study of the elastic properties of wound materials, is better carried out continuously, in view of the greater sensitivity to the loading conditions.

In torsion tests of tubular specimens, they are usually fastened and loaded on the outer surface. Under these conditions, participation of the entire cross section in work must be ensured and slipping of the ends of the specimen prevented. For this reason, special plugs are inserted into the ends of the specimen, through which metal rods are passed which rigidly fasten the plug to the test specimen. Under loading, both ends of the specimen may be fastened in the grips of the testing machine, or one end may be fastened and the other left free. The latter scheme is especially convenient where restriction of axial deformation must be avoided.

To eliminate stress concentration in the transmission of the torque to the specimen, fastening of its end parts must not be passive, i.e., a gradual increase in stress must be ensured over the length of the fixed part of the specimen. In [194], for example, such gradual transmission of the torque was complished by means of an epoxy resin disk molded together with the specimen.

In tests of tubular specimens, as in torsion of solid bars, the angles of twist of two sections and the torque are measured over the linear section of the stress-strain curve. Angles of twist can be measured with a Martens mirror strain gage, by Aistov's strain gage, or by means of special devices. Torque is measured by means of a special dynamometer or directly from the machine scale.

The disadvantages of measuring twist angles by the Martens scheme consist not only of technical difficulties, but also of the extreme sensitivity of the specimen to the quality of fabrication. Therefore, methods of measurement are used which enable the twist angle to be averaged over the cross section, for example, with the use of an Aistov's strain gage or induction displacement sensors.

In thin laminated (cloth reinforced) specimens, when plugs are not used, mutual slippage of the layers is possible, and measurements on the surface can be erroneous. In such cases, a simple procedure is frequently used in which tubes are drilled at the ends of the gage length through the diameter. Metal rods are inserted into the opening, and during torsion the linear displacement of their ends is measured. To prevent slippage of layers, special end caps on the specimen can be used. To eliminate loss of stability in thin-walled specimens, light fillers of foam plastics are used, which press tightly against the inner surface of the specimen.

4.3. PANEL SHEAR

4.3.1. Loading Methods

There are several loading methods for producing pure shear in the plane of thin plates. The best known loading methods are shown in Fig. 4.3.1; the

Loading scheme			a 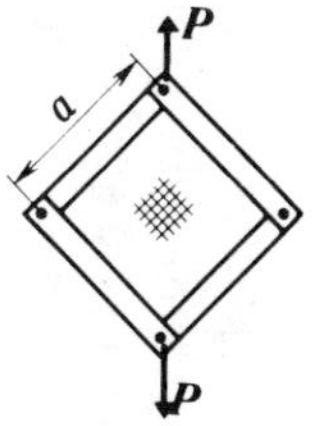	b 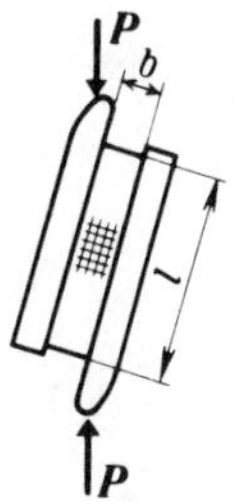	c 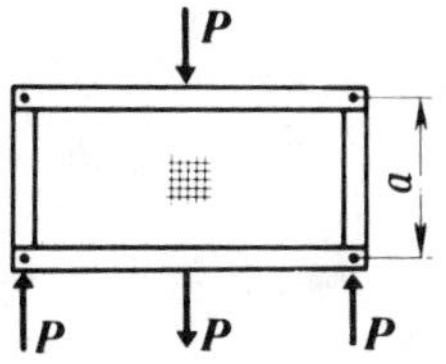
Determinable characteristics			G_{xy}, τ_{xy}^{u}	G_{xy}, τ_{xy}^{u}	G_{xy}
Measurable values			P, ϵ	P, ϵ	P, w
Geometrical sized			a, h	b, h, l	a, h
Limitations	Structural	Layup	0° ; 90° ; 0/90°		
		Orientation	0°, 90°		
	Physical		Linear range "load displacement;" buckling excluded; layup symmetrical about midplane		
	Geometrical		Limited by testing machine power by the buckling of the gage section and the stress distribution in the specimen	$l/b > 10$	Limited by the buckling of the gage section

Fig. 4.3.1. Loading schemes for determination of shear characteristics in the plane of reinforcement layup: (a) four-part frame; (b) rail shear; (c) thin-walled beam.

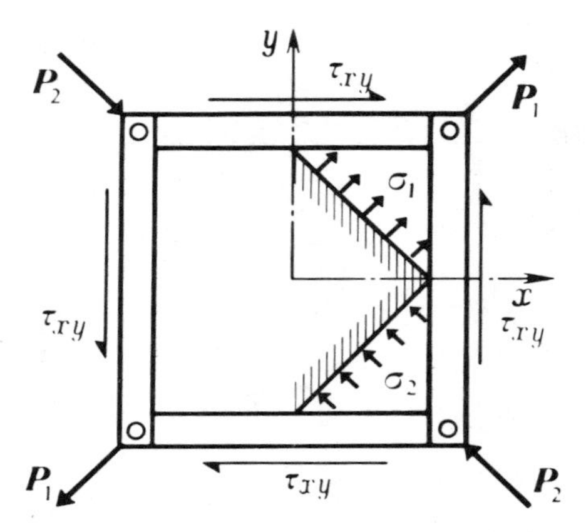

Fig. 4.3.2. Loading scheme of a balanced isotropic material in a four-link rig [216].

magnitudes to be experimentally evaluated, force factors, and the variable geometrical dimensions of specimens are also shown.

The most popular method is the use of a four-part test jig ("picture-frame") [216; 222, p. 105]; initially, this method was used for testing plywood. The fixture consists of a massive double frame connected by pin hinges, in which a plate of the material tested is fixed by means of bolts, rivets, or an adhesive. In loading, the four-part frame converts tensile forces acting along one of the diagonals of the plate P_1 (tensile forces are applied to the two opposite hinges) into shear forces (Fig. 4.3.2). The links of the fixture, between which the specimen is clamped, should have sufficiently high rigidity in bending and in tension-compression, i.e., $\varepsilon_{x'} = \varepsilon_{y'} \approx 0$ (Fig. 4.3.3) and it should not change its configuration in the loading process. Assuming that the edges of the gage section of a square specimen remain rectilinear in loading and the square is transformed into a rhomboid, the elastic constants and strength are easily calculated. The diagonals of the four-part frame are elongated and shortened by one and the same amount. Therefore, longitudinal deformations of a specimen of a balanced material (isotropic or anisotropic) in these directions are equal in magnitude but opposite in sign. Hence it follows that the normal stresses σ_1 and σ_2, acting on the areas, located at an angle 45° to the x and y axes (see Fig. 4.3.2), are numerically equal to:

$$\sigma_1 = -\sigma_2 = \tau_{xy} \tag{4.3.1}$$

The forces applied to the pin hinges are: $p_1 = p_2$, i.e., the specimen is in a

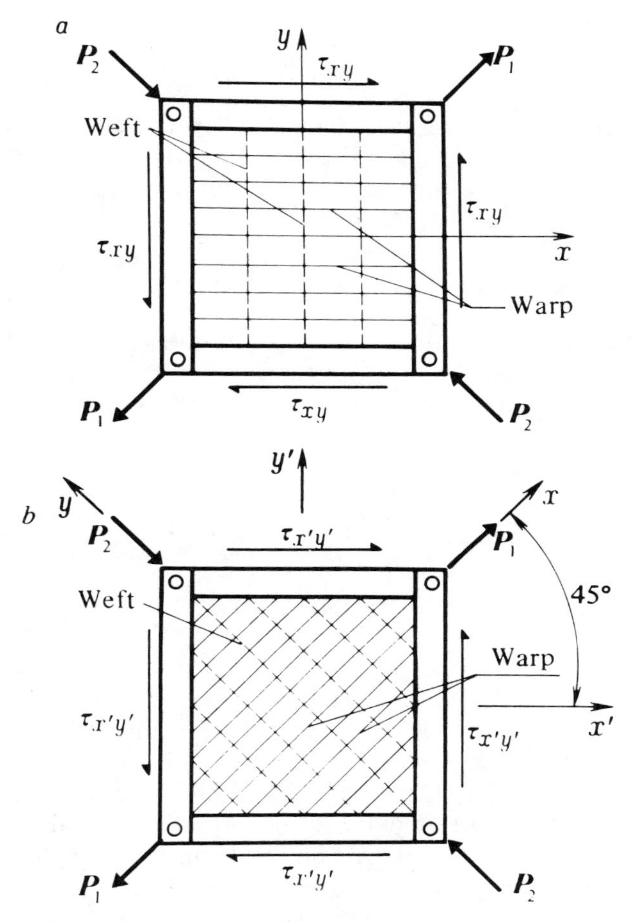

Fig. 4.3.3. Loading scheme of an orthotropic material with reinforcement layup parallel to the arms of the four-part frame (a) and parallel to the x and y axes (b) [216].

state of pure shear ($\sigma_x = \sigma_y = 0$, $\tau_{xy} \neq 0$). Also in this state of stress is a plate made of an unbalanced orthotropic material, in which the principal axes of symmetry x' and y' are located parallel to the links [Fig. 4.3.3(a)], since in this case $\sigma_{x'} = -\sigma_{y'}$.

If the principal axes of symmetry of a plate of an unbalanced orthortropic material coincide with the x and y axes [Fig. 4.3.2(b)], from the condition $\varepsilon_x = -\varepsilon_y$ it follows that

$$\sigma_y = -\frac{1 - \nu_{xy}}{1 - \nu_{yx}} \frac{E_y}{E_x} \sigma_x \tag{4.3.2}$$

i.e., the forces acting along the diagonal in tension are not equal to the forces acting along the diagonal in compression, and the edges of the plate, besides tangential stresses equal to

$$\tau_{x'y'} = -\frac{\sigma_x - \sigma_y}{2} = \frac{\sigma_x}{2}\left(1 + \frac{1 - \nu_{xy}}{1 - \nu_{yx}} \frac{E_y}{E_x}\right) \tag{4.3.3}$$

are loaded with normal stresses as well. An orthotropic plate loaded in this manner is no longer in a state of pure shear, and actually it is not the shear modulus of the material which is determined from experiment, but the elastic constant $a_{6'6'}$:

$$\tau_{x'y'} = a_{6'6'}\gamma_{x'y'} \tag{4.3.4}$$

where

$$a_{6'6'} = \frac{1}{4(1 - \nu_{xy}\nu_{yx})}[E_x(1 - \nu_{yx}) + E_y(1 - \nu_{xy})]. \tag{4.3.5}$$

Besides these limitations, the standard four-part frame has still other deficiencies, also due to its design. In standard pin-hinged four-link frames, the axes of rotation of the joints, as a rule, do not coincide with the corners of the gage section of the specimen, i.e., the basic principle of the force flow is violated: transmission of the shear forces directly to the edges of the specimen gage section. Such deviation results in irregular distribution of strains and stresses in the gage section of the specimen, with a clearly defined concentration of strains and stresses at its corners (Fig. 4.3.4). Besides, the numerical values of the maximum stresses and their coordinates depend on the dimensions of the gage section of the specimen. An analysis shows that the load, absorbed by the lateral pins (along the y axis) of the four-part frame also depend on the dimensions of the specimen; thus, with smaller specimens [Fig. 4.3.4(a)] the lateral pins absorb 12.1% and with larger sizes [Fig. 4.3.4(b)] 20.6% of the tensile load. Disregarding the stress concentration results in experimentally determined values of the shear modulus and strength are 1.5–2.5 times greater than the actual values [170].

The error due to irregularity of strain distribution decreases with increase in plane dimensions of the specimen, i.e., with increase in the ratio of the dimensions of the gage section of the specimen to the distance between the

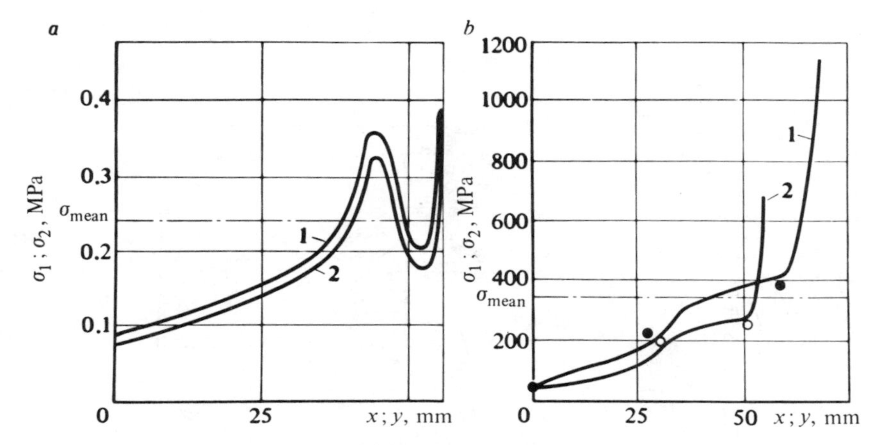

Fig. 4.3.4. Stress distribution in a 6-ply (a) and 12-ply (b) boron composite, with fiber layup at an angle $\pm 45°$ to the axes of the four-part frame [82]. For a 6-ply boron composite the stresses σ_1 (σ_2) are given for a unit load, i.e., $P = 1$. Dimensions of the gage section of the specimen 38 × 38 mm (a) and 114 × 114 mm (b). Load: (1) tension (stress σ_1); (2) compression (stress σ_2).

hinges of the fixture. However, the increase in the specimen dimensions is limited by the availability of the material, the testing machine power, and mainly by the possibility of buckling of the gage section of the specimen due to forces acting along the diagonals in compression. Experience in testing boron composites [82] has shown that it is not possible to completely eliminate buckling of the gage section of the specimen by either an increase in thickness of the material or the use of sandwich-type specimens.

The standard four-part frame also has the shortcoming that the specimen is weakened in the zone of fastening the fixture (in the case of fastening by means of bolts or rivets) and it undergoes strong edge squeezing, i.e., the edge effect zone is large. This practically eliminates the possibility of correct determination of the in-plane shear strength. Efforts to eliminate the shortcomings of the standard four-part frame have resulted in the creation of a series of new fixtures. The devices proposed in [169, 170] consist of a pair of pin-hinged frames between which the specimen is clamped. The axes of rotation of the hinges of each four-link frame extend above the surface of the specimen and they are made to coincide with the corners of the gage section of the specimen. In this way the concentration of stresses near the corners and the systematic experimental error connected with this are eliminated. The other shortcomings of the four-part frame—edge squeezing, interrelation of the strains ε_1 and ε_2—still remain with this fixture.

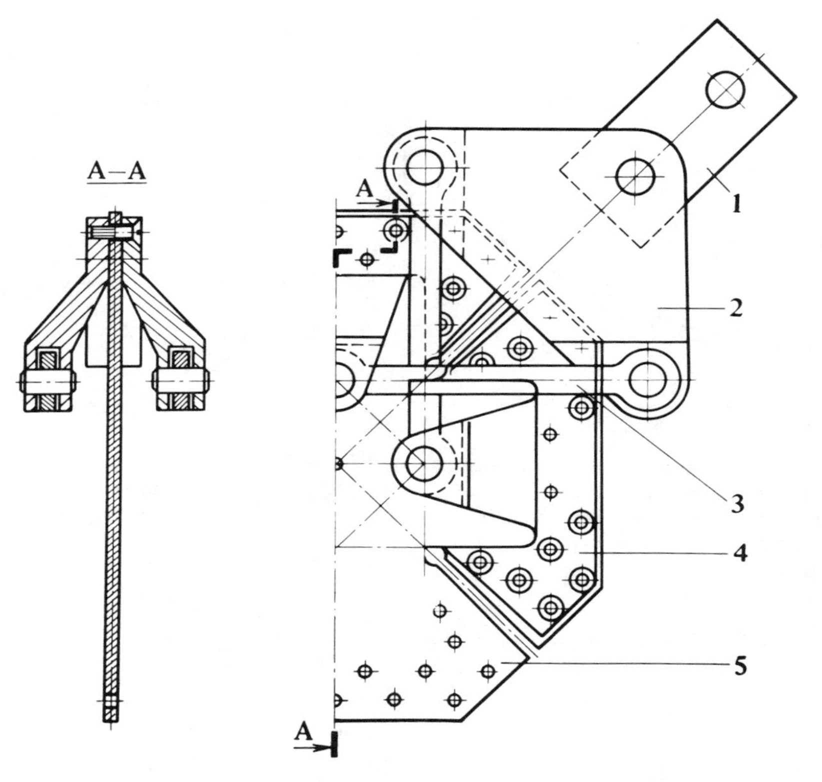

Fig. 4.3.5. An improved four-part frame [219]: (1) pull rod; (2) cross-piece; (3) pull rod; (4) arm of four-part frame; (5) specimen.

Another method of load application is used in the device shown in Fig. 4.3.5. In this device, the tensile load is applied through a system of independent levers at a distance from the gage section of the specimen. On condition that the rigidity of the specimen fastening is sufficient, this device is free from all the shortcomings inherent in the standard four-part frame. This device gives satisfactory results in testing glass fiber and high-modulus composites with varying reinforcement layups and it permits the shear strength of relatively rigid plates to be determined. In the case of testing thin or less rigid plates, buckling of the gage section of the specimen is observed.

A less rigid loading system is exemplified by the fixture in Fig. 4.3.6, in which shear forces on the specimen are transmitted through a rigid hinged frame and calibrated adjustable loading links [234].

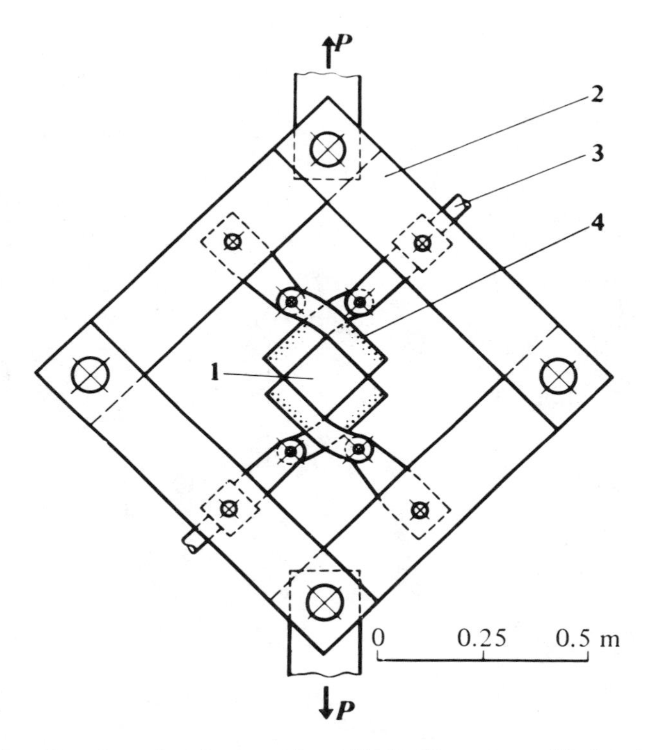

Fig. 4.3.6. Loading scheme in a four-part frame [234]: (1) specimen; (2) pinned steel frame; (3) calibrated adjustable loading links; (4) steel edge members for fastening of specimen.

The devices described above are intended for study of shear resistance or strength in specimens of thin sheet materials. The method can be extended to thicker materials, if the specimens tested are in the form of rectilinear parallelepipeds [99]. Phenomena connected with loss of stability are avoided in thick specimens; however, to eliminate of possible misalignment of specimens, assembly of the test fixture must be more precise. The loading devices usually are attached by means of intermediate metal plates bonded to the specimen.

Shear testing of a relatively narrow strips clamped between two independent rigid bars, so-called rail shear, is also a method of panel shear (Figs. 4.3.7 and 4.3.8). This method of testing was proposed relatively recently (in 1967) and evaluation of it is sometimes controversial [82, 217, 260]. However, this is a consequence of incomplete consideration of the features of the method; it should be noted that this method has been used successfully in fatigue

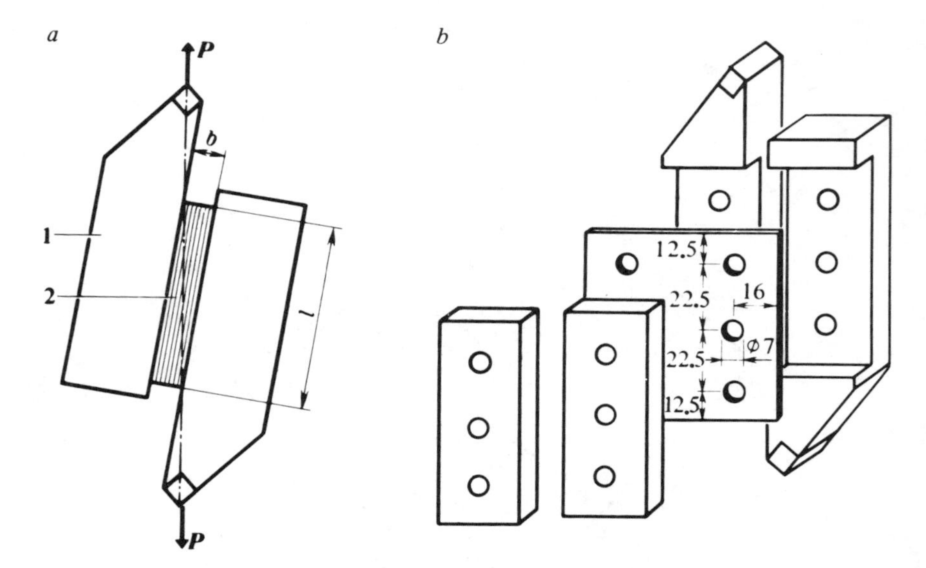

Fig. 4.3.7. Schematic (a) and example of structural realization (b) of an ordinary device for in-plane shear testing of plates [194]: (1) link of the device; (2) specimen.

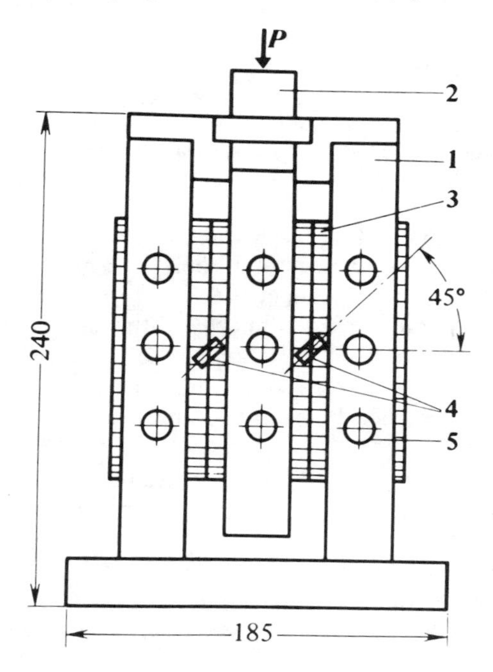

Fig. 4.3.8. Dual device for in-plane shear testing of plates [217]: (1) frame; (2) plunger; (3) specimen; (4) resistance strain gages; (5) specimen fastening bolts.

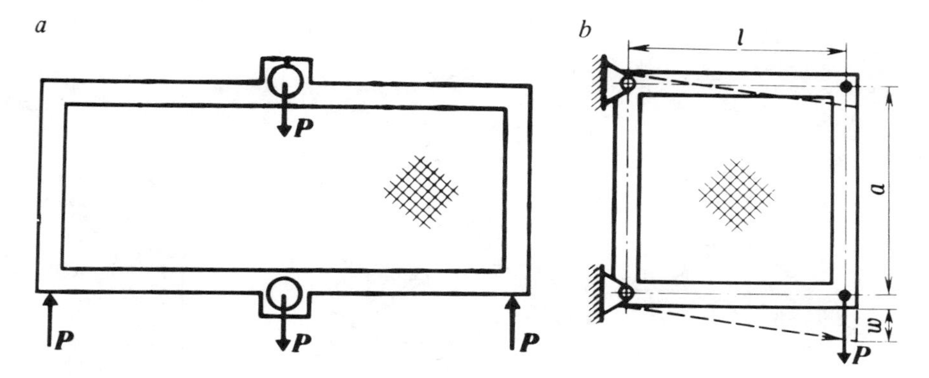

Fig. 4.3.9. Thin-walled beams: (a) on two supports; (b) cantilever.

testing [162]. Two types of fixture are used for loading of specimens in rail shear: one- and two-piece (Fig. 4.3.8). The advantages of the latter are higher output and a purer state of stress in the specimen. One-piece devices are not installed vertically in the testing machine, but at an angle, so that the specimen is subjected to biaxial stress. Therefore, in testing unidirectional materials (fiber layup parallel to the rails) premature failure is observed. Because of restriction of the deformation of the specimen, deformation at failure is somewhat overestimated. For more detail on selection of specimen dimensions see Section 4.3.2.

For the determination of in-plane shear characteristics, thin-webbed beams (panels) are also used, consisting of a rigid frame and a thin web of the material being tested (Fig. 4.3.9). In distinction from thin-webbed beams, used in structures, loss of web stability in this case is inadmissible. These specimens have very large dimensions, for example, a panel 650 × 234 mm [130]. This limits their use as a consequence of the consumption of a large amount of test material, as well as because of possibility of buckling of the specimen. Some difficulties may arise in processing the experimental results because the rigidity of the frame must be taken into account.

4.3.2. Shape and Dimensions of Specimens

In panel shear by a four-part frame, thin plates with a square gage section and projections for fastening are used [Fig. 4.3.10(a)]. The in-plane dimensions of the gage section are usually set at 100 × 100 mm. It is more difficult to use strain-measuring devices in plates of smaller sizes and the measured strains of the gage section can prove to be distorted as a consequence of squeezing of the plate edge. With very large in-plane dimensions of the gage

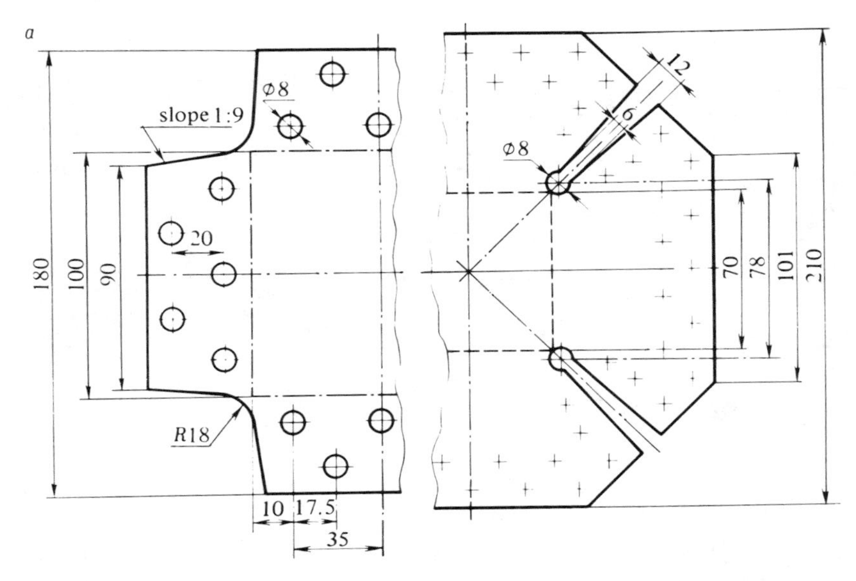

Fig. 4.3.10. A specimen for testing of glass fiber composites: (a) in a standard four-part frame [222]; (b) in a modified four-part frame [219].

section of the specimen buckling can occur, which is sometimes unnoticeable to the eye, and the experimental results turn out to be inaccurate.

The thickness of the specimen is usually assumed to be equal to thickness of the sheet from which the specimen is cut. The necessary breaking power of the testing machine depends on the dimensions of the gage section of the specimen. The dimensions of the projections are selected on structural grounds, based on considerations of reliability of specimen fastening and transmission of forces. The specimen can be fastened in the four-link frame by means of bolts, rivets, or adhesive. In the latter two methods, the construction cannot be demounted without causing specimen failure; therefore, their use is desirable only in repeated testing of the same specimen.

The specimens used in the modified four-part frame differ from those described above only by the shape of projections, which are separated from each other by diagonal cuts to the corners of the specimen gage section [Fig. 4.3.10(b)]. The width of the cuts is selected so that the edges of the cuts are not in contact and do not restrict deformation of the plate under load. There are no recommendations on selection of the dimensions of rectangular parallelepipeds for plane shear tests. In [99], in determination of the in-plane shear modulus, 20 × 20 × 20 mm cubes were used; such small specimens require very careful preparation and testing.

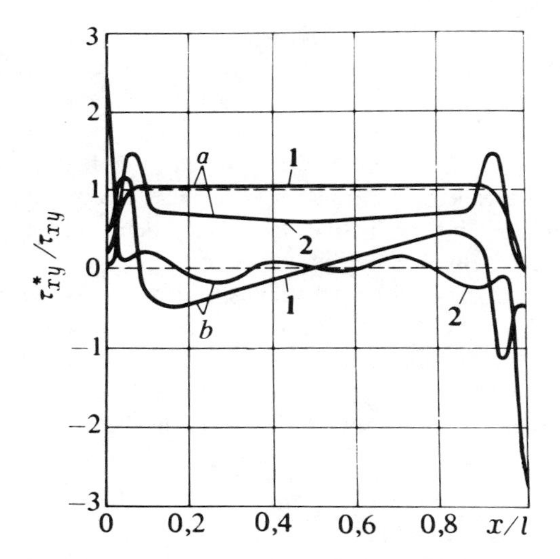

Fig. 4.3.11. Distribution of tangential stresses in the gage section of a carbon composite (Thornel 50) specimen with fiber layup at angles $0/\pm45°$ (1) and $\pm45°$ (2): (a) in the middle of the specimen ($y = b/2$); (b) at the side of the specimen ($y = b/100$); the ratio $l/b = 12$ [260]; τ_{xy} is the average value; τ_{xy}^* is the calculated value of tangential stresses.

In selection of specimen dimensions for rail shear testing (see Figs. 4.3.7 and 4.3.8), the distribution of tangential stresses in the specimen should be taken into account. Around the free edges a state of stress differing from pure shear is observed. These are the so-called transition or edge effect zones. Analysis shows that the edge effect zones and distribution of tangential stresses depend on the length-to-width ratio l/b of the specimen gage section and on the ratios between the elastic constants of the material. For example, with $G_{xy}/E_y > 1$, a uniform distribution of tangential stresses over almost the entire length of the gage section of the specimen can be obtained even with small l/b ratios (for example, at $l/b = 2$). For fibrous polymeric composites the effect of the edge zones is small with $l/b > 10$, exception in the case where $\nu_{xy} \approx \nu_{yx} \approx -1$ (or $A_{12}/A_{11} \approx A_{12}/A_{22} \approx 1$), i.e., in materials with greater Poisson's ratios. In this case, irregular distribution of the tangential stresses over the entire length of the gage section of the specimen is observed over a wide range of values of l/b ($2 \leqslant l/b \leqslant 14$). Fibrous composites with a $\pm45°$ fiber layup correspond to this case. The distribution of tangential stresses also changes over the width of the specimen.

Figure 4.3.11 shows the calculated distribution of tangential stresses in the gage section of a specimen of epoxy-carbon composite reinforced with Thornel 50 fibers with $l/b = 12$ over two longitudinal sections: $y = b/2$ and

$y = b/100$. Irregularity of the distribution of tangential stresses over the width of the specimen increases with squeezing of a relatively narrow (about 15 mm) gage section by the links of the device.

There are no recommendations on selection of dimensions of thin-webbed beams. In selection of the dimensions of thin-webbed beams, the design peculiarities of these beams should be taken into account: the web absorbs only the shear force, while the frame absorbs the bending moment. Moreover, loss of stability in the web must be prevented.

It should be noted in conclusion that in studying in-plane shear characteristics the symmetry of the material about its midplane must be assured. Since this requirement is practically unrealizable, material symmetry is assured by selection of a sufficiently large number of reinforcement layers. In [39] it was established that in determining the in-plane shear modulus the number of reinforcement layers should be no less than eight.

4.3.3. Elastic Constants

In determining the in-plane shear modulus of a plate in a four-part frame, the tensile force P_1 and strains ε_1 and ε_2 in the directions of gage section diagonals are measured (the deformation scheme is shown in Fig. 4.3.12). Resistance strain gauges or mechanical extensometers are used to measure the strains.

If a plate is in a state of pure shear, the in-plane shear modulus G_{xy} is determined by the formula

$$G_{xy} = \tau_{xy}/\gamma_{xy}. \tag{4.3.6}$$

The tangential stress τ_{xy} and shear strain γ_{xy} are equal to

$$\tau_{xy} = \frac{P}{\sqrt{2}ah} = \frac{0.707P_1}{ah} \tag{4.3.7}$$

$$\gamma_{xy} = \frac{\varepsilon_1 + \varepsilon_2}{1 + \varepsilon_1 - \varepsilon_2} \tag{4.3.8}$$

where a is the length of a side of the gage section of the plate; h is the thickness of the plate; ε_1 and ε_2 are the absolute values of the strains of the plate along the diagonals in tension and compression, respectively.

If the axes of rotation of the hinges of the four-part frame do not coincide with the corners of the gage section of the specimen, i.e., stress concentration (see Section 4.3.1) is observed in the specimen and a correction should be

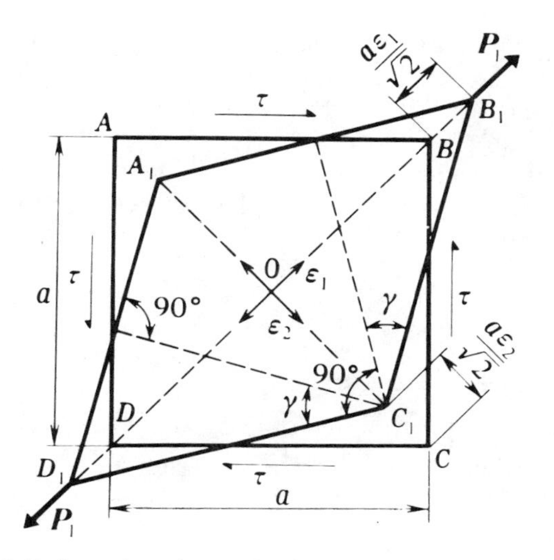

Fig. 4.3.12. Deformation scheme of a plate in a four-part test frame [219].

introduced into Equation (4.3.6). The refined value of the shear modulus is determined by the formula

$$G_{xy}^{*} = G_{xy} - \frac{0.354\Delta P}{ah\varepsilon_{\text{exp}}}. \tag{4.3.9}$$

The technique of determination of the correction $\Delta P/\varepsilon_{\text{exp}}$ is reported in [170].

In the rail shear test, the in-plane shear modulus G_{xy} is determined using (4.3.6) by substitution into it the following values τ_{xy} and γ_{xy}:

$$\tau_{xy} = \frac{P}{F} \tag{4.3.10}$$

$$\gamma_{xy} = 2\varepsilon_{45^\circ} \tag{4.3.11}$$

where $F = lh$ is for a one-piece fixture, $F = 2lh$ for a two-piece fixture; l and h are the length and thickness of the gage section of the specimen, respectively; and ε_{45° is the experimentally measured strain.

In tests of unidirectional composites with fiber layup parallel to the arms of the fixture, the modulus of interlaminar shear [see Equation (4.6.2)] can also be determined.

Results of determination of the elastic constants by the rail shear method are insensitive to the relative dimensions of the specimen l/b, since the measurements are carried out in the center of the specimen, where the state of stress is most uniform.

Comparison of test results shows that the values of the shear moduli obtained by rail shear testing are somewhat greater than those obtained by testing in a four-part frame [17]:

	G_{xy}, MPa	
	FOUR-PART FRAME	RAIL SHEAR
Fiberglass composite AG-4S (0°)	5,240	5,480
Fiberglass composite AG-4S (0/90°)	5,470	5,530
Boron composite (0/90°)	10,400	11,000
Carbon composite (0/90°)	3,920	4,100
Epoxy binder ED-16	152	169

In tests of thin-webbed beams, the bending of a cantilever beam, consisting of the material studied in a form of a thin web and a link [see Fig. 4.3.9(b)] is the simplest case. In this case, the in-plane shear modulus of a web is determined using (4.3.6) with the substitutions

$$\tau_{xy} = Q/F = P/ah \qquad (4.3.12)$$

$$\gamma_{xy} = w/l \qquad (4.3.13)$$

where h is the web thickness [see Fig. 4.3.9(b) for the remaining dimensions] and the web height a is measured between the centers of gravity of the cantilever flanges.

4.3.4. Strength

In tests of a plate in the four-part frame the in-plane shear strength τ_{xy}^u is calculated by the formula

$$\tau_{xy}^u = \frac{P^u}{F} = \frac{P^u}{\sqrt{2}ah} = \frac{0.707P^u}{ah} \qquad (4.3.14)$$

where P^u is the breaking force.

Stress concentrations (when the pin axes extend above the gage section of the specimen) are taken into account by the formula

$$(\tau_{xy}^u)^* \approx \tau_{xy}^u - \frac{0.707\Delta P'}{ah} \tag{4.3.15}$$

The method of determination of $\Delta P'$ has been described in [170]. Equation (4.3.15) is applicable if the failure takes place in the calculated section along the diagonal in compression. With the four-part frame, because of squeezing of the specimen edge failure frequently takes place along the edge of the gage section of the plate. In processing experimental data, results from such samples are rejected.

The strength τ_{xy}^u in rail shear is determined from (4.3.10) by substitution of $P = P^u$. The in-plane shear strength or interlaminar shear strength can be determined, depending on the orientation of the material relative to the load direction.

Values of in-plane shear strength obtained by rail shear are somewhat lower than values obtained in a four-part frame; however, they are within the range of experimental scatter (deviation does not exceed 8%) [17]:

	τ_{xy}^u, MPa	
	FOUR-PART FRAME	RAIL SHEAR
Fiberglass composite AG-4S (0°)	48.0	44.4
Fiberglass composite AG-4S (0/90°)	50.2	48.2
Boron composite (0/90°)	49.5	48.2
Carbon composite (0/90°)	44.2	43.5
Epoxy binder ED-16	42.0	40.3

4.4. SQUARE PLATE TWIST

4.4.1. Basic Methods

Shear modulus determination by the square plate twist method is well known and was initially used to study relatively rigid isotropic materials, i.e., metals. The method is not used in practice for determination of shear strength. The method assumes small deflections: up to $0.1h$ (h is plate thickness). In [35] it has been noted that a linear theory is suitable up to $w \approx h$. Therefore its extension to fibrous polymeric composites, whose shear stiffness is much less than that of metals, requires increased accuracy of load and deflection measurement. If the deflection of the plate exceeds the value assumed by the linear theory, the shape of the bent specimen surface changes due to instability and the equations given below become invalid.

The formulas for large deflections presented in [35] have so far not found widespread use.

The possibilities of studying the strength of fibrous polymeric composites by square plate twist are rather limited because of difficulties in processing the test results. The relative plate thickness h/l must be small enough that it would be possible to neglect the effect of shear on deflection. Completing the general description of the method, let us note that the relationships presented below, are applicable only to materials uniform through the thickness and orthotropic in the axes of the plate. Testing of materials, for example, with a $\pm 45°$ fiber layup results in large errors [194].

Two methods of determining the in-plane shear modulus by square plate twist are known: anticlastic (or two-point) and modified anticlastic (or three-point). The term "anticlastic" refers to the shape of the bent surface which the plate acquires during loading. A bent surface is called anticlastic if its principal curvatures have opposite signs. It is more convenient to use terms based on the supporting scheme of the plates being tested: two-point and three-point methods.

The two-point method is known from the work of A. Nadai [147, p. 39] and M. Bergsträsser [23]. It was initially used for studying metals. Later, the method was extended to orthotropic plates of fibrous polymeric composites [92, 242, 261]. A square plate is supported on two supports, located at opposite corners (quadrants 1 and 3 in Fig. 4.4.1), while at the other two corners (quadrants 2 and 4) the plate is loaded with a concentrated force P. Under the effect of uniformly distributed bending and torsion moments M_x, M_y, and M_{xy} the plate assumes the shape of a surface of second order, which can be described by the equation [242]:

$$wh^3 = 6M_x(a_{11}x^2 + a_{12}y^2 + a_{16}xy) + 6M_y(a_{21}x^2 + a_{22}y^2 + a_{26}xy) + 6M_{xy}(a_{61}x^2 + a_{62}y^2 + a_{66}xy) \quad (4.4.1)$$

where h is the plate thickness; w is deflection of the plate measured over a preset measurement base relative to the center of the plate;* a_{ij} are elastic constants (see Section 1.3).

Let the axes of coordinate system ξ, η (Fig. 4.4.1) be the principal axes of symmetry of the plate ($a_{16} = a_{26} = 0$). In this system of coordinates, the elastic constants are designated by A_{ij}. In the x, y coordinate system the

*In [261] deflection of a plate is measured relative to a stationary reference system; in that case the equations are different.

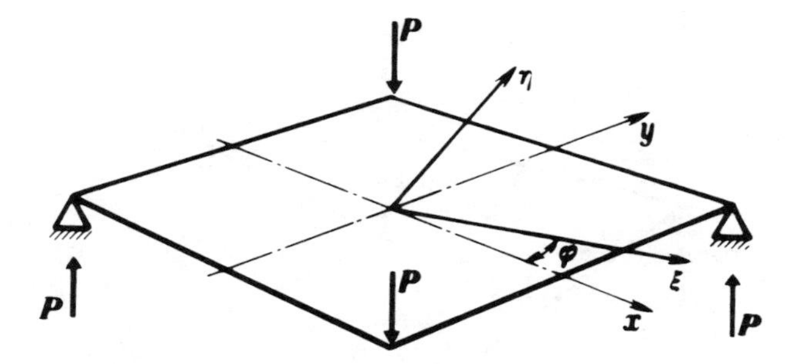

Fig. 4.4.1. Square plate twist on two fixed supports [92].

plate is in a state of pure twist, so that

$$M_x = M_y = 0; \qquad M_{xy} = \frac{P}{2}. \tag{4.4.2}$$

In the ξ, η coordinate system

$$M_\xi = -M_\eta = 0.5P \sin 2\varphi, \qquad M_{\xi\eta} = 0.5P \cos 2\varphi. \tag{4.4.3}$$

Then Equation (4.4.1) takes the form

$$wh^3 = 3P \sin 2\varphi[(A_{11} - A_{12})\xi^2 + (A_{12} - A_{22})\eta^2] + 3P \cos 2\varphi A_{66}\xi\eta. \tag{4.4.4}$$

By selection of the angle φ and the measurement base coordinates ξ and η from square plate twist test results, not only the in-plane shear modulus of the plate, $G_{\xi\eta} = G_{xy}$, but also elastic constants A_{11}, A_{12}, and A_{22} can be determined.

The three-point method differs from the two-point plate loading scheme in that the plate is supported at three corners and loaded with force P only at one corner (Fig. 4.4.2). Deflection of a plate is determined from the following equation [239]:

$$wh^3 = -3P\left[\left(x^2 - \frac{l^2}{4}\right)a_{61} + \left(y^2 - \frac{l^2}{4}\right)a_{62} + \left(xy - x\frac{l}{2} + y\frac{l}{2} - \frac{l^2}{4}\right)a_{66}\right]. \tag{4.4.5}$$

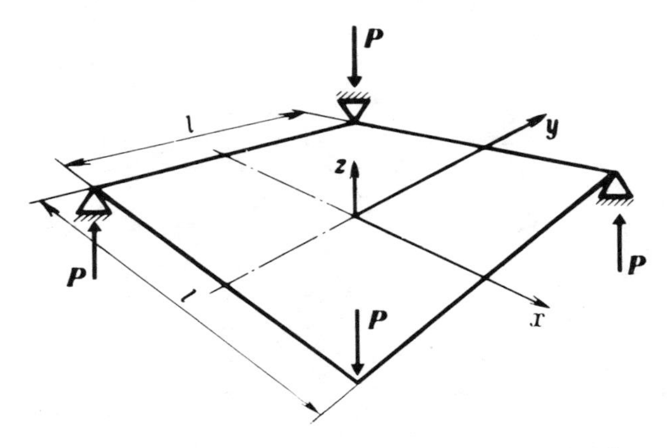

Fig. 4.4.2. Square plate twist on three fixed supports [239].

Deflection of the plate at the center, w_c, is found by

$$w_c h^3 = \frac{3Pl^2}{4}(a_{61} + a_{62} + a_{66}). \tag{4.4.6}$$

Hence, the sum of elastic constants is

$$A_G = a_{61} + a_{62} + a_{66} = \frac{4w_c h^3}{3Pl^2} \tag{4.4.7}$$

i.e., pure shear tests yield a sum of three elastic constants. Moreover, when the principal axes of elastic symmetry coincide with the coordinate axes of an orthotropic material, we obtain the following relationship

$$A_G = -2mn(m-n)^2 A_{11} - 8m^2n^2 A_{12} + 2mn(m+n)^2 A_{22} + (m^2 - n^2)^2 A_{66} \tag{4.4.8}$$

where $m = \cos\varphi$, $n = \sin\varphi$.

At angles $\varphi = 0°$ and $\varphi = 45°$ of rotation of the coordinate system, the elastic constants A_{11}, A_{12}, A_{22}, and A_{66} (see Fig. 4.4.3) can be determined from Equations (4.4.4) and (4.4.8).

In [35] an equation is presented which takes into account large deflections and the weight of the plate

$$\alpha\bar{w}^3 + \beta\bar{w} - \bar{P} - \frac{\bar{\rho}}{4} = 0 \tag{4.4.9}$$

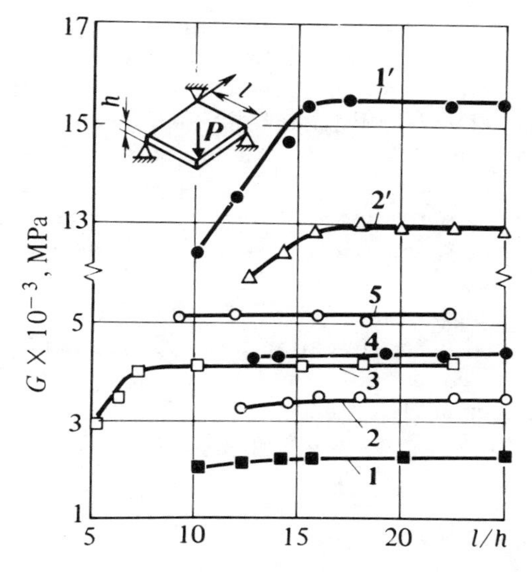

Fig. 4.4.3. Dependence of values G_{xy} (1–5) and G_{45° (1′ and 2′) of carbon (1–4) and glass fiber composites (5) on the l/h ratio in square plate twist by the three-point scheme [273]: Reinforcement layup: (1 and 1′) 0/90°; (2 and 2′) ±30°; (3) 2:2; (4) 0°; (5) 0/90°.

where

$$\alpha = \frac{128}{\pi^8}\left(1 + \frac{E_x}{E_y} + \frac{E_x}{G_{xy}} - 2\nu_{xy}\right)^{-1},$$

$$\beta = \frac{1}{3}\frac{G_{xy}}{E_x}, \qquad \bar{P} = \frac{Pl^2}{E_x h^4}, \qquad \bar{\rho} = \frac{\rho l^4}{E_x h^4}, \qquad \bar{w} = \frac{w_P}{4}.$$

Equation (4.4.9) can be used for determination of the shear modulus G_{xy} if the elastic moduli E_x and E_y, Poisson's ratio ν_{xy}, and density ρ of the specimen material are known.

The advantage of the three-point method over the two-point method consists in the simplicity of determination of the in-plane shear modulus, for which it is enough to measure the force P and deflection w_p of the plate at the point of application of the force.

It should be noted that under actual conditions cases are possible where, for example, due to defects in fabrication, the plate does not form the surface described by Equations (4.4.1) or (4.4.5). This naturally has a negative effect on the accuracy of experimental data processing. In practice, it is possible to estimate these deviations by measuring the radii of curvature of the deformed surface of the plate [20]. Further, it should be kept in mind that supporting the plate at the corners is impossible. In practice, it is necessary to allow some free distance from the edge or make projections at the corners of the plate. This introduces some inaccuracies into the measurements. It has been established in [273] that these errors can be neglected if the distance between the supporting (or loading) point and the peaks of corners of the plate does not exceed $(1.5\text{–}2)h$. Increasing the distance beyond this value results in excessive measured values of the shear moduli.

A common deficiency of methods of plate twist for studying shear resistance of the material is the limitation imposed on the allowable value of deflection. As a consequence of this limitation, both methods require high precision of measurement to estimate the elastic constants, and their use for determination of the in-plane shear strength is limited.

4.4.2. Specimen Dimensions

In square plate twist, specimens in the form of thin square plates are used. The dimensions of these plates have not been standardized. In selection of the dimensions, it should be taken into account that deflection or the applied load depends on the rigidity of the plate, which is also determined by its dimensions. At low plate rigidity the allowable load (determined by the deflection of the plate, which must not exceed $0.1h$, where h is the thickness of the plate) can be difficult to measure with sufficient accuracy. The specimen should be strictly planar, without initial waviness, and its thickness should be constant (in the equation, the thickness h enters in the third and fourth orders!).

Taking into account local effects (local bending at points of loading; supporting and loading at a distance from the corner of the plate; bearing of support surfaces), the following in-plane plate dimensions are suggested [89, p. 735] for testing of boron composites: 240×240 mm at room temperature and 115×115 mm at elevated temperature. However, in practice, plates of considerably smaller size, for example, 50×50 mm are used. This requires a more sophisticated test technique. From a theoretical point of view, the main geometrical characteristic of the square plate is the ratio of the side of the square, l, to plate thickness h. According to data given in [89], this ratio must be within the range $25 \leqslant l/h \leqslant 100$. However, in [273] it has been established from test results for glass fiber, carbon, and boron composites with various fiber layups that reliable readings are obtained at $l/h \geqslant 15$ (see Fig. 4.4.3).

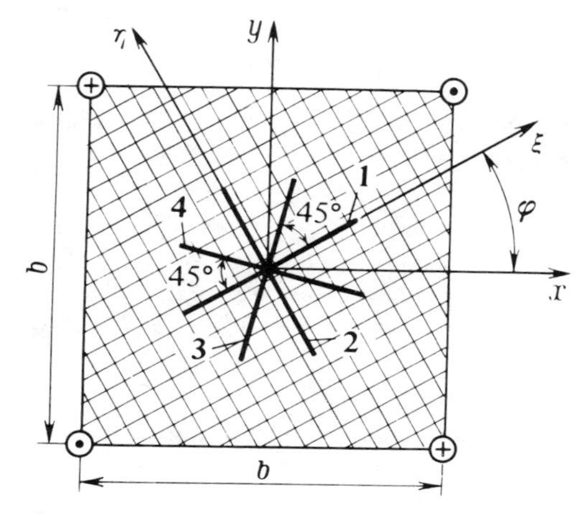

Fig. 4.4.4. Location of a measurement base in square plate twist tests by the two-point scheme [242] (ξ, η are the principal axes of elastic symmetry of the material). The direction of the load: ⊕ into the figure; ⊙ away from the figure.

4.4.3. Elastic Constants

In testing by the two-point method, measurements are made of load P and deflection w from the center of the plate of the a preset measurement baseline of length l. The location of the measurement baseline is shown in Fig. 4.4.4, where the coordinates of the measurement baseline may be equal to $\eta = 0$, $\xi = \text{const}$; $\xi = 0$, $\eta = \text{const}$; $\xi = \eta$; or $\xi = -\eta$.

For determination of the elastic constants Equation (4.4.4) is used. At a given angle φ, it is possible to obtain not only an expression for the determination of the in-plane shear modulus $G_{\xi\eta} = 1/A_{66}$, but also interrelations between the elastic constants $A_{11} = 1/E_\xi$, $A_{22} = 1/E_\eta$, $A_{12} = -\nu_{y\xi}/E_y$ and A_{66}. By solving them jointly, it is possible to determine all the mentioned constants, for which the tests must be conducted at angles $\varphi = 0$, 22.5, and 45° (i.e., there must be three series of specimens with different orientations of the principal axes of elastic symmetry of the material). The equations for $\varphi = 0$, 22.5, and 45° are given in Table 4.4.1.*

In testing according to three-point method measurements are made of the load P, deflection w_p at the point of application of the load, and deflection w_p at the center of the plate relative to a fixed reference system. For determination of the elastic constants, Equations (4.4.7) and (4.4.8) are used. The

*A variant method of determination of elastic constants is described in [221].

Table 4.4.1. Formulas for Determination of Elastic constants* in square plate twist [239, 242].

TWO-POINT METHOD			THREE-POINT METHOD	
φ	MEASUREMENT BASELINE	FORMULA	φ	FORMULA
0°	3 or 4	$A_{66} = wh^3/(3P\xi\eta)$	0° or 90°	$A_{66} = 4w_0h^3/(3Pl^2)$
22.5°	1	$A_{11} - A_{12} = \sqrt{2}wh^3/(3P\xi^2)$	45°	$A_{22} - A_{12} = 2w_0h^3/(3Pl^2)$
22.5°	2	$A_{12} - A_{22} = \sqrt{2}wh^3/(3P\eta^2)$	−45°	$A_{11} - A_{12} = 2w_0h^3/(3Pl^2)$
22.5°	3	$A_{11} - A_{22} + A_{66} = \sqrt{2}wh^3/(3P\eta^2)$		
22.5°	4	$A_{11} - A_{22} - A_{66} = \sqrt{2}wh^3/(3P\eta^2)$		
45°	1	$A_{11} - A_{12} = wh^3/(3P\xi^2)$		
45°	2	$A_{12} - A_{22} = wh^3/(3P\eta^2)$		

* A_{ij} are the elastic constants in the ξ, η coordinate system, which are the principal axes of elastic symmetry.

equations for $\varphi = 0$ or 90° and $\varphi = \pm 45°$ are presented in Table 4.4.1. As can be seen from the table, in testing at $\varphi = 0$ and $\varphi = \pm 45°$ one obtains three equations in four unknown elastic constants. To determine all the elastic constants, A_{11} and A_{22} are still needed. They are obtained from individual bending tests or tension of bars cut from a similar plate at 0° and 90° angles.

Deflection in the corner of the plate under the load is equal to

$$w_P = \frac{3Pl^2}{h^3} A_{66}. \tag{4.4.10}$$

Hence

$$A_{66} = \frac{w_P h^3}{3Pl^2} \tag{4.4.11}$$

i.e., by measuring the load and deflection of the point of loading it is possible to directly determine the elastic constant A_{66}. This method of determining A_{66} can be used for verification of the value obtained from Equation (4.4.6).

In processing of plate twist test results one must make sure of the linearity of P versus w relationship. Experience shows that deviation from linearity can be observed at different levels of load on the plate, i.e., at different w/l ratios, depending on the material studied and the dimensions of the plate (h/l ratio) (Fig. 4.4.5). In processing experimental results, neglecting

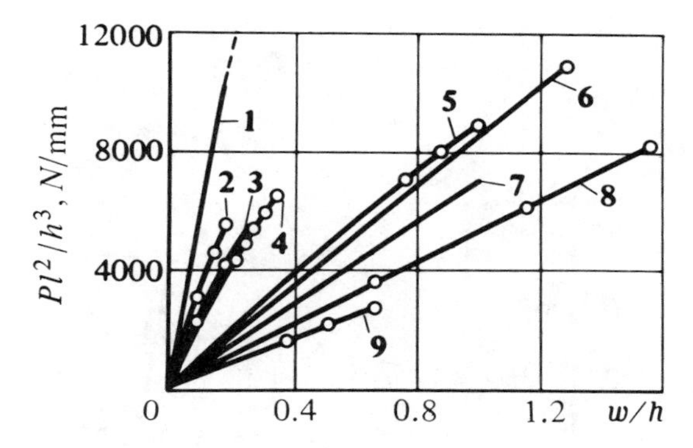

Fig. 4.4.5. Curves $Pl^2/h^3 = f(w/h)$ in square plate twist by two- (curves 1, 2, 4–7) and three-point (curves 3, 8, 9) methods. Materials: (1) 0/90° carbon composite at an angle $\pm 45°$; $l/h =$ 17 (linear up to $w = 0.722h$); (2, 3) three-dimensional glass fiber composite, $l/h = 9.2$ and 18.5, respectively; (4) carbon composite of complex structure; $l/h = 12.0$; (5, 6, 9) unidirectional boron composite, $l/h = 15.5$, 17.5, 13.2, respectively; (7) unidirectional carbon composite, $l/h = 17.0$; (8) balanced glass cloth composite, $l/h = 18.6$.

deviation from linearity of the P versus w relation can result in appreciable errors.

4.4.4. Interlaminar Shear Strength

In plate twist tests according to a two-point scheme, the interlaminar shear strength can be determined. The strength is calculated from the formula:

$$\tau_{xz}^{u} = \frac{3}{2}\frac{P^{u}}{h^{2}} \tag{4.4.15}$$

where P^u is the breaking load of the plate.

Studies of epoxy glass cloth reinforced composites show [274] that the test results depend slightly on the relative dimensions of the plate l/h:

$h = 3$ mm		$h = 5$ mm		$h = 8$ mm	
l/h	τ_{xz}^u, MPa	l/h	τ_{xz}^u, MPa	l/h	τ_{xz}^u, MPa
13.5	68	8.3	77	5.2	79
10.0	71	7.0	79	4.4	80
8.0	70	6.0	78	3.8	82
6.5	71				
4.8	76				

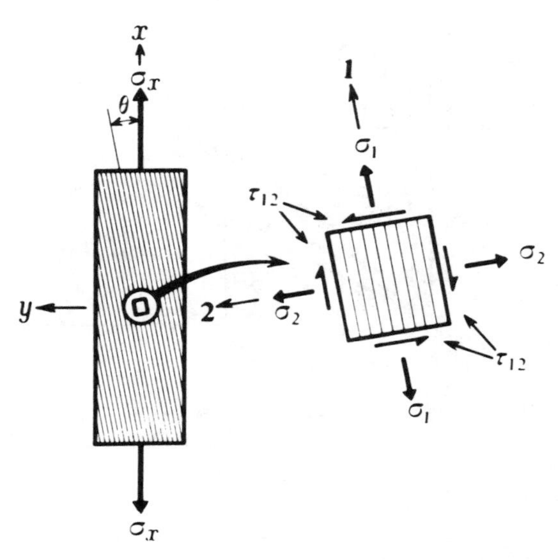

Fig. 4.5.1. Loading scheme of an anisotropic strip used in determination of the in-plane shear modulus [33].

The overrated values of strength τ_{xz}^{u} of a plate of thickness 5 and 8 mm and a plate of thickness 3 mm with a ratio $l/h = 4.8$ in [274] have been explained by nonlinearity of the p versus w diagram in the indicated cases.

4.5. TENSION OF AN ANISOTROPIC STRIP

For determination of the in-plane shear modulus, the method of tension of anisotropic strips with a variety of fiber layup schemes is used. The simplest case is a strip with a fiber layup at an angle θ to the longitudinal axis (Fig. 4.5.1). Reference [33] presents a determination of the magnitude of the angle θ at which the ratio of shear deformation γ_{12} to the longitudinal strain ε_x is greatest and exceeds the strain ratios $\varepsilon_1/\varepsilon_x$ and $\varepsilon_2/\varepsilon_x$ (for designation of axes, see Fig. 4.5.1). At this value of θ, only τ_{12} among stresses σ_1, and σ_2 and τ_{12} is near its critical value. For this purpose, the following equations are used:

$$\tau_{12} = \tfrac{1}{2}\sigma_x \sin 2\theta \tag{4.5.1}$$

$$\gamma_{12} = (\varepsilon_y - \varepsilon_x)\sin 2\theta + \gamma_{xy}\cos 2\theta \tag{4.5.2}$$

$$\varepsilon_x = \frac{\sigma_x}{E_x} \tag{4.5.3}$$

$$\varepsilon_y = -\nu_{xy}\varepsilon_x \tag{4.5.4}$$

$$\gamma_{xy} = a_{16}\sigma_x. \tag{4.5.5}$$

The ratio $\gamma_{12}/\varepsilon_x$ has a maximum at the following angle:

$$\theta = \pm\tan^{-1}\left[(B + D^{1/2})^{1/3} + (B - D^{1/2})^{1/3} - \frac{\delta_1}{3}\right]^{1/2} \tag{4.5.6}$$

where

$$B = \frac{1}{2\alpha} - \frac{\delta_1^3}{27} - \frac{\delta_1\delta_2}{6},$$

$$D = \frac{1}{4\alpha^2} \cdot \frac{\delta_1}{3\alpha}\left(\frac{\delta_2}{2} + \frac{\delta_1^2}{g}\right) - \frac{\delta_2}{27}\left(\frac{\delta_1^2}{4} + \delta_2\right),$$

$$\alpha = \frac{E_1}{E_2}, \qquad \delta_1 = 3 - \frac{E_2}{G_{12}} + 2\nu_{12},$$

$$\delta_2 = 3\frac{E_2}{E_1} - \frac{E_2}{G_{12}} + 2\nu_{12}.$$

For the unidirectional carbon composites studied in [33], the optimal value of the angle θ is equal to:

Modmor I/ERLA 4617	10°
Thornel T-300/PR 288	11°
S-glass/PR 288	15°

For isotropic materials, $\theta = 45°$

An analysis shows [33] that relations of stresses σ_1/σ_x, σ_2/σ_x and τ_{12}/σ_x are considerably more sensitive to variation in θ (Fig. 4.5.2) than are the strain ratios $\varepsilon_1/\varepsilon_x$, $\varepsilon_2/\varepsilon_x$, and $\gamma_{12}/\varepsilon_x$ (Fig. 4.5.3). On this basis, a tolerance of $\pm 1°$ has been specified for specimen cutting angle, strain gage position, and the load alignment. Relatively narrow specimens ($l/b \approx 14$) are used to ensure a uniform state of stress.

For experimental evaluation of shear deformation γ_{12} the readings of rosettes of resistance strain gages at 0°/120°/240° or 0°/45°/90° are used.

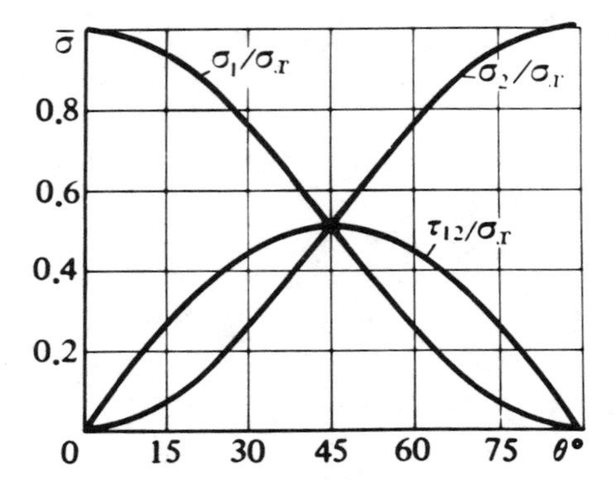

Fig. 4.5.2. Change in relative stresses $\bar{\sigma} = \sigma_1/\sigma_x$, σ_2/σ_x, and τ_{12}/σ_x, depending on the angle θ for a unidirectional carbon composite [33].

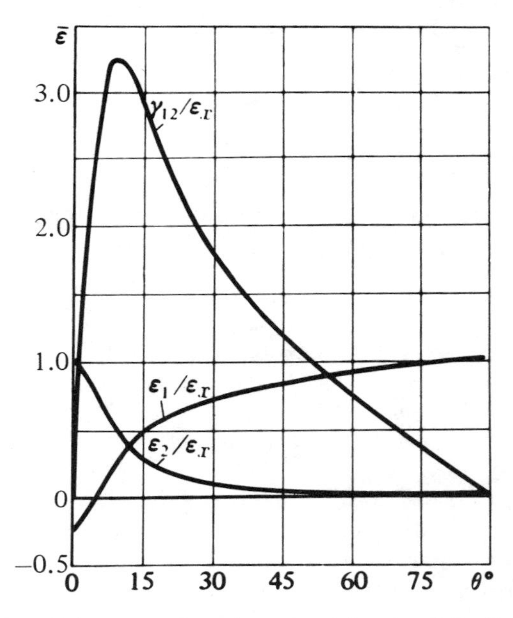

Fig. 4.5.3. Change in relative strains $\bar{\varepsilon} = \varepsilon_1/\varepsilon_x$, $\varepsilon_2/\varepsilon_x$ and $\gamma_{12}/\varepsilon_x$, depending on angle θ for a unidirectional carbon composite [33].

For a 60° delta-shaped rosette (0°/120°/240°):

$$\varepsilon_x = \varepsilon_{0^\circ}, \qquad \varepsilon_y = \frac{1}{3}(\varepsilon_{120^\circ} + \varepsilon_{240^\circ} - \varepsilon_{0^\circ}),$$

$$\gamma_{xy} = \frac{2}{\sqrt{3}}(\varepsilon_{240^\circ} - \varepsilon_{125^\circ}). \tag{4.5.7}$$

For a rectangular rosette (0°/45°/90°):

$$\varepsilon_x = \varepsilon_{0^\circ}$$

$$\varepsilon_y = \varepsilon_{90^\circ} \tag{4.5.8}$$

$$\gamma_{xy} = -\varepsilon_{0^\circ} + 2\varepsilon_{45^\circ} - \varepsilon_{90^\circ}.$$

At $\theta = 10°$, it follows from Equations (4.5.1) and (4.5.2) that

$$\tau_{12} = 0.1716_x = 0.171\frac{p}{F} \tag{4.5.9}$$

$$\gamma_{12} = -0.456\varepsilon_{0^\circ} - 0.875\varepsilon_{120^\circ} + 1.313\varepsilon_{240^\circ} \quad \text{(delta rosette)} \tag{4.5.10}$$

or

$$\gamma_{12} = -1.282\varepsilon_{0^\circ} + 1.879\varepsilon_{45^\circ} - 0.598\varepsilon_{90^\circ} \quad \text{(rectangular rosette)} \tag{4.5.11}$$

A stress-strain (τ_{12} versus γ_{12}) curve is plotted according to the experimentally measured and calculated stresses, and strains γ_{12} and the tangential shear modulus $G_{12} = \tau_{12}/\gamma_{12}$ are calculated from this curve.

In tension of a strip with a nonsymmetrical fiber layup relative to its axis, it should be taken into account that the initially rectangular strip under the load turns into a parallelogram. Consequently, the design of the testing machine grips must ensure free rotation of the end sections of the specimen. To eliminate premature failure (splitting parallel to the reinforcement direction), specimens with nonsymmetrical fiber layups are covered with a thin layer of cold-cure epoxy resin. Equations (4.5.1) and (4.5.2) can be also used in the case of tension of a strip with a fiber layup of $\pm 45°$ for determination of the in-plane shear modulus of constituents of a unidirectional lamina [83, 202]. For this reason, the stress $\sigma_x = P/F$ and strains of a strip in the axial (ε_{0°) and transverse (ε_{90°) directions are measured.

The in-plane shear modulus G_{xy} of a material with a general type anisotropy can be determined by tension or compression testing of three series of specimens, cut from this material in two mutually perpendicular directions x and y and at an angle 45° to these directions. In this experiment, moduli of elasticity E_x, E_y, and E_{45° and Poisson's ratios ν_{yx} and ν_{45° are determined. The shear modulus G_{xy} is determined from the following formula [273]:

$$G_{xy} = \frac{1}{\dfrac{4\nu_{45^\circ}}{E_{45^\circ}} + \dfrac{1 - 2\nu_{yx}}{E_x} + \dfrac{1}{E_y}}. \tag{4.5.12}$$

The index x of the Poisson's ratio designates the direction of the load.

Reference [178] proposes a method of determining all elastic constants of a unidirectional material (E_1, E_2, ν_{12} and G_{12}) from the testing of flat specimens—strips with fiber layups of 0, 90, and ±45°—in uniaxial tension or tension-compression (if the elastic characteristics in tension and compression differ). Characteristics of a unidirectional material are determined from individual experiments: moduli of elasticity E_1 and E_2, Poisson's ratios ν_{12} and ν_{21} (on the principal axes of the material). Then the value U_1 and the tangential shear modulus G_{12} are calculated:

$$U_1 = \frac{1}{8(1 - \nu_{12}\nu_{21})}(E_1 + E_2 + 2\nu_{21}E_1) \tag{4.5.13}$$

$$G_{12} = \frac{2U_1 E_{\pm 45^\circ}}{8U_1 - E_{\pm 45^\circ}} \tag{4.5.14}$$

where

$$\nu_{21} = \nu_{12}\frac{E_2}{E_1}.$$

The modulus of elasticity $E_{\pm 45^\circ}$ is determined from tension tests of a specimen with a fiber layup of ±45° (the percent fiber content should be the same in specimens with fiber layups of 0, 90, ±45°! Since the relation of $G_{\pm 45^\circ}$ to $\varepsilon_{\pm 45^\circ}$ is nonlinear, the obtained curve is partitioned into segments for which this relation is assumed to be linear. The tangential modulus of elasticity $E_{\pm 45^\circ}$ is found for each of these segments of the experimentally

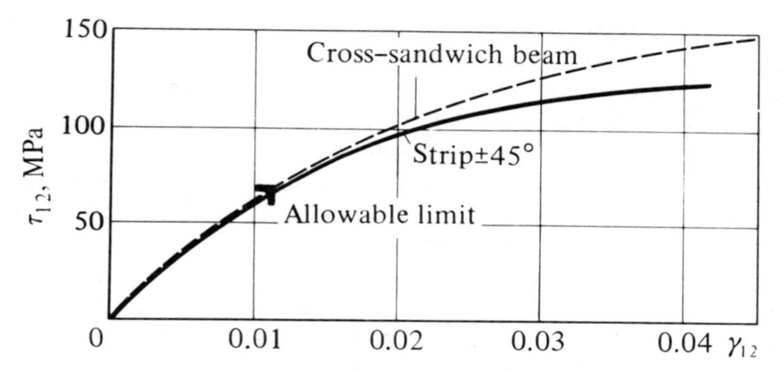

Fig. 4.5.4. γ_{12} versus τ_{12} diagram for a boron composite, obtained in testing of a strip with a reinforcement layup at an angle of $\pm 45°$ and of a cross-sandwich beam [178].

obtained curve, and from Equation (4.5.16) the shear modulus G_{12} is found. Further, for each portion of the $\sigma_{\pm 45°}$ versus $\varepsilon_{45°}$ curve the shear deformation and the tangential stresses are determined from the experimentally obtained values of $\varepsilon_{\pm 45°}$ and $\nu_{\pm 45°}$:

$$\Delta\gamma_{12} = (1 + \nu_{\pm 45°})\varepsilon_{\pm 45°} \tag{4.5.15}$$

$$\Delta\tau_{12} = G_{12}\gamma_{12}. \tag{4.5.16}$$

After this, the τ_{12} versus γ_{12} curve is plotted.

In the development of this method some assumptions have been made which somewhat limit the area of its application. First, under tension a specimen with a fiber layup of $\pm 45°$ is not in the state of pure shear ($\varepsilon_x^{\pm 45°} \neq \varepsilon_y^{\pm 45°}$ and $\nu_{xy} < 1$, i.e., the center of Mohr's circle does not coincide with the origin of the system of coordinates), and small normal stresses σ_x and σ_y also act in it over the shear plane. This results in stresses in the τ_{12} versus γ_{12} plot which are somewhat lower than the test results for a cross-ply sandwich beam with a fiber layup of $\pm 45°$ (Fig. 4.5.4). Therefore, the range of application of the method using τ_{12} is limited; however, this range is extended in practice. It should be noted that this is why the tension and compression strengths for strips with a fiber layup of $\pm 45°$ are different.

Secondly, in tension of a specimen with a fiber layup at an angle of $\pm 45°$ there is a zone of edge effect (see Section 2.1), as a result of which actual distribution of stresses through the width of a specimen differs from the calculated value. This limitation can be reduced to a minimum by proper selection of the specimen width. In [178], a specimen 25 mm wide is considered to be a compromise—between a very narrow specimen in which it is easy to attain a uniform distribution of stresses throughout its width but with a relatively large edge effect zone, and a very wide specimen with negligible edge effect but in which it is hard to assure a uniform distribution of stresses throughout the specimen width.

This method is not laborious, regardless of the fact that a set of five curves ($\sigma^t_{0°}-\varepsilon^t_{0°}, \sigma^t_{90°}-\varepsilon^t_{90°}, \sigma^c_{0°}-\varepsilon^c_{0°}, \sigma^c_{90°}-\varepsilon^c_{90°}$, and $\sigma^t_{\pm 45°}-\varepsilon^t_{\pm 45°}$) is needed for complete evaluation of the material.

Anisotropic strips are not used for determination of in-plane shear strength

The dimensions of anisotropic strips have not been standardized, and their effect on the characteristics to be determined has not been systematically studied. Selection of dimensions must assure uniform stress in the gage length of the strip and eliminate deformations perpendicular to its plane (skewness, loss of stability). In [38] strips of dimensions 300 × 25 × 2 mm with aluminium tabs were used, while in [33] strips 254 × 13 mm in planar dimensions, 8- layers thick, with Micarta tabs.

4.6. TENSION-COMPRESSION OF NOTCHED SPECIMENS

4.6.1. Shapes and Dimensions of Specimens

In tension or compression tests of notched specimens, the interlaminar shear modulus and strength are determined. In selecting the shape of a grooved specimen, a cross section must be chosen which guarantees that only tangential stresses will act and that failure of the specimen in interlaminar shear will take place.

Specimens with two notches [Fig. 4.6.1(b) and Fig. 4.6.2(a)] or with notches and a hole [Fig. 4.6.1(c) and 4.6.2(b)] are tested in tension or compression. In the first type of specimen, the calculated cross section F is located between the notches and parallel to its longitudinal axis; in the second case it lies between the hole and the notches, and the shear takes place in two planes as well.

In selection of the dimensions and the test techniques for notched specimens, bending of the specimen and stress concentration should be taken into account. In tension tests on a specimen with asymmetrically located notches, a bending moment appears in the specimen (the analogy with an adhesive lap joint is shown in Fig. 4.6.3(b)], equal to $M = Pt/2$, where P is the tensile force and t is the width of the solid part of the specimen in a weakened section (see Figs. 4.6.1 and 4.6.3). As a result the breaking load P^u decreases. The value ME/I is assumed to measure the effect of bending [141] (M is the bending moment, E is the modulus of elasticity of

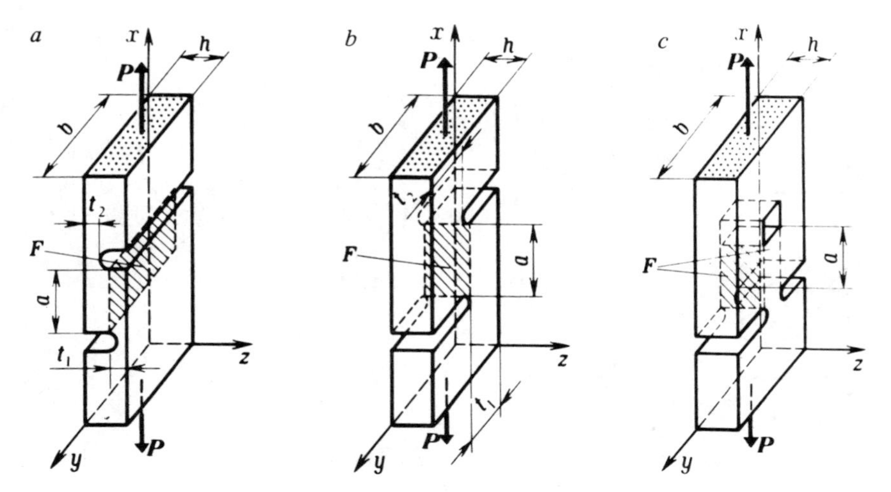

Fig. 4.6.1. Loading schemes and basic dimensions of specimens for interlaminar shear tests under tension or compression (F is the area through which shear takes place) [141]: (a) grooves in the reinforcement layup plane; (b) grooves in the transverse plane; (c) grooves and an opening.

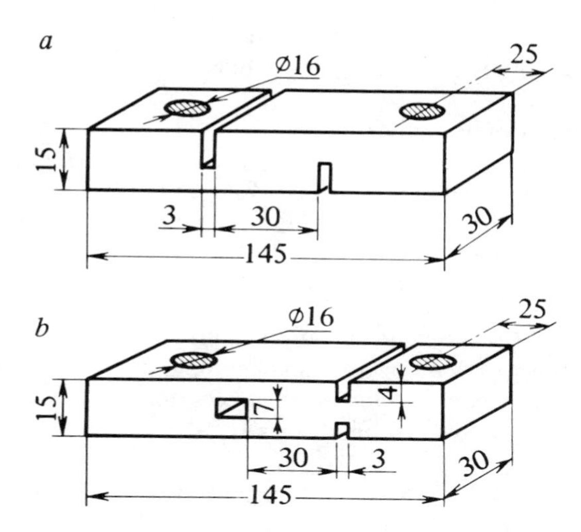

Fig. 4.6.2. Specimens for interlaminar shear tests [222]: (a) with grooves in two sections; (b) with grooves in one section and a central opening.

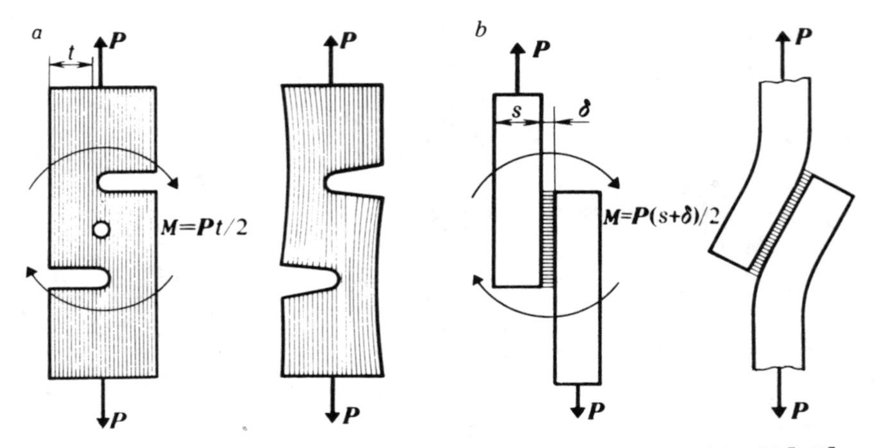

Fig. 4.6.3. Deformation scheme for grooved specimens (a) and adhesive joint (b) [141].

the material of a specimen, I is the inertia moment of the whole part of the weakened section of the specimen). For the specimen shown in Fig. 4.6.1(a) $I = bt^3/12$; in Fig. 4.6.1(b), $I = ht^3/12$.

The effect of bending on the measured interlaminar shear strength in tension of glass fiber composites is shown in Fig. 4.6.4, where the measured interlaminar shear strength has been plotted as a function of ME/I. The maximum values of interlaminar shear strength in tension have been obtained at $ME/I \to 0$, i.e., on specimens which have been installed in guides which preventing specimen bending. There are no data on the effect of the notch spacing and the length of the entire specimen on the bending strength of specimens.

Stress concentration is also observed in notched specimens, i.e., distribution of tangential stresses along the length of the section between the notches is not constant. The maximum tangential stresses of specimens with unequal notches [see Fig. 4.6.2(a)] can be determined by the following formula [136]:

$$\tau_{\max} = \tau\theta \coth \theta. \tag{4.6.1}$$

Here $\tau = P/ab$ is the mean value of tangential stresses (where P is the tensile or compressive load, a is the groove spacing, and b is the specimen width), and $\theta = 2ak/h$ where is the specimen thickness and $k = \sqrt{G_{xz}/2E_x}$, with E_x and G_{xz} moduli of elasticity and shear of the material of a specimen, respectively).

It follows from (4.6.1) that the concentration of tangential stresses increases with the increasing notch spacing a and shear modulus G_{xz} and with decreasing specimen thickness h and modulus of elasticity E_x. Diagrams of tangential stresses along the length of a section between notches at different

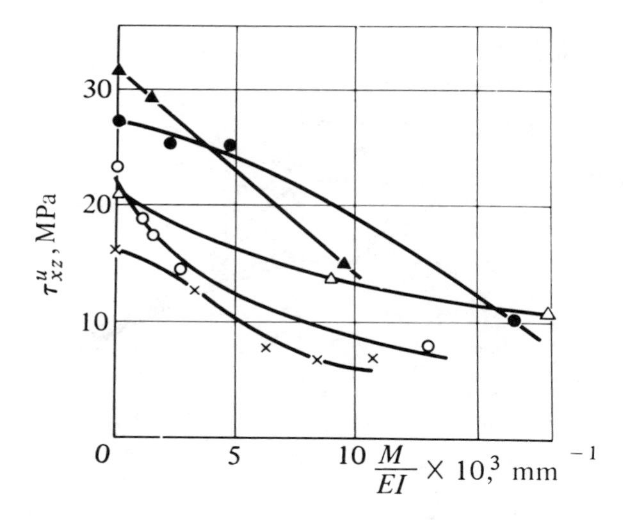

Fig. 4.6.4. The effect of bending a grooved specimen on the measured interlaminar shear strength of glass fiber composites [141]. The points on the ordinate, obtained experimentally for specimens installed in guides ($M/EI \to 0$). Characteristics of specimens (material, distance a between grooves):

▲ a strand, $V_f = 0.45$, epoxy resin, $a = 13$ mm;
● roving, $V_f = 0.46$, epoxy resin, $a = 20$ mm;
× roving, $V_f = 0.58$, epoxy resin, $a = 20$ mm;
○ roving, $V_f = 0.53$, polyester resin, $a = 20$ mm;
△ polyester fibers, $V_f = 0.45$, epoxy resin, $a = 20$ mm.

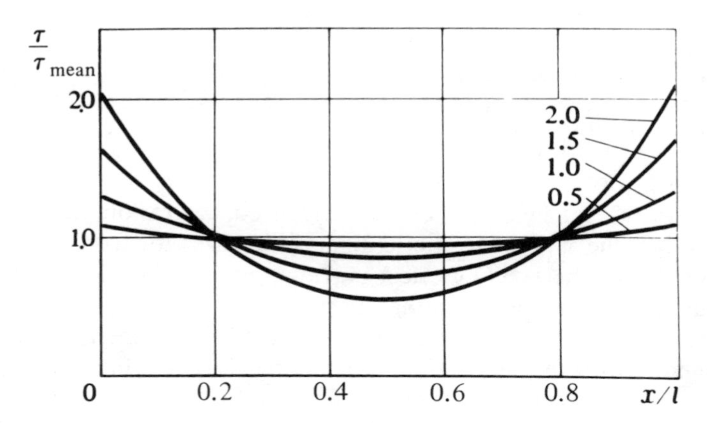

Fig. 4.6.5. Distribution of tangential stresses over the length of a section between grooves depending on the parameter θ (a number on the curves) [136].

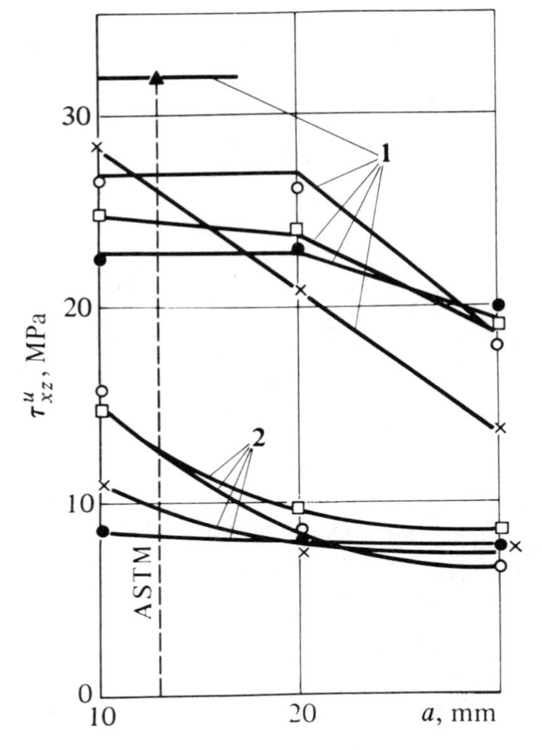

Fig. 4.6.6. Effect of the distance between grooves and bending of specimens on the measured interlaminar shear strength of fiberglass composites [133]: (1) specimens, installed in guides (bending eliminated); (2) freely supported specimens. Materials:

- ▲ a strand, $V_f = 0.45$, epoxy resin;
- × polyester fibers $V_f = 0.44$, epoxy resin;
- ○ a strand, $V_f = 0.44$, epoxy resin;
- □ a strand, $V_f = 0.52$, polyester resin;
- ● roving, $V_f = 0.53$, polyester resin.

θ are shown in Fig. 4.6.5. The effect of stress concentration, i.e., the notch spacing a, on the measured interlaminar shear strength with and without consideration of the bending effect is shown in Fig. 4.6.6. It can be seen from this figure that for obtaining optimal results in determination of the interlaminar shear strength by tension of a notched specimen, the specimen should be installed in guides which prevent bending and the notch spacing must be no more than 10 mm [141]. The specimens recommended in the ASTM standard (Fig. 4.6.7) answer the latter requirement; the notch spacing $a = 30$ mm of the specimens shown in Fig. 4.6.2(a) is too large.

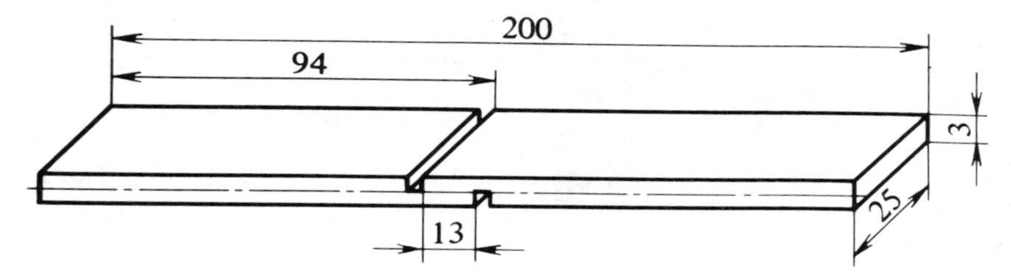

Fig. 4.6.7. A specimen for determination of interlaminar shear strength according to ASTM D 2733–70 (1976) [11].

The effect of bending is excluded in specimens of the type shown in Fig. 4.6.2(b). Deformation occurs in two planes. Stress concentration in these specimens has not been investigated, but its presence around the hole and the notches is irrefutable. The merit of these specimens consists in that it is possible to determine both the shear modulus and interlaminar shear strength. The specimens are loaded in tension or compression in testing machines or by means of special devices.

Besides the shapes described above, a variety of notched specimen shapes have been developed for investigation of interlaminar shear characteristics [21, 102, 129]; researchers on wood have acquired vast experience in this field [21]. Experience has shown that a change in the shape of the notches and of the specimen itself does not qualitatively affect the character of stress distribution over the gage section. However, the search for new specimen shapes for this type of tests continues [12].

There are no recommendations as to selection of dimensions (length, width, thickness) of notched specimens, except those specified in the ASTM standard. Specimen dimensions are frequently determined by considerations of fabrication and structure. However, some precautions may be in order because of the peculiarities of tension and compression of specimens of fibrous polymeric composites (force transmission, the effect of specimen width, scale effect, etc.).

Let us note in conclusion that the quality of cutting of the notches has a significant effect on the test results of specimens. Undercutting (where the notch does not reach the midplane of the specimen) leads to an increase in the measured strength τ_{xz}^{u}; overcutting (beyond the midplane of the specimen) leads to its decrease [38]. High-quality cutting instruments (diamond wheels) must be used to cut the notches.

Table 4.6.1 [38] presents data on the effect of notch depth on the interlaminar shear strength of unidirectional organic-epoxy composites (Kevlar-49, $V_f = 65\%$) during testing of notched specimens (V is the coefficient of variation; for k, see Section 4.6.3).

Table 4.6.1. Effect of Notch Depth on Interlaminar Shear Strength of Unidirectional Organic Epoxy Composite [38].

NOTCH DEPTH	$\frac{2a}{h}$	MEAN SHEAR STRESS τ, MPa	V, %	k	MAXIMUM SHEAR STRESS τ_{max}, MPa	MEAN VALUE OF τ_{max}, MPa
Notches cut to the center ply or slightly less deep	5.43	15.2	5.8		14.03	
	8.12	16.7	10.0		23.05	
	11.94	14.0	22.0	0.17	28.42	23.46
	12.58	10.9	12.3		23.31	
	15.67	10.7	8.1		28.50	
Notches cut through the center ply or slightly deeper	3.94	14.14	6.2		12.26	
	5.87	11.73	6.6		15.15	
	7.87	8.49	7.8	0.22	14.70	15.52
	9.73	8.18	4.3		17.51	
	10.98	7.44	11.0		17.97	
Overcut notches	4.41	5.20	13.5		4.82	
	6.03	5.42	7.2		6.86	
	8.21	5.05	6.4	0.21	8.71	7.71
	10.40	4.12	2.8		9.00	
	11.61	3.76	4.2		9.17	

4.6.2. Interlaminar Shear Modulus

For determination of the interlaminar shear modulus specimens of the type shown in Fig. 4.6.2(b) are used. In this case, the obtained characteristic is usually called the *apparent* interlaminar shear modulus G_{xz} because of several assumptions [222, p. 117]: interlaminar shear is caused by shear of a binder layer located between two reinforcement layers; the thickness of the reinforcement is small and it practically does not resist shear of a layer of binder; the displacement Δ, measured in the process of testing is the sum of the tensile deformation Δl^t and the displacement Δl^s.

The apparent interlaminar shear modulus is determined from the formula

$$G_{xz} = \frac{\Delta P}{b l_g \Delta \gamma} \tag{4.6.2}$$

where ΔP is a load increment; $\Delta\gamma = \Delta l^s/h_1$ is the average shear strain increment, corresponding to load increment ΔP; $\Delta l^s = \Delta - \Delta l^t = \Delta - (\Delta P l_e/bhE_x)$ $h_1 = h/n$ is the average thickness of a layer of the binder; h is the thickness of a specimen; n is the number of reinforcement layers in the specimen; b is the specimen width; l_e is the reference length of the extensometer; l_g is the average gage length of the specimen; E_x is the elastic modulus of the material under tension.

Strains are measured with the help of a Martens mirror tensometer. Speed of free movement of testing machine grips (without a specimen) is 10 mm/min [222, p. 119].

4.6.3. Interlaminar Shear Strength

In tensile tests, the interlaminar shear strength of grooved specimens is determined from Equation (4.6.1) by substitution of $\tau_{xz}^u = \tau_{max}^u$ and $\tau^u = P^u/F$ (where P^u is the breaking force and F is the shearing area).

The factor $\theta \coth \theta$ is unknown a priori. In [136] the following method of successive approximation is proposed for processing the experimental results:

1. The curve $\tau^u = f(a)$ is plotted from to test data of a series of specimens of various dimensions a [Fig. 4.6.8(a)].
2. By extrapolating the curve $\tau^u = f(a)$ to $a = 0$, τ_{max}^u is determined.
3. Using the value of τ_{max}^u, the coefficient k is determined from the derivation $(2ak/h)\tau^u = \tau_{max}^u$ (here $\theta \coth \theta \to \theta$ is approximated, which is valid at $2a/h > 10$).
4. Using the coefficient k (for fibrous polymeric composites k is within the range of $k = 0.2$–0.4) the straight line $1/\tau^u = f(\theta \coth \theta)$ is drawn [Fig. 4.6.8(b)].
5. The point of intersection of the straight line $1/\tau^u = f(\theta \coth \theta)$ with the straight line $\theta \coth \theta = 1$ gives a value of τ_{max}^u, which is used for refining the coefficient k; then calculation is repeated until the assumed and obtained values of k coincide completely.
6. The value of the coefficient k is used for calculation of τ_{max}^u from all experimentally determined stresses τ^u; for a correctly conducted experiment and proper processing of the results, $\dot{\tau}_{max}^u$ should be constant and independent of the ration $2a/h$.

The described algorithm for determination of $\tau_{max}^u = \tau_{xz}^u$ is quite laborious; however, experience has shown [136] that after careful specimen preparation it yields more reliable results than the method of determination τ_{xz}^u from three-point bending tests.

Interlaminar shear strength can also be evaluated on ring specimens. In this case, the same type specimen is used as in tensile tests, but of greater width (up to 50 mm). Two rectangular notches are cut into the ring surface

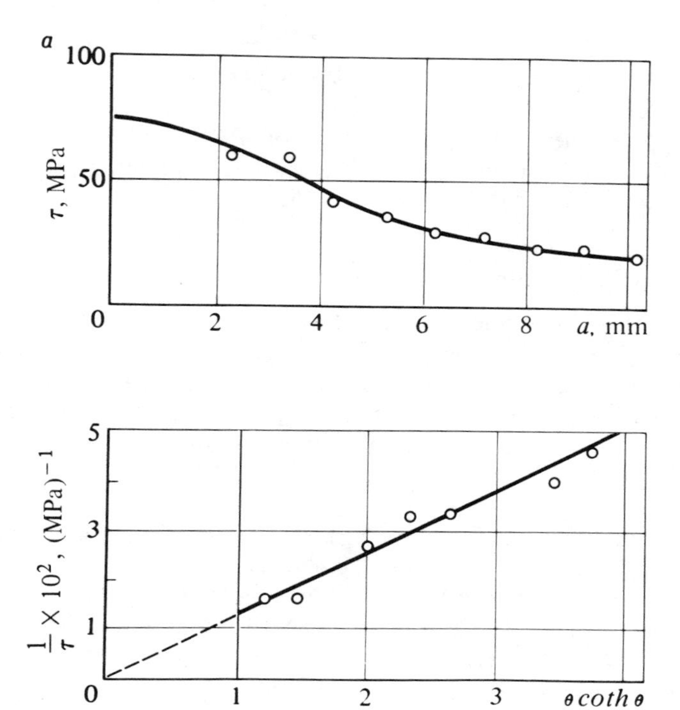

Fig. 4.6.8. Schemes of determination τ_{max} (a) and coefficient k (b) [136].

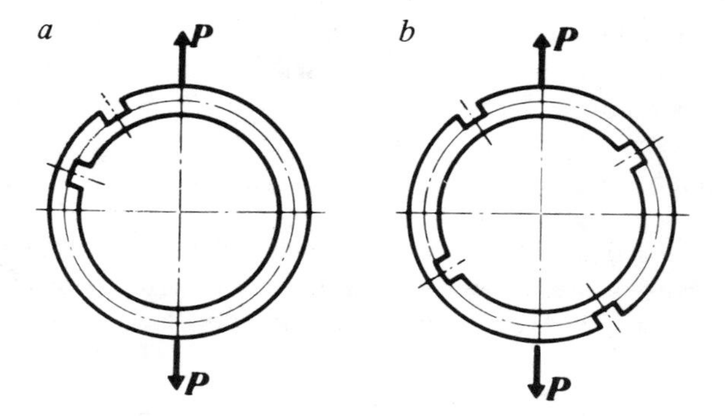

Fig. 4.6.9. Ring specimens for interlaminar shear tests (over circumference): (a) specimen with two notches; (b) specimen with four diametrically opposed notches.

across the entire width of the ring (Fig. 4.6.9). Usually a notch is 5 mm wide, and its depth somewhat exceeds $h/2$. Notches can be cut [Fig. 4.6.9(a)], but two pairs of diametrically opposed grooves can also be cut [Fig. 4.6.9(b)]. Diametrical opposition of notches is allowable only in loading of the specimen with uniform internal pressure. In tension by a split disk, the notches must be located in areas of the ring situated at a 30–45° angle to the midplane of split disk. In tension by split disk, bending of the specimen in this section is avoided, but in other respects the method has the same disadvantages as in tension-compression of flat notched specimens (see Section 4.6.1). The strength $\tau_{\theta r}^u$ is determined from Equation (4.6.1) by substituting $\tau_{\theta r}^u = \tau_{\max}^u$, $\tau^u = P^u/2F$, where P^u is the breaking force and F is the area of shear. The recommended loading rate is 1.3 mm/min [225].

4.7. BIAXIAL TENSION-COMPRESSION

For determination of shear characteristics in biaxial tension-compression the relationships between normal and tangential stresses and the corresponding strains are used. The in-plane shear modulus is determined from the formula:

$$G_{xy} = \tau_{xy}/\gamma_{xy} \tag{4.7.1}$$

where

$$\tau_{xy} = \frac{|\sigma_x| + |\sigma_y|}{2} = |\sigma_x| \quad \text{or} \quad \tau_{xy} = \frac{|\sigma_x| + |\sigma_y|}{2} = |\sigma_y| \tag{4.7.2}$$

$$\gamma_{xy} = 2|\varepsilon_x| = 2|\varepsilon_y| = \sqrt{2}\varepsilon_{45^\circ}. \tag{4.7.3}$$

The strength τ_{xy}^u is determined from the formula:

$$\tau_{xy}^u = |\tau_x^u| \quad \text{or} \quad \tau_{xy}^u = |\sigma_y^u|. \tag{4.7.4}$$

Two practically feasible ways of static loading of fibrous composites under biaxial tension-compression are known. In testing of a cross-sandwich beam (Fig. 4.7.1) the tensile and compressive stresses are produced by bending of arms of the beam according to four-point scheme. Stresses σ_x or σ_y are calculated according to Equation (5.3.12), but the strain is measured by means of resistance strain gages glued in the center of the beam at a 45° angle to one of the axes of the beam (Fig. 4.7.2). The remaining four

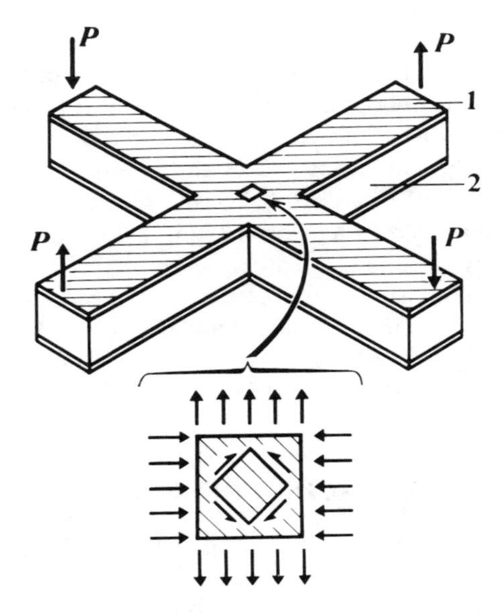

Fig. 4.7.1. Loading scheme for a cross-ply sandwich beam [104].

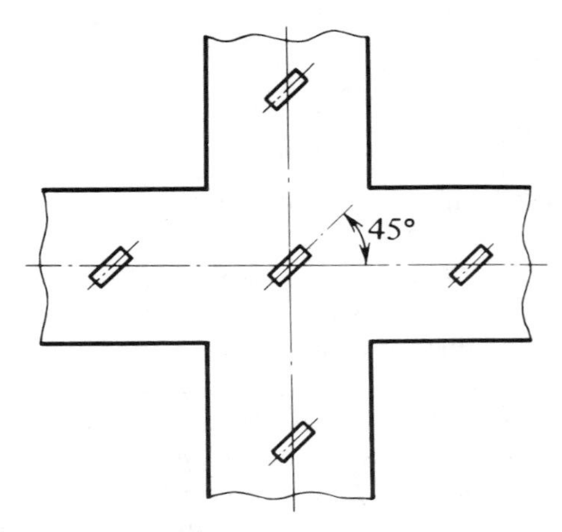

Fig. 4.7.2. Gluing of resistance strain gages in testing of a cross-sandwich beam.

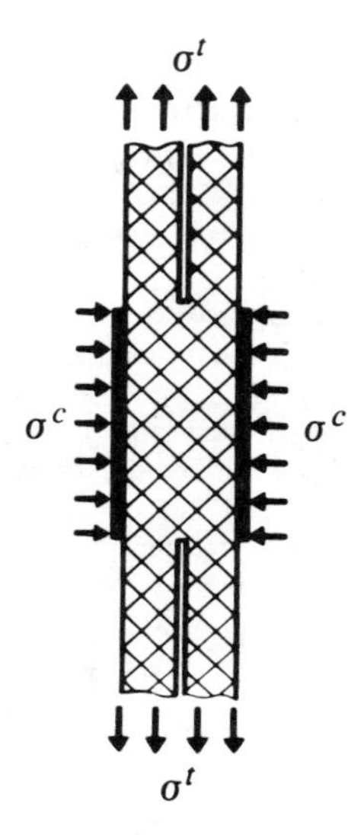

Fig. 4.7.3. Biaxial loading of a slotted specimen [66].

resistance strain gages, glued at a small distance from the edge of the gage length of the specimen, serve to control the accuracy of loading of the beam. To ensure a state of pure shear in the gage length of the beam, the condition $|\varepsilon_x| = |\varepsilon_y|$ must be satisfied.

In the material tested, the reinforcement is laid symmetrically about the axes of the beam. In the case of correct fabrication and loading of the beam (equality of tensile and compressive stresses must be ensured), the method yields good results; however, this method is uneconomical, because specimens of large dimensions (of the order 500 × 500 mm in plane) and of complex structure are needed to satisfy the loading conditions. The advantage of this method is the simple realization of biaxial loading. Structurally, the cross-ply sandwich beam is made like the sandwich beams described in Section 5.1.3.

Another method for biaxial tension-compression testing is suggested in [66]. Here, specimen strips with axial slots (Fig. 4.7.3) are used, which ensure purity of the state of stress in the gage length. The loading scheme is shown from the figure. The system of controlling the loads must ensure fulfillment of the condition $\sigma^t = \sigma^c$. For reliable transmission of transverse compressive load, soft inserts are used (at small loads, thin lead sheets; at large loads, glass fiber composite sheets). Specimen dimensions are not given in [66]; all that is mentioned is that the distance between slots is twice the specimen

Table 4.7.1. Shear Resistance of Various Materials under Biaxial Tension-Compression in Comparison with Other Methods [66].

MATERIAL, LAYUP	METHOD	MODULUS, GPa	STRENGTH, MPa
Fiberglass composite ($V_f = 40\%$, discontinuous fibers, polyphenylene sulfide matrix)	Slotted-tension	3.55	—
	Tube torsion	3.45	—
Grapite composite T 300/E 788; $[0^\circ_8]_s$ (interlaminar shear)	Slotted-tension	6.21	65.0
	Modified lap-shear	6.20	—
	Rail shear	5.17	42.9
Graphite composite T 300/E 788; $[0/90^\circ]_{4s}$ (interlaminar shear)	Slotted-tension	6.90	132.2
	145° Shear test	5.52	93.4
	Rail shear	6.27	118.7
Graphite composite T 300/E 788; $[0^\circ/+45^\circ/90^\circ/45^\circ]_s$ (50% of fibers intersect the shear planes at 45°)	Slotted-tension	21.58	356.6
	Rail shear	18.69	209.6
Graphite composite T 300/E 788 $[0/90^\circ]_{4s}$ (100% of fibers intersect the shear planes at 45°)	Slotted-tension	36.75	Fixture force limit exceeded
	Rail shear	31.37	
Aluminium alloy 6061-T4	Slotted-tension	27.30	87.9
	Rail shear	23.79	65.0 (yield)
Graphite/aluminium GY 70/201 (lamina)	Slotted-tension	16.15	57.5
	Off-axis (10°) shear	18.41	38.1
	Tube torsion	17.25	—

width, i.e., $P^c = 2P^t$. The main difficulty of the method lies in ensuring synchronous tensile and compressive loading. The method has been checked on various materials and it yields good results, especially in the determination of shear strength, as follows from Table 4.7.1.

4.8. SHEARING (TRANSVERSE SHEAR)

Shearing (transverse shear) tests are conducted to determine material resistance in structural elements subjected to the effect of shearing forces—in bolted and riveted joints, in axles, etc. The main difficulty in shearing tests is to ensure uniform distribution of tangential stresses in the gage section of the specimen. In practice, not one of the existing shearing test methods satisfies this requirement. In the case of the most widely used test method in single and double shearing (Fig. 4.8.1), accomplished through mutual displacement of a pair of cutting edges in the apparatus, the distribu-

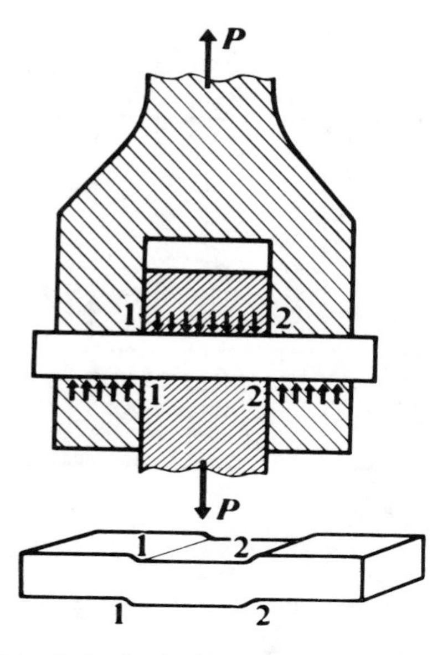

Fig. 4.8.1. Diagram of the device for double shearing tests in planes 1-1 and 2-2 [222].

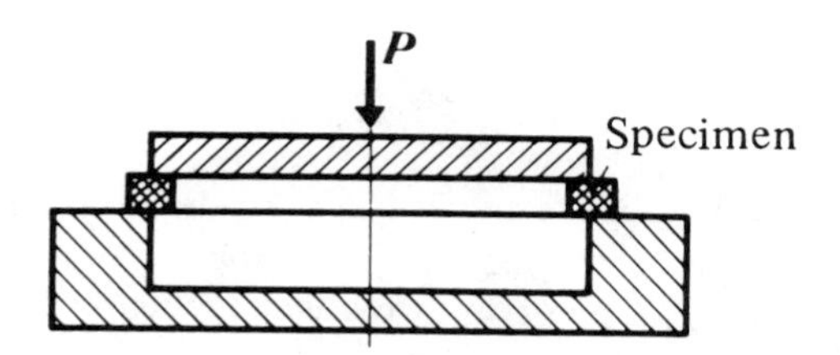

Fig. 4.8.2. Shearing (interlaminar shear) tests of rings [201].

tion of deformations in the material is the same as under compression and, consequently, there is no state of pure shear in the gage section of the specimen. Besides tangential stresses, normal stresses are also acting and failure of the material takes place simply as a consequence of action of the normal contact stresses under high stress concentration. In improperly made or worn fixtures the specimen is also subject to bending.

Due to these deviations, transverse and interlaminar shear tests in theoretically equivalent schemes of action of tangential stresses yield numerically different results. Therefore, shearing tests are inapplicable for the study of shear and serve only for qualitative comparison of materials. Specimens of different shape and their estimates are presented in works [21, 102].

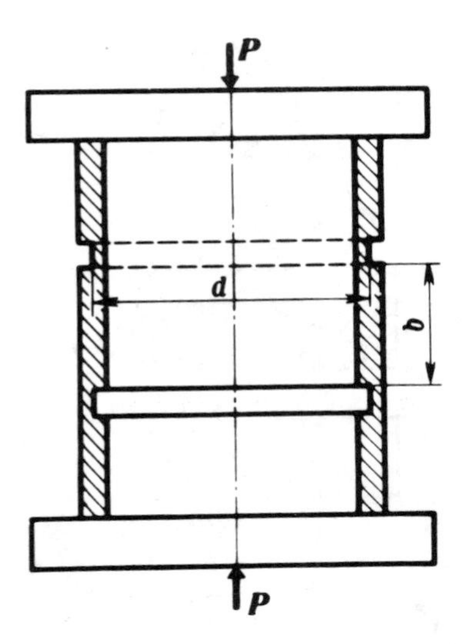

Fig. 4.8.3. Shearing (interlaminar shear) tests of tubular specimens.

Transverse shear strength is calculated according to formula:

$$\tau^u_{\text{shear}} = \frac{P^u}{F_{\text{shear}}} \tag{4.8.1}$$

where P^u is the breaking force and F_{shear} is the shearing area.

Equation (4.8.1) was derived under the assumption of uniform distribution of tangential stresses in the section over which shearing takes place. Since the actual distribution of these stresses is nonuniform, the determinable characteristic τ^u_{shear} is arbitrary.

Shearing strength of filament wound materials is studied on rings or short cylinders. Ring specimens are installed into a special fixture and loaded under compression, as shown in Fig. 4.8.2. Axial shearing strength is determined from the formula

$$\tau^u_{\text{shear}} = \frac{P^u}{\pi D b} \tag{4.8.2}$$

where D is the diameter of the sheared surface and b is the ring width.

A short cylinder with two circumferential grooves (one on the outer surface of the cylinder, another on the inner surface) is loaded with compressive forces over the end faces (Fig. 4.8.3). The depth of the grooves somewhat exceeds $h/2$. Axial interlaminar shearing strength is equal to:

$$\tau^u_{\text{shear}} = \frac{P^u}{\pi d b}. \tag{4.8.3}$$

The disadvantage of testing with grooved specimens is the requirement of high accuracy in the cutting of the grooves to ensure failure of the specimen in the gage section. General disadvantages of this type of testing for study of the shear strength are indicated in Section 4.6.

4.9. TORSION OF STRAIGHT BARS

4.9.1. Testing Characteristics

In torsion of solid bars of circular and rectangular cross sections it is possible to determine shear moduli and strengths. However, possibilities for determining the shear strength of composites in torsion experiments of solid bars are rather limited, since the formulae for calculation of stresses in anisotropic bars (particularly at the instant of failure) are complicated, failure of the material may take place in the nonlinear region of the M_T versus φ torsion curve, for which the calculation relationships are unknown, and the failure mechanism in torsion of composites with various fiber layups has not yet been investigated. At the present time, only attempts to determine interlaminar shear strength from torsion experiments are known; this method is described in Section 4.9.4.

The method of determination of the shear moduli from torsion tests on solid bars of circular and rectangular cross sections has been treated in detail elsewhere in this book. This method has been sufficiently well substantiated by experiment, and is being used for a wide variety of composites.

The complexity of the method of determining the shear moduli from torsion experiments consists in the impossibility of directly determining the shear moduli from experimentally measured torque M_T, angle of twist φ, and the geometrical dimensions of the specimen by simple substitution in a set of equations. Torsional stiffness C is determined from experimentally measured data using an analytical expression which contains, as a rule, two shear moduli, i.e., two unknown values, besides the geometrical dimensions of the specimen cross section. Therefore, to determine the shear moduli from torsion experiments it is necessary to have two values of torsional stiffness C evaluated on specimens with different cross-sectional dimensions. It is unreasonable to independently determine one of the shear moduli with the aim of simplifying the calculation operations, because of additional errors. It should be noted that the method of determination of the shear moduli from torsion experiments is quite sensitive to experimental errors, so that to obtain reliable data, a high order of test accuracy is required. The method of determining the shear moduli from torsion experiments on straight bars is described in Sections 4.9.2 and 4.9.3, and that from torsion experiments on intact and split rings in Section 4.10.

4.9.2. Torsional Stiffness

The shear moduli are calculated from the experimentally determined torsional stiffness C, which is related to the torque M_T and the relative angle of twist $\Delta\varphi$ by

$$C = \frac{M_T}{\Delta\varphi}. \tag{4.9.1}$$

For determination of $\Delta\varphi$ the absolute angles of twist φ_1 and φ_2 for two sections of the specimen and the distance between these sections (gage length l) are measured. The relative angle of twist $\Delta\varphi$ is determined from

$$\Delta\varphi = \frac{\varphi_1 - \varphi_2}{l}. \tag{4.9.2}$$

Experience shows that the shape of the M_T versus φ torsion curve depends on the type of fiber layup and the specimen shape. The relationship of M_T to φ can remain linear right up to failure of the material, but it may also have a break point with a subsequent nonlinear section (Fig. 4.9.1). Angles of twist are measured, as a rule, within the range of initial linear section of the torsion curve. Therefore, before the experiment it is necessary to obtain a complete torsion curve and set the allowable angle of twist.

In torsion of an orthotropic bar around the principal axis of orthotropy z, the corresponding torsional stiffness C_z depends on the two shear moduli G_{xz} and G_{yz} (as well as on the cross-sectional dimensions (analogously, C_y depends on G_{xy} and G_{yz}, and C_x depends on G_{xy} and G_{xz}).

For bars of circular cross section with diameter d:

$$C_x = \frac{\pi d^4}{16} \frac{G_{xy}G_{xz}}{G_{xy} + G_{xz}}, \tag{4.9.3}$$

$$C_y = \frac{\pi d^4}{16} \frac{G_{xy}G_{yz}}{G_{xy} + G_{yz}}, \tag{4.9.4}$$

$$C_z = \frac{\pi d^4}{16} \frac{G_{xz}G_{yz}}{G_{xz} + G_{yz}}. \tag{4.9.5}$$

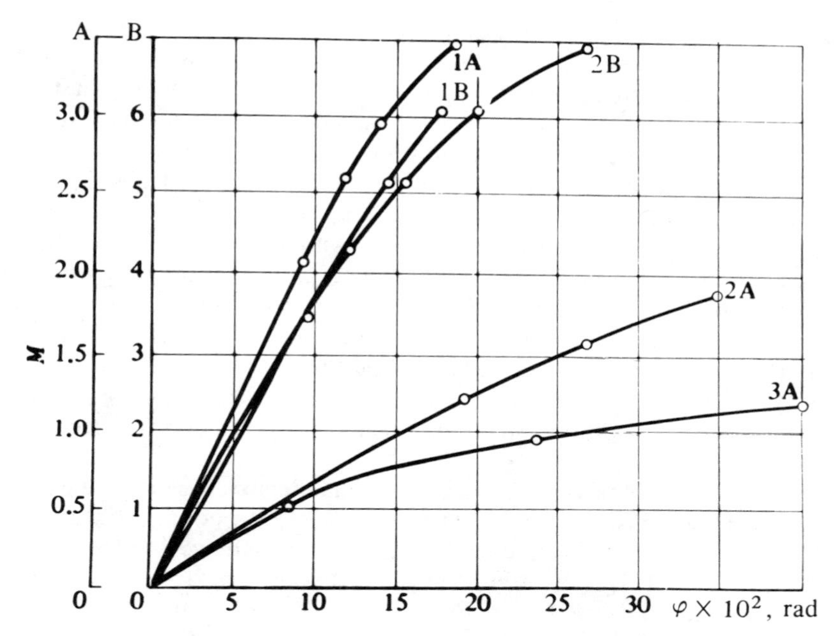

Fig. 4.9.1. M_T versus φ torsion curve of composites with various structure: (1A) carbon composite of complex structure; (2A) boron composite (0°); (3A) carbon composite (0°); (1B) carbon composite (0/90°); (2B) three-dimensionally reinforced glass fiber composite.

For bars of rectangular cross section the relationship between shear moduli, cross-sectional dimensions, and torsional stiffness is more complex:

$$C_x = G_{xy} b h^3 f(\eta) \tag{4.9.6}$$

where

$$f(\eta) = \frac{32\eta^2}{\pi^4} \sum_{k=1,3,5\ldots}^{\infty} \frac{1}{k^4}\left(1 - \frac{2\eta}{k\pi}\tanh\frac{k\pi}{2\eta}\right);$$

$$\eta = \alpha\beta, \qquad \alpha = \frac{b}{h}, \qquad \beta = \sqrt{\frac{G_{xz}}{G_{xy}}};$$

and b and h are the cross-sectional dimensions. Direction of axes: x is the longitudinal axis; the y axis coincides with the width b; the z axis with the thickness (height) h. Analogous formulas for C_y and C_z can be obtained by permutation of subscripts.

4.9.3. Determination of Shear Moduli

The method of determination of the shear moduli of an orthotropic material from torsion experiments has not been standardized. Moreover, at the present time there are no recommendations on selection of the shape and dimensions of specimens. Specimens are cut from blanks (plates, bars) so that their longitudinal axes coincide, in accordance with the purpose of the tests, with one of the principal axes of elastic symmetry of the material being studied. Solid bars of circular and rectangular cross section are used. The equations for bars of noncircular cross section are complex. In practice, there is a tendency to test bars of narrow rectangular cross section, one dimension of which is considerably larger than the other (for example, $b \gg h$). As will be shown later, in this case the equations are greatly simplified; however, testing of strip specimens also involves some technical difficulties.

Torsion tests of bars are conducted in special devices. The torque can be produced by suspension of a weight from a lever, but continuous loading of specimens is more reasonable. For this purpose, however, special test machines are needed. It must be noted that the formulas for processing of experimental results were derived under the assumption of free deformation of the specimen cross section. This must be taken into account in the design of the grips or supports of the device, since restriction of deformation in the axial direction leads to overstated values of torsional stiffness.

To avoid the problem of end effect, angles of twist φ_1 and φ_2 must be measured in sections, which are at a distance from the points of application of the torque ($z > z_{\min}$). The following manner of estimating the length of end effect zones may be recommended [150, 158]:

$$z_{\min} \approx c\sqrt{\frac{G_{\alpha z}}{G_{\beta z}}} \tag{4.9.7}$$

where c is the greater of dimensions b and h (see Fig. 4.9.1), and $G_{\alpha z}$, $G_{\beta z}$ are the greater and the smaller of the shear moduli G_{xz} and G_{yz}, respectively.

Analytical evaluation of the length of the end effect zone and suitability of specimen grips of various constructions is impossible. However, this information is important when the angle of twist is determined from the rotation of the bar grips. In [148] a simple method of experimental determination of the length of the end effect zone is given. In this technique, the same bar is repeatedly loaded with the same torque M_T but with different values of length l_1 between the grips, and the angle of twist is measured for each bar length. From these data $\Delta\varphi$ is plotted versus l_1 (Fig. 4.9.2). The

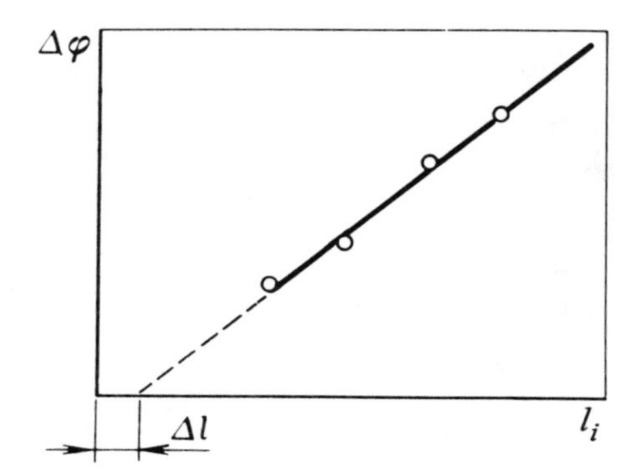

Fig. 4.9.2. Determination of the end effect zone in bar torsion by means of the l_i, $\Delta\varphi$ diagram [148].

point of intersection of straight line drawn through the experimental points with the abscissa gives the correction value Δl. This correction may be negative or positive, depending on whether the attachment stiffens the bar or not.

It is recommended that the angle of twist be measured by means of mirror extensometers (even in cases where the torsion machine has automatic angle of twist registration). The baseline for mounting of mirrors for a specimen length between the grips of 200–250 mm is usually 50–100 mm, i.e., it amounts to one-third of the free bar length. Recently, low-power lasers [224] have been used as light source which ensures high accuracy of measurements and safe operation in a normally lit laboratory.

Torsional stiffness of the specimen is determined from experimentally measured values according to the formula

$$C = \frac{\Delta M_T}{\Delta\varphi} = \frac{2ls}{\Delta k_1 - \Delta k_2}\Delta M_T \tag{4.9.8}$$

where ΔM_T is an increment of torque; l is the gage length; s is a distance from the measuring rule to the mirror (Fig. 4.9.3); Δk_1 and Δk_2 are readings on the scale of the device at angles of twist φ_1 and φ_2.

In the derivation of (4.9.8) $\tan 2\varphi = 2\varphi$ is assumed (the error is less than 5% for $\varphi < 11°$) and variation in the distance s in rotation of the mirrors is ignored.

The torsional stiffness C determined in the experiment is described by a ratio of two shear moduli. This does not allow direct determination of shear moduli from test data for one bar. The shear moduli G_{xy}, G_{xz}, and G_{yz} can

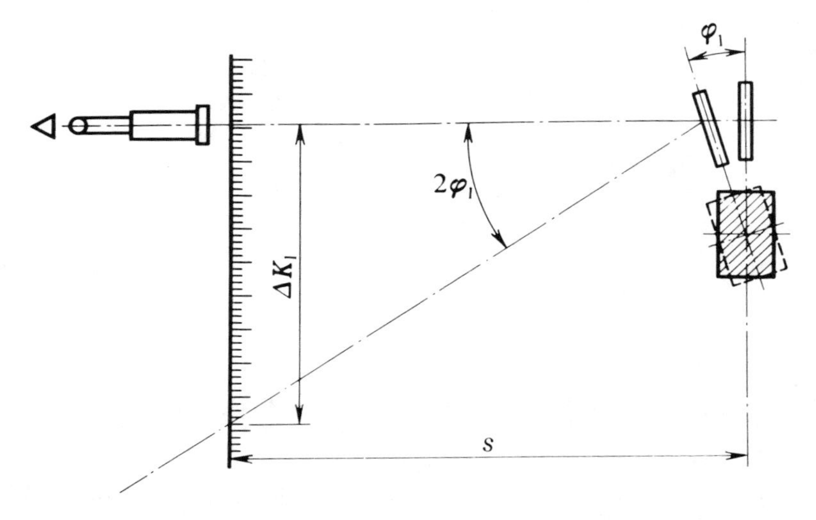

Fig. 4.9.3. Measurement scheme for the angle of twist.

be found if torsional stiffnesses C_x, C_y, and C_z of bars of circular cross section have been determined. For bars having a diameter d at already known C_x, C_y, C_z the shear moduli are found as follows:

$$\frac{1}{G_{xy}} = \frac{\pi d^4}{32}\left(\frac{1}{C_x} + \frac{1}{C_y} - \frac{1}{C_z}\right), \tag{4.9.9}$$

$$\frac{1}{G_{xz}} = \frac{\pi d^4}{32}\left(\frac{1}{C_x} - \frac{1}{C_y} + \frac{1}{C_z}\right), \tag{4.9.10}$$

$$\frac{1}{G_{yz}} = \frac{\pi d^4}{32}\left(-\frac{1}{C_x} + \frac{1}{C_y} + \frac{1}{C_z}\right). \tag{4.9.11}$$

However, it is not always possible to determine all three torsional stiffnesses of bars of circular cross section, since the material is very often supplied in the form of thin sheets of insufficient thickness to allow cutting of specimens whose axis is perpendicular to the reinforcement plane. Therefore, several specimens are used which have been cut along axes which lie in the reinforcement plane. Thus, shear moduli G_{xz} and G_{yz} can be found from test results of one circular bar and one bar of rectangular cross section. In this case, the system of equations of G_{xz} and G_{yz} is made up of Equations (4.9.6) and (4.9.11).

It is possible to avoid the computational difficulties arising from use of Equation (4.9.6) by replacing bars with strips in which width b is considerably greater or smaller than thickness h. When

$$\frac{h}{b}\sqrt{\frac{G_{yz}}{G_{xz}}} < \frac{\pi}{4} \tag{4.9.12}$$

or

$$\frac{h}{b}\sqrt{\frac{G_{yz}}{G_{xz}}} > 4\pi \tag{4.9.13}$$

the torsional stiffnes C can be respectively determined from the following approximate relationships [128]:

$$\frac{3}{bh^3}C_z = G_{yz} - 0.63025\frac{h}{b}G_{yz}\sqrt{\frac{G_{yz}}{G_{xz}}} \tag{4.9.14}$$

or

$$\frac{3}{bh^3}C_z = G_{xz} - 0.63025\frac{b}{h}G_{xz}\sqrt{\frac{G_{xz}}{G_{yz}}}. \tag{4.9.15}$$

Designation of axes in Equations (4.9.12)–(4.9.15) is as follows: the x axis is directed through the thickness of the strip; the y axis runs along the width of the strip; the z axis is the longitudinal axis of the strip.

For determination of the shear moduli G_{xz} and G_{yz}, torsional stiffness C_z is experimentally measured at various values of h/b, and the reduced stiffness $(3/bh^3)\ C_z$ is plotted versus h/b (Fig. 4.9.4). At $h/b = 0$, it follows from (4.9.14) that

$$G_{yz} = \frac{3}{bh^3}C_z \tag{4.9.16}$$

i.e., the linear plot of $(3/bh^3)C_z$ versus h/b intersects the ordinate at the point G_{yz}. Given G_{yz} we obtain from (4.9.14)

$$G_{xz} = \frac{0.3972\ G_{yz}^3}{k_1^2} \tag{4.9.17}$$

where k_1 is the slope of the $(3/bh^3)C_z$ line.

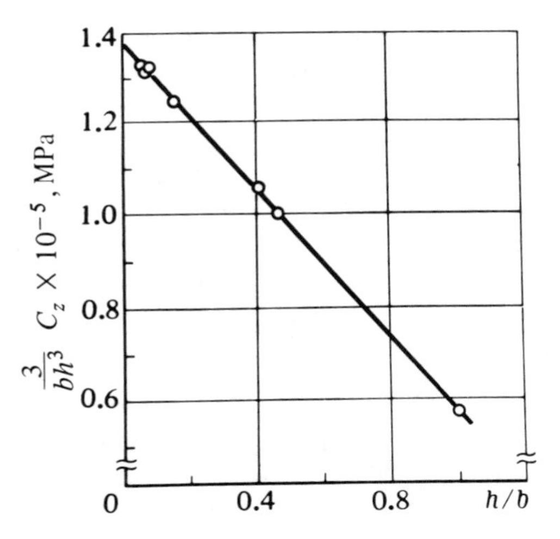

Fig. 4.9.4. Determination of shear modulus G_{yz} by means of a reduced stiffness versus h/b diagram [128].

It can be seen from (4.9.17) that this method requires very high accuracy in the determination of the shear modulus G_{yz} and the slope k_1, since the error in determination of the shear modulus G_{xz} increases as the square (or the cube) of the error in determination of k_1 or G_{yz}). It should be noted that stringent demands on accuracy in the determination of specimen cross-sectional dimensions and of one shear modulus (the first one) present a great difficulty in processing the results of torsion of bars of noncircular cross section. It is possible to use computers to solving the system of equations (4.9.14) and (4.9.15).

In processing test results for bars of almost quadratic cross section, the curve $C_i/b_ih_i^3 = f(b_i/h_i)$ is plotted from the experimentally determined values of torsional stiffnesses C_i by the method of least squares (Fig. 4.9.5). For a reliable plot, it is necessary to have no less than 4–5 experimental values of C_i within a sufficiently wide range of b_i/h_i ratios. By means of the $C_i/b_ih_i^3 = f(b_i/h_i)$ curve two values of torsional stiffness C_1 and C_2 are selected from which the shear moduli G_{xy} and G_{xz} are determined.

1. From the known values of b_1, h_1, C_1 and b_2, h_2, C_2 the following relation is determined:

$$F(\beta) = \frac{f(\eta_1)}{f(\eta_2)} = \frac{b_2}{b_1}\frac{h_2^3}{h_1^3}\frac{C_1}{C_2}. \tag{4.9.18}$$

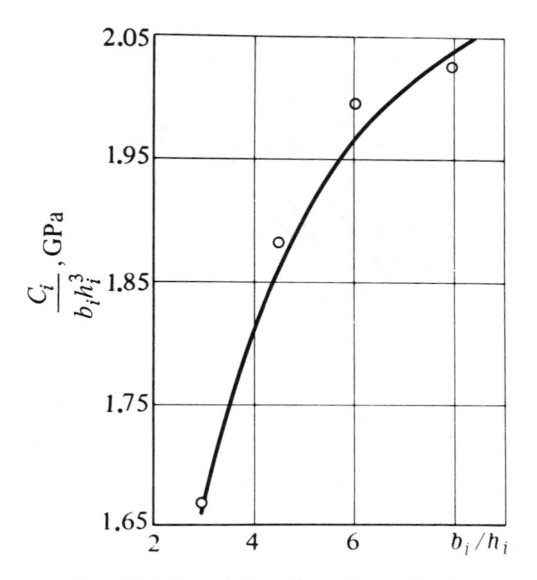

Fig. 4.9.5. Plot of $C_i/(b_i h_i^3) = f(b_i/h_i)$ [224].

2. From the known values of b_1, h_1, b_2, h_2 and given $\beta = \sqrt{G_{xz}/G_{xy}}$ (the range of β depends on the material studied; it is usually $\beta \leqslant 1$) the following function is derived by computer from Equation (4.9.6):

$$F(\beta) = \frac{f(\alpha_1 \beta)}{f(\alpha_2 \beta)}, \quad \text{where} \quad \alpha_1 = \frac{b_1}{h_1}, \qquad \alpha_2 = \frac{b_2}{h_2} \tag{4.9.19}$$

3. Equating (4.9.18) and (4.9.19), the ratio β, the function $f(\alpha_1 \beta)$, and finally the shear moduli are determined:

$$G_{xy} = \frac{C_1}{b_1 h_1^3 f(\alpha_1 \beta)} \tag{4.9.20}$$

$$G_{xz} = \beta^2 G_{xy} \tag{4.9.21}$$

When it is impossible to use electronic computers, the procedure for determination the shear moduli is somewhat different:

1. $F(\beta)$ is determined by Equation (4.9.18).
2. From the plot given in Fig. 4.9.6, β is determined for known α_1, α_2 and $F(\beta)$.
3. The function $f(\alpha_1 \beta)$ and the shear moduli [Equations (4.9.20) and

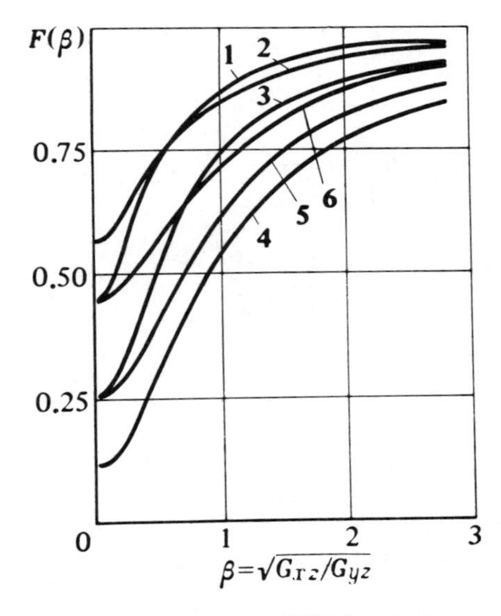

Fig. 4.9.6. A graph for determination of $\beta = \sqrt{G_{xz}/G_{yz}}$ with a known function $F(\beta)$ [158].

Curve	α_1	α_2	α_2/α_1
1	2	3	1.5
2	1.5	2	1.33
3	1.5	3	2
4	1	3	3
5	1	2	2
6	1	1.5	1.5

(4.9.21)] are determined from the graph shown in Fig. 4.9.7, or from Table 4.9.1.

4.9.4. Interlaminar Shear Strength

In torsion of bars of rectangular cross section the interlaminar shear strength is determined by

$$\tau_{xz}^{u} = \frac{k_1 M_T^u}{hb^2} \quad (\text{at } h > b) \tag{4.9.22a}$$

or

$$\tau_{xz}^{u} = \frac{k_2 M_T^u}{hb^2} \frac{1}{\sqrt{G_{xz}/G_{xy}}} \quad (\text{at } h < b) \tag{4.9.22b}$$

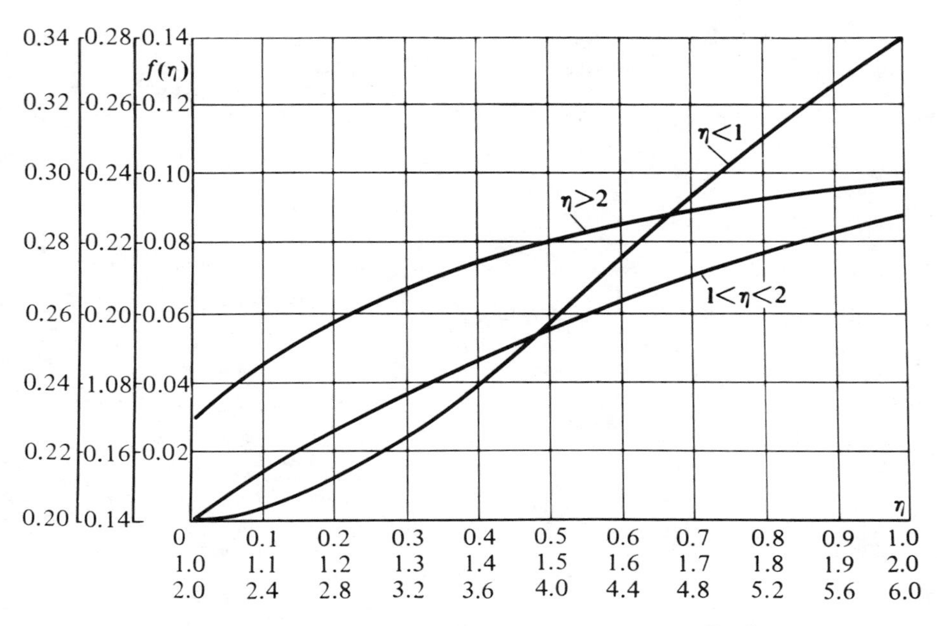

Fig. 4.9.7. A graph for determination of the function $f(\eta)$ [150].

where M_T^u is the torque at specimen failure; b, h are width and thickness of the specimen, respectively; G_{xy}, G_{xz} are the shear moduli, determinable in an independent experiment.

Coefficients k_1 and k_2 are selected [125, p. 58] depending on $(h/b)\sqrt{G_{xy}/G_{xz}}$:

$\frac{h}{b}\sqrt{\frac{G_{xy}}{G_{xz}}}$	k_1	k_2	$\frac{h}{b}\sqrt{\frac{G_{xy}}{G_{xz}}}$	k_1	k_2
1	4.804	4.804	10	3.202	2.379
1.5	4.330	3.767	20	3.098	2.274
2	4.068	2.975		3.000	0
5	3.430	2.548			

Investigations on glass cloth epoxy composites [274] show that experimental results are quite sensitive to changes in the relative dimensions of the specimen b/h and l/b (l is the gage length of the specimen). It can be seen from Fig. 4.9.8 that a change in b/h ratio of thick specimens ($h = 20$ mm) affects the strength less than a change in thin specimens ($h = 10$ mm). It also follows from the figure that there is a b/h range over which the materials

Table 4.9.1. Values of $F(\beta) = f_1(\alpha_1 \beta)/f_2(\alpha_2 \beta)$ at given parameters α_1 and α_2 [158].

β	$f(\beta)$	$F(\beta; 1; 1.5)$	$F(\beta; 2; 3)$	$F(\beta; 1; 2)$	$F(\beta; 1.5; 3)$	$F(\beta; 1; 3)$	$F(\beta; 1.5; 3)$
0	—	0.444	0.444	0.250	0.250	0.111	0.562
0.033	—	—	—	—	0.258	—	—
0.050	0.0008	—	—	0.258	—	—	—
0.067	—	—	—	—	0.265	—	—
0.100	0.0031	—	0.490	0.265	—	—	—
0.167	—	—	—	—	0.306	—	—
0.200	0.0118	—	—	0.296	—	—	0.603
0.250	0.0176	—	—	0.307	—	—	—
0.333	0.0292	0.510	0.619	0.335	0.406	0.207	—
0.400	0.0398	—	—	0.362	0.452	—	0.710
0.500	0.0573	—	0.719	0.406	—	—	—
0.533	—	—	—	—	0.539	—	—
0.572	0.0699	—	—	—	—	—	—
0.667	0.0870	0.617	—	—	0.616	0.380	—
0.800	0.1100	—	—	0.542	—	—	—
1.000	0.1410	0.719	0.870	0.616	0.746	0.537	0.846
1.250	0.1720	—	0.898	0.691	—	—	—
1.500	0.1960	—	0.921	0.745	—	0.683	0.910
1.750	0.2140	—	—	0.783	—	—	—
2.000	0.2290	0.872	—	0.815	0.882	0.768	0.938
2.500	0.2490	—	0.945	0.857	—	0.817	—
3.000	0.2630	0.916	—	0.882	0.926	—	0.956
4.000	0.2810	0.942	0.978	0.916	0.949	0.895	0.977
5.000	0.2910	—	—	0.933	—	—	—
10.000	0.3120	—	—	0.967	—	—	—
20.000	0.3230	—	—	—	—	—	—
∞	0.3330	—	—	—	—	—	—

with $\tau_{xy}^u < \tau_{xz}^u$ can not be tested. It follows from Fig. 4.9.9 that increasing the relative length of the specimen l/b results in a decrease in the measured strength τ_{xz}^u.

In torsion of specimens of circular cross section, the interlaminar shear strength is determined by

$$\tau_{xz}^u = \frac{16}{\pi} \frac{M_T^u}{d^3} \tag{4.9.23}$$

where d is a diameter of the gage length of the specimen.

Investigations performed on glass cloth epoxy composites [274] show that in tests of specimens with circular cross section the measured strength does

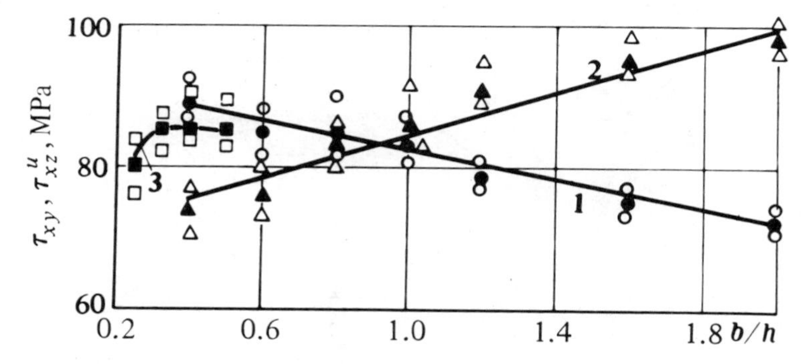

Fig. 4.9.8. The dependence of τ_{xz}^{u} (1, 3) and τ_{xy} (2) on the relative width of the specimen with a rectangular cross section. The specimen heights 10 mm (1, 2) and 20 mm (3) [274]. Open circles show scatter.

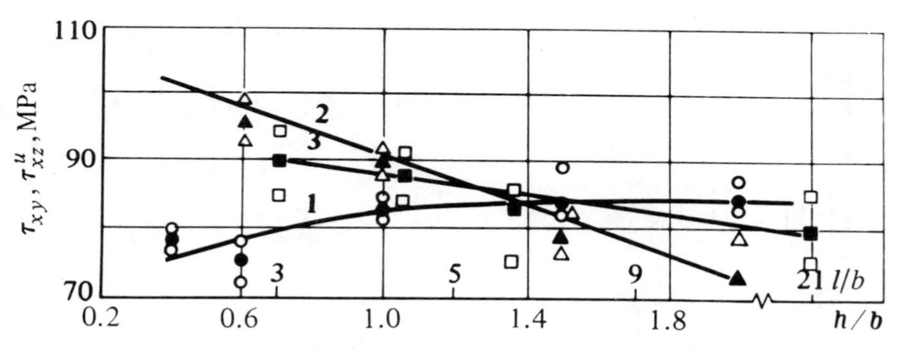

Fig. 4.9.9. Dependence of τ_{xz}^{u} (1, 3) and τ_{xy} (2) on the relative height h/b (1, 2) and gage length l (3) of the specimen with a rectangular cross section; the specimen width 10 mm [274]. Open circles show scatter.

not directly depend on the relative gage length l/d, but there is a certain optimum range of the diameter d over which the measured strength reaches its maximum (Fig. 4.9.10).

4.9.5. Torsion of Waisted Specimens

Consideration of the behavioral peculiarities of multidirectionally reinforced composites in shear has resulted in the development of waisted specimens, which are used for the determination of interlaminar shear strength. There are two types of these specimens. Reference [274] gives test results on specimens with a relatively narrow circular groove [Fig. 4.9.11(a)]. Two

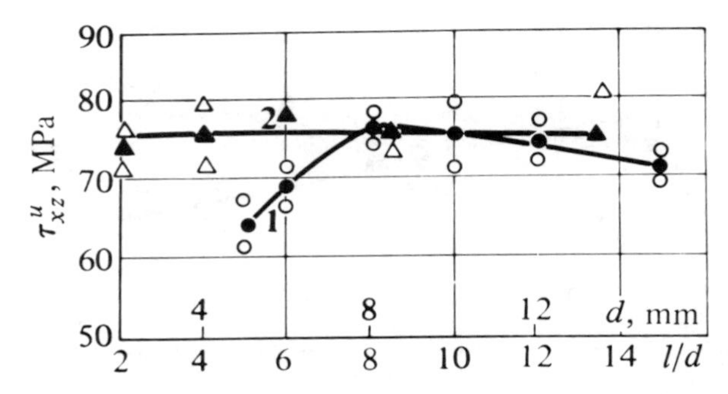

Fig. 4.9.10. Dependence of τ_{xz}^u on the diameter d (1) and relative gage length of the specimen l/d (2) with a circular cross section [274]. Open circles show scatter.

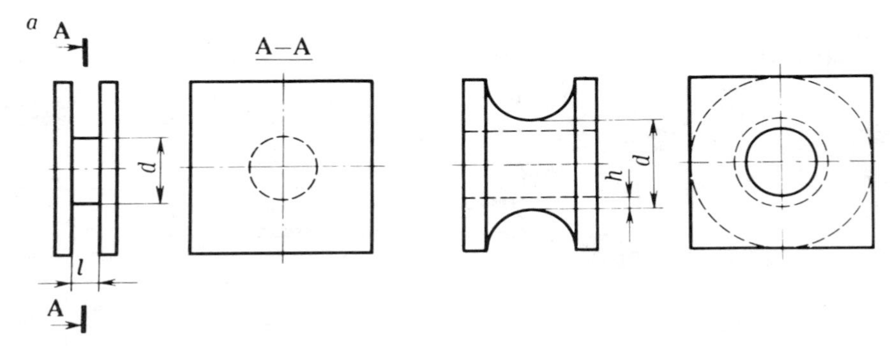

Fig. 4.9.11. Specimen with a circular groove: (a) without central opening; (b) with a central opening.

geometrical dimensions are of importance for this type of specimen—the gage length l and the diameter d. Investigations show that over the range $l/d = 0.2$–1.0 the gage length l does not affect the strength τ_{xz}^u (Fig. 4.9.12). Increasing the diameter of the gage section d from 5 to 10 mm also does not affect the value of the measured strength τ_{xz}^u, but on further increase in the diameter d a sharp drop occurs in the measured strength; this is connected with the increase in nonhomogeneity of the state of stress in the specimen.

The waisted specimens studied in [140] differ from those described above mainly in that there is a central opening [Fig. 4.9.11(b)], i.e., they are actually in the form of short tubes. The advantage of these specimens is the more uniform state of stress in the gage length. The chief variable geometric parameter of specimens of this type is wall thickness h in the gage length of the specimen. For the polyester and epoxy fiberglass com-

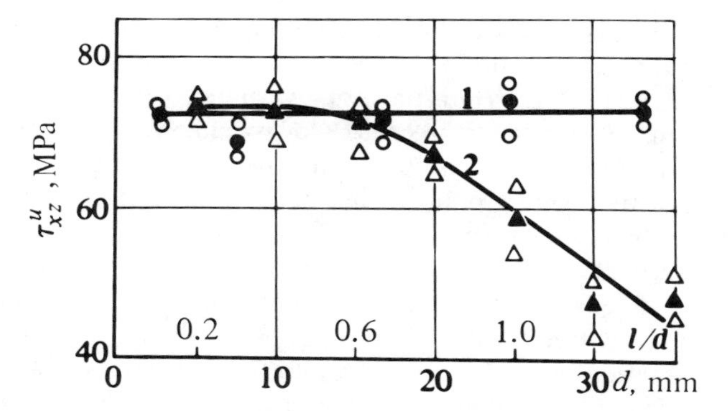

Fig. 4.9.12. Dependence of the strength τ_{xz}^{u} on the width of a circular groove (1) and diameter of the specimen gage length (2) [274]. Open circles show scatter.

posites studied in [140] the most acceptable wall thickness was $h = 2$ mm (however, this cannot be extended to all types of composites).

Interlaminar shear strength τ_{xz}^{u} is determined from the following formulas:

for specimens without a central opening:

$$\tau_{xz}^{u} = \frac{16}{\pi} \frac{M_T^u}{d^3} \tag{4.9.24}$$

for specimens with a central opening

$$\tau_{xz}^{u} = \frac{2}{\pi} \frac{M_T^u}{d_{\text{mean}}^2 h} \tag{4.9.25}$$

where M_T^u is the torque at failure of the specimen; d is the diameter of the gage length of the specimen; $d_{\text{mean}} = d - h$ is the mean diameter of the gage length of the specimen; h is the wall thickness.

4.10. TORSION OF RINGS

4.10.1. Purpose and Specific Features of Tests

In torsion testing of intact and split rings it is possible to determine torsional stiffness C and, consequently, shear moduli in two mutually perpendicular planes θ_r and θ_z, i.e., shear moduli $G_{\theta r}$ and $G_{\theta z}$. Directions of axes are as follows: r is along the radius, z is perpendicular to the plane of the ring, θ is along the circumference of radius R. Dimensions of specimens are: R is the mean radius, b is the width (measured along the z axis of the specimen; h is the height (measured along the radius of the specimen); $F = bh$ is the cross-sectional area; $I = hl^3/12$ is the moment of inertia in the cross section.) For this reason, no less than three series of specimens with various ratios of cross-sectional dimensions are tested. In the experiment the load P and displacement of the loading point w_P are measured; torsional stiffnesses C_i are calculated from the formula for displacement w_P and then, by the procedure described in Section 4.9.3, shear moduli $G_{\theta r}$ and $G_{\theta z}$ are calculated. Loading of specimens, as a rule, is accomplished only within the linear range of the P versus w_P curve.

For determination of shear moduli $G_{\theta r}$ and $G_{\theta z}$ from two known torsional stiffnesses C_1 and C_2, obtained from the graph of $(C_i/b_i h_i^3) = f(b_i/h_i)$, and from the relative dimensions of sections α_1 and α_2, the formulas for prismatic rods are used. The limits of applicability of this methodology for anisotropic ring specimens are not specified; for specimens of isotropic materials, this approach is acceptable at $R/h > 5$. The few known analytical solutions to problems in the torsion of bars of anisotropic material with a circular axis [26, 36] are complex and unsuitable for processing of experimental results.

The formulas presented below were derived under the assumption that the specimen has been fabricated from a homogeneous material, and is subject to small deflections and zero displacement in the plane of the ring. Nonhomogeneity is a specific feature of composites, which can be reduced only by eliminating manufacturing defects. Satisfaction of the remaining two requirements and avoidance of loss of stability of the specimen must be ensured by proper selection of specimen dimensions (R, b, h).

In torsion of circular specimens with concentrated force P, in the general case torque M_T and bending moment M_b vary around the circumference, while a constant shearing force Q also acts on the specimen. Consequently, a vertical (parallel to the z axis) displacement w at any point (except a support point) on the specimen circumference is the sum of the displacements due to the torque (w_T), bending moment (w_b), and to the shearing force (w_Q). However, formulas for displacements w_P at the points of application load can be used for processing of experimental results (i.e., the corresponding loading scheme can be used for experimental determination of torsional stiffness C) only in two cases:

1. When displacements w_b and w_Q are negligibly small compared to placement w_T
2. When displacement w_Q is negligibly small and the circumferential modulus of elasticity E_θ, which is needed for processing of experimental results, is determined from individual tests.

The limits of applicability of individual loading schemes and formulas for processing of experimental results are presented below.

4.10.2. Torsion of Intact Rings

Three loading schemes for torsion of intact rings are known; ways of loading and supporting are shown in Fig. 4.10.1. Formulas for displacement w_P at the point of load application take the form [114]:

For scheme (a) in Fig. 4.10.1:

$$w_P = w_T + w_b + w_Q = \frac{PR^3}{2C}\left(\frac{3\pi}{4} - \frac{1}{\pi} - 2\right) + \frac{PR^3}{2E_\theta I}\left(\frac{\pi}{4} - \frac{1}{\pi}\right) + \frac{3}{20}\frac{\pi PR}{G_{\theta z}F}. \tag{4.10.1}$$

For scheme (b):

$$w_P = w_T + w_b + w_Q = \frac{PR^3}{C}(\pi - 3) + \frac{PR^3}{E_\theta I}\left(\frac{\pi}{2} - 1\right) + \frac{3}{5}\frac{\pi PR}{G_{\theta z}F}. \tag{4.10.2}$$

For scheme (c):

$$w_P = w_T + w_b + w_Q = \frac{PR^3}{2C}\left(\pi - \frac{8}{\pi}\right) + \frac{4}{\pi}\frac{PR^3}{E_\theta J}\frac{1}{1 + \dfrac{C}{E_\theta I}} + \frac{3}{5}\frac{\pi PR}{G_{\theta z}F}. \tag{4.10.3}$$

It follows from analysis* [114] of Equations (4.10.1)–(4.10.3), with displacements w_b ($\leqslant 0.03\ w_T$) and w_Q ($\leqslant 0.03\ w_T$) assumed negligible, that: in tests under loading schemes (b) and (c) the displacements w_b are negligibly small only for specimens of materials with high anisotropy ($E_\theta/G_{\theta z} > 20$) and of narrow cross section ($b/h > 5$); displacements w_Q are negligibly small if at $b/h = 0.5$–10 the condition $R/h > 12$ is satisfied for loading scheme (b) and $R/h > 8$ for loading scheme (c). Consequently, loading schemes (b) and

*The following ranges of parameters were included in the analysis: anisotropy $E_\theta/G_{\theta z} = 5$–$10$; shear anisotropy $G_{\theta r}/G_{\theta z} = 0.5$–$2$; relative dimensions $R/h = 5$–20 and $b/h = 0.5$–10.

Loading scheme			a	b	c
Determinable characteristics			See Section 4.10.2	$G_{\theta r}$, $G_{\theta z}$	$G_{\theta r}$, $G_{\theta z}$
Measurable values			—	P, w_p	P, w_p
Geometrical sizes			—	$R/h, b/h$	$R/h, b/h$
Limitations	Structural	Layup	—	0°; 90°: 0/90°	
		Orientation	—	0°, 90°	
	Physical		—	Linear range $P \sim w_p$ independently determined E_θ	
	Geometrical		—	$\frac{R}{h} > 12$ at $\frac{b}{h} = 0.5 - 10$	$\frac{R}{h} > 8$ at $\frac{b}{h} = 0.5 - 10$

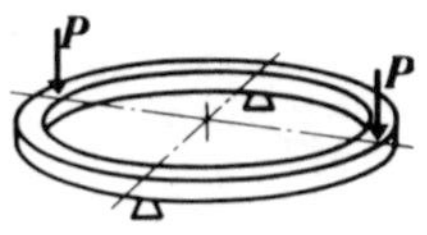

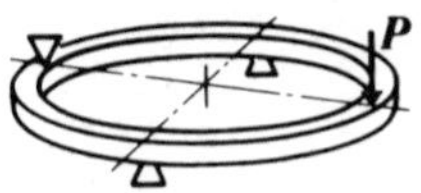

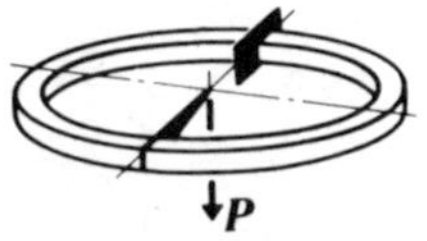

Fig. 4.10.1. Loading schemes of intact rings under torsion: (a) two-point; (b) three-point; (c) loading in the ring center.

(c) are applicable only when the circumferential modulus of elasticity E_θ is determined from independent experiments. Then the formulas for calculation of torsional stiffness C take the following forms:

For scheme (b):

$$C = \frac{\dfrac{3\pi}{4} - \dfrac{1}{\pi} - 2}{\dfrac{2w_P}{PR^3} - \dfrac{1}{E_\theta I}\left(\dfrac{\pi}{4} - \dfrac{1}{\pi}\right)} = \frac{0.037885}{\dfrac{2w_P}{PR^3} - \dfrac{0.46709}{E_\theta I}}. \tag{4.10.4}$$

For scheme (c):

$$C = \frac{\pi - 3}{\dfrac{2w_P}{PR^3} - \dfrac{1}{E_\theta I}\left(\dfrac{\pi}{2} - 1\right)} = \frac{0.14159}{\dfrac{2w_P}{PR^3} - \dfrac{0.57080}{E_\theta I}}. \tag{4.10.5}$$

Loading scheme (a) is essentially unsuitable for experimental determination of torsional stiffness C, since in this case displacements w_b and w_Q cannot be neglected in processing the experimental results.

To establish the range of applicability of loading schemes (b) and (c) under other conditions imposed on displacement ratios w_b/w_T and w_Q/w_T, the graphs $w_b/w_T = f(E_\theta/G_{\theta z}, b/h)$ and $w_Q/w_T = f(b/h, R/h)$ must be plotted.

4.10.3. Torsion of Split Rings

At the present time, three loading schemes in torsion of split rings are known; the modes of load application and support are shown in Fig. 4.10.2. Formulas for displacement in the point of load application w_P take the following forms [114]:

For scheme (a) in Fig. 4.10.2:

$$w_P = w_T + w_Q = \frac{PR^3}{C}\theta + \frac{6}{5}\frac{PR}{G_{\theta z}F}\theta \qquad (w_b = 0). \tag{4.10.6}$$

For a scheme (b) (mutual displacement of free ends of the specimen):

$$w_P = w_T + w_b + w_Q = \frac{3\pi PR^3}{C} + \frac{\pi PR^3}{E_\theta I} + \frac{12}{5}\frac{\pi PR}{G_{\theta z}F}. \tag{4.10.7}$$

Loading scheme			a 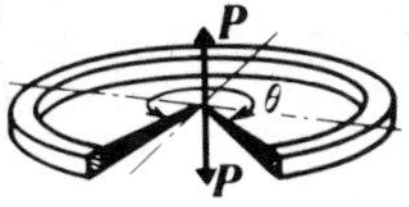	b 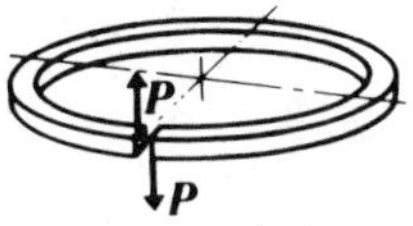	c 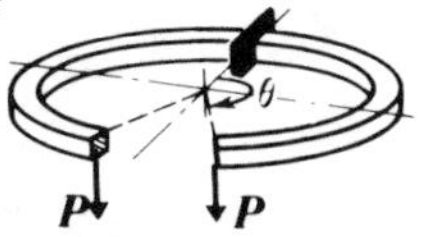
Determinable characteristics			$G_{\theta r}, G_{\theta z}$	$G_{\theta r}, G_{\theta z}$	$G_{\theta r}, G_{\theta z}$
Measurable values			P, w_p	P, w_p	P, w_p
Geometrical sizes			$R/h, b/h$	$R/h, b/h$	$R/h, b/h$
Limitations	Structural	Layup	0°; 90°; 0/90°		
		Orientation	0°, 90°		
	Physical		Linear range of the curve $P \sim w_p$		
	Geometrical		—	$\frac{12}{E_\theta hb^3} \to 0$ at $\frac{b}{h} > 3$	$\theta > 70°$

Fig. 4.10.2. Loading schemes of split rings under torsion: (a) in the ring centre with two opposite forces; (b) the same at the point of splitting; (c) at the point of splitting with two equally directed forces.

For scheme (c):

$$\begin{aligned} w_P = w_T + w_b + w_Q &= \frac{PR^3}{2C}[3\theta - (4 - \cos\theta)\sin\theta] \\ &+ \frac{PR^3}{2E_\theta I}[\theta - \sin\theta\cos\theta] + \frac{6}{5}\frac{PR}{G_{\theta z}F}\theta. \end{aligned} \tag{4.10.8}$$

Analysis of Equations (4.10.6)–(4.10.8) leads to the following conclusions.

The most rational loading scheme for split rings is scheme (a), in the realization of which $w_b = 0$ but displacement w_Q is negligibly small ($w_Q/w_T < 0.015$) within the entire range studied in [114].

In loading according to scheme (b), displacement w_Q is also negligibly small; numerical values of w_Q/w_T are one-third smaller than in loading according to scheme (a). The range of values of the relative width b/h within which displacement w_b is negligibly small depends on the anisotropy of the material, $E_\theta/G_{\theta z}$. It follows from the condition $w_b/w_T \leqslant 0.03$ that at $E_\theta/G_{\theta z} =$ 5–10 it must be $b/h \geqslant 3$; at $E_\theta/G_{\theta z} = 10$–$30$ and $E_\theta/G_{\theta z} > 30$, $b/h \geqslant 2$ and $b/h \geqslant 1$, respectively. In comparison to loading schemes (a) and (b), scheme (c) has the disadvantage of constant length of specimen gage section, which makes evaluation of the end effect zone impossible.

Loading scheme (c) is the least reasonable of the three loading schemes for split ring specimens. At $\theta = \pi$, the values of w_b/w_T and w_Q/w_T for loading schemes (b) and (c) are equal; however, at $\theta < \pi$, the ratios for scheme (c) increase rapidly and at $\theta < 70°$ become greater than for the loading scheme shown in Fig. 4.10.1(a) (i.e., than for an intact ring). Consequently, except the case when $\theta = \pi$, the range of applicability of scheme (c) is smaller than that of scheme (b) (see Fig. 4.10.2).

Equation (4.10.8) is applicable also to a simpler loading scheme, where instead of the entire split ring only a small circular section is used. In this case, for the purpose of more reliably fastening the specimen as well as to satisfy the assumptions underlying the calculation procedure, it is more reasonable to load the specimen symmetrically with respect to the restrained section.

By assuming $w_Q = 0$ we obtain the equations for the calculation of torsional stiffness:

For scheme (a) in Fig. 4.10.2:

$$C = \frac{PR^3}{w_P}\theta. \tag{4.10.9}$$

For scheme (b):

$$C = \frac{3}{\dfrac{w_P}{\pi P R^3} - \dfrac{1}{E_\theta I}}. \tag{4.10.10}$$

For scheme (c):

$$C = \frac{3\theta - (4 - \cos\theta)\sin\theta}{\dfrac{2w_P}{PR^3} - \dfrac{\theta - \sin\theta\cos\theta}{E_\theta I}}. \tag{4.10.11}$$

4.10.4. Experimental Technique

In accordance with the calculation scheme, supporting and loading of specimens as well as measuring of displacements w_P must be realized exactly along the circumference of mean radius R. These conditions practically exclude the possibility of testing thin specimens ($h < 2$ mm) according to loading schemes (a) and (b) in Fig. 4.10.1.

Since the required load is small, specimen loading is accomplished by suspension of a weight or with the help of a calibrated loading screw.

Displacements w_P can be measured by any sufficiently accurate instrument —clock-type indicator, calibrated metallic strips with resistance strain gages, variable reluctance pickup, optical cathetometer, etc. When mechanical displacement measuring instruments are used their internal resistance (friction) must be taken into account, since this can turn out to be comparable with the load to be applied to the specimen, particularly in testing split ring specimens.

In practice, realization of the loading schemes shown in Fig. 4.10.1(a) and (b) presents no difficulties. In loading according to the scheme of Fig. 4.10.1(c) the specimen fastening and the load cantilever must be sufficiently rigid and not allow mutual rotation of the coupled elements. Loading according to the scheme of Fig. 4.10.2(a) also presents no difficulties. In practice, usually one of the cantilevers is fixed rigidly and it forms a base of the specimen, while the other is used for loading. Bonding of cantilevers to the specimen should be secure and not allow arbitrary rotation of the cross section of the specimen under load.

Experimental technique in loading according to the scheme of Fig. 4.10.2(b) is simpler than in loading according to Fig. 4.10.2(a), but it requires care. Both forces P must be applied exactly in one section of the specimen, otherwise the specimen will rotate under load and the loading scheme will be distorted. Therefore, one of the free ends of the specimen is usually fixed during the experiment, without restriction of its deformation. Fabrication of a loading unit for metallic specimens is easy: the specimen is drilled in the necessary section and then cut. In a filament wound specimen, the cut ends of the specimen are displaced because of initial stresses, and the loading unit must be adjusted into place.

In loading according to the shceme of Fig. 4.10.2(b), reliable fastening of the specimen must be ensured, which does not allow its arbitrary rotation under the load.

* * *

Summary Table II is based on the following test methods: in-plane shear, interlaminar shear, shearing, torsion of bars with a straight and circular axis, and torsion of thin-walled tubes. In general, the techniques used in research practice are presented. Shear tests have not been standardized in practice, with the exception of interlaminar shear strength.

Summary Table II. Methods for Determination

METHOD	STANDARDS	CHARACTERISTICS DETERMINED: ELASTIC CONSTANTS	CHARACTERISTICS DETERMINED: STRENGTH	EXPERIMENTALLY MEASURED VA[LUES]: IN ELASTIC CONSTANT DETERMINATION	IN ST[RENGTH] DETERM[INATION]
Torsion of thin-walled tubes	—	G_{xy}	τ_{xy}^u	M_T (applied torque), $\Delta\varphi$ (angle of twist over the length of a measurement base)—or $\varepsilon_{45°}$, $\varepsilon_{-45°}$ (strains)	M_T^u (applie failure of men fron
Panel shear test	—	G_{xy}	τ_{xy}^u	p (load), ε_1, ε_2 (strains along tension and compression diagonals)	p^u (load at the speci
Square plate twist two-point scheme	—	G_{xy} (indirectly and other elastic constants)	τ_{xz}^u	p (load), w (relative deflection over the length of a measurement base)	The same
three-point scheme	—	G_{xy} (indirectly and other elastic constants)	—	p (load), w_p (deflection at the point of loading)	—
Tension of an anisotropic strip	—	G_{xy}	—	p (load), ε (strain; orientation depends on the selected method)	—
Tension-compression of a bar or ring with grooves	—	G_{xz}	τ_{xz}^u	p (load), Δ (displacement)	p^u (load at specimen
Biaxial loading	—	G_{xy}	τ_{xy}^u	p (load), $\varepsilon_{45°}$ (strain)	The same
Shearing of prismatic or circular specimens	—	—	τ_{xz}^u	—	p^u (load at specimen
Torsion of circular cross section bars	—	G_{xy}, G_{xz}, G_{yz}	—	M_T (applied torque), $\Delta\varphi$ (angle of twist over the length of measurement base)	—
Torsion of rectangular cross section bars	—	G_{xy}, G_{xz}, G_{yz}	τ_{xz}^u	The same	M_T^u (the a torque a specimen
Torsion of waisted specimens	—	—	τ_{xz}^u		The same
Torsion of intact and split rings	—	$G_{\theta r}$, $G_{\theta z}$	—	p (load), w_p (displacement in the point of loading)	—

haracteristics of Fibrous Polymeric Composites

SPECIMEN SHAPES:				
FOR DETERMINATION)F ELASTIC CONSTANTS	FOR DETERMINATION OF STRENGTH	TESTING EQUIPMENT	DEFICIENCIES OF METHOD	SECTION OF THIS BOOK
'ilament or fabric wound tubes		Machine or device for shear tests	In determination of strength it is difficult to assure failure from in-plane shear	4.2
hin square plate with projections for fastening in the device		Tensile machine, a four-part frame	Limited use for non-equilibrium orthotroic material and for strength determination; large edge effect	4.3
hin square plate		Device for plate support and loading	Usable for balanced materials; requires high load and deflection measurement accuracy; sensitive to specimen shape and dimensions (projections for support, etc.)	4.4
hin square plate	—	Device for plate support and loading		4.4
'hin rectangular strip		Tensile testing machine	Approximate method	4.5
ar with two notches in one section and a :entral opening	Bar with one or two notches in two sections or a ring with notches	Tensile testing machine	Experimentally obtained characteristics are averaged; method usable for qualitative comparison of materials	4.6
'ross-ply sandwich beam or slotted specimen		Device for four-point bending of a cross; biaxial loading system	Testing a cross requires a large specimen of complex construction; in tension-compression of a strip synchronous loading is necessary	4.7
-	Short rod or ring	Tensile testing machine and a special fixture	Experimentally obtained characteristics are averaged; method usable for qualitative comparison of various materials	4.8
'ircular cross section bar	—	Torsion testing machine	Not used for strength determination; material in the form of a sufficiently thick plate is necessary for fabrication of three identical specimens along the x, y, and z axes	4.9
ectangular cross section bar		The same	Complicated processing of experimental results	4.9
-	Waisted specimen or a tube	Torsion testing machine	The method is somewhat approximate	4.9
ing (relative dimen- ;ions depend on oading scheme)	—	Device for loading of specimens	Complicated processing of experimental results; relative dimensions depend on anisotropy of the properties of test material	4.10

Chapter 5
Flexure Testing (BENDING)

5.1. BARS WITH A STRAIGHT AXIS

5.1.1. Loading Methods

Owing to their simplicity, bending (flexural) tests of bars with a straight axis are very widespread. At the same time, bending tests of composites have not so far been thoroughly standardized. There is a British standard BS 2782 (Part 10, Method 1005), as well as several ISO recommendations, but they are far from satisfactory for use in research and analysis. Therefore, in practice, standards for testing rigid plastics in flexure (GOST 4648-71, GOST 9550-71, ASTM D 790-71, DIN 53452, DIN 53457, etc.) are frequently used. Such an approach is permitted only when scrupulous account is taken of the specific features of composites, which are treated in later sections of this chapter. Otherwise, gross errors are inevitable.

In bending tests, the modulus of elasticity E_x^b, interlaminar shear modulus G_{xz}^b, strength according to normal stresses σ_x^{bu}, and interlaminar shear strength τ_{xz}^{bu} can be determined. Despite the large amount of information obtained, bending tests are frequently considered secondary, and many authors do not recommend their use in the calculation of composite structures. There are several reasons for this distrust: limited capabilities of proper processing of the test results, peculiarities of the test materials, and the state of stress in bending. At the same time, bending tests of composites are, for example, an effective and simple means of studying environmental effects [258], when as a result of partial destruction of the material a change occurs in the ratio of resistances of its components—the reinforcement and matrix. In this case, loading schemes in bending permit variations in the ratio of normal to tangential stresses.

The main difficulty in bending tests is connected with processing of the test results. In bending tests, the load and deflection or strain of the outer layers of the specimen are measured, and the properties of the materials are calculated from these parameters. Analysis of bending test results for anisotropic materials is not so simple and obvious as, for example, in the test of uniaxial tension. The quantities measured in testing (load, deflection, strains) are connected with the material characteristics under study (strength and elastic constants) by analytical functions whose accuracy is determined by their underlying assumptions.

The elementary (technical) theory of bending outlined in strength of materials courses uses a number of simplifying assumptions. The material of a bar is assumed to be isotropic, homogeneous, and of equal tensile and compression strengths; deflection w of the bar is assumed to be small compared with the span l (i.e., $w \ll l$), which permits the quantity $(dw/dx)^2$ in the differential equation of the bent axis of a bar to be disregarded. In the majority of cases, the effect of transverse shear stress is also disregarded, i.e., the theory of plane cross sections is used. This is equivalent to provision of material of infinite shear ($G_{xz}^b \to \infty$) and transverse ($E_z^b \to \infty$) stiffnesses; the uneven stress distribution over the width of the bar is not taken into account. The errors introduced by these assumptions essentially increase with height-to-span ratio and anisotropy of the mechanical properties.

The applicability of these assumptions to homogeneous, isotropic materials has been tested over many years of experience. The applicability of an assumption to composites depends on the anisotropy of the material and the state of stress, i.e., on the dimensions and the scheme of loading and supporting of the specimen. Unfortunately, for the materials examined the limits of applicability of the assumptions have not been clearly formulated. Therefore, difficulties can arise in selection of specimen dimensions and loading conditions even in the simplest cases. At the same time, experience shows that unjustified application of the formulas of the elementary theory of bending in processing test results for bars of anisotropic nonhomogeneous materials, i.e., composites, leads to gross errors in interpretation of the experimental results and to underestimation of the capabilities of bending test methods. A more detailed analysis of the applicability of the mentioned assumptions is presented in the discussions of specimen shape and dimensions and the analyses of various loading schemes. The effect of nonhomogeneity of the materials and cross-sectional dimensions of the bar are discussed in Section 5.2. The hypothesis of plane cross sections is discussed in Section 5.2.1. Large deflections are discussed in Sections 5.2.3 and 5.3.1.

Consideration of the peculiarities of the mechanical properties of composites has led to the development and experimental evaluation of a number of loading schemes in bending. The specimen supporting and loading schemes used in current practice for testing are shown in Fig. 5.1.1. The characteristics to be determined and the variable specimen dimensions are also presented in the figure. For testing of specimens of isotropic materials, the so-called three-point loading method [Fig. 5.1.1(a)], i.e., a simply supported bar loaded with a concentrated force P in the middle of span l, is used almost exclusively. This loading method is the most widely used in composite testing, but in this case the three-point scheme is complicated: the state of stress of the specimen varies over the length, height, and width of the specimen and a bending moment and shearing force, i.e., normal and tangential stresses, develop on the specimen. The capabilities of three-point loading are expanded in tests of composites: it is also used for the determination of interlaminar shear characteristics.

Efforts to eliminate the effect of tangential stresses have led to a pure bending scheme [Fig. 5.1.1(b)] in which, in the idealized case, the specimen is loaded by moments applied to its end faces. The advantage of this loading method is the uniform state of stress over the entire length of the specimen and the absence of shear deformation. Loading with moments is technically very difficult. Therefore, in practice varieties of the pure bending schemes are more widespread—four-point loading, in which the bending moments are on the ends of the bar due to concentrated forces P either within the span l [Fig. 5.1.1(c)] or outside the span [Fig. 5.1.1(d)]. The deficiency of these loading methods compared with the pure bending scheme is that a shearing forces (the tangential stresses) act on the sections of the specimen of length a or c [see Fig. 5.1.1(c, d)]. The five-point loading scheme [Fig. 5.1.1(b)] is a modification of three-point loading. The five-point method is used relatively rarely, since the stress on the specimen is complicated. However, its possibilities have been underestimated. The most complicated method from a state-of-stress viewpoint the bending of bars with fixed ends [Fig. 5.1.1(f)]. This scheme is used very rarely because of indeterminacy of the fastening conditions of the specimen.

Transverse loads in bending tests of materials are usually created by one or more concentrated forces or moments. Loading of specimens with a uniform distribution of load over the span is difficult because of the complexity of the apparatus used, although it is useful in verification of theoretical solutions.

Bars with a straight or circular axis (rings and their segments) and constant rectangular cross section, as well as sandwich beams, are used in bending tests. Although in tests of bars with a circular axis (ring segments) the equations used have usually been derived for straight bars, the behaviors of straight and circular bars made of anisotropic materials differ markedly.

Loading scheme	*a* 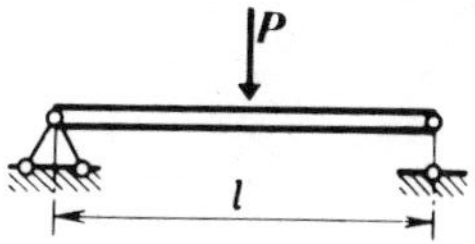	*b* 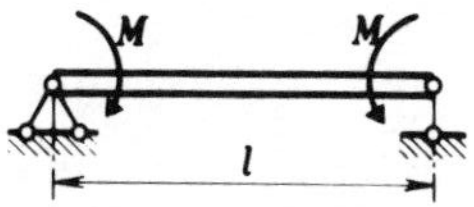	*c* 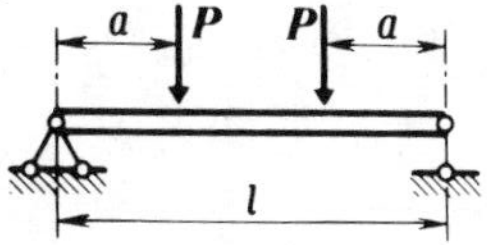
Determinable characteristics	E_x^b, G_{xz}^b, σ_x^{bu}, τ_{xz}^{bu},	$E_x^{t(c)}$, σ_x^{bu}	E_x^b, σ_x^{bu}
Measurable values	p, w	M, w	P, w
Geometrical sizes	$l/h, b, h$	$l/h, b, h$	$l/h, a, b, h,$
Limitations — Structural — Layup	0°; 90°; 0/90°		
Limitations — Structural — Orientation	0°, 90°		
Limitations — Physical	For E_x^b and G_{xz}^b linear range of the curve $P \sim w$		
Limitations — Geometrical	For E_x^b: $l/h \geqslant 40$ For E_x^b and G_{xz}^b $8 \leqslant \frac{l}{h} \leqslant 1.095\sqrt{\frac{1-K_\tau E_x^b}{K_\tau G_{xz}^b}}$	$\frac{l}{h} \geqslant 5$	For σ_x^{bu}: $\frac{a}{h} > \frac{\sigma_x^{bu}}{4\tau_{xz}^{bu}}$

Fig. 5.1.1. Loading schemes of prismatic bars in bending: (a) three-point bending; (b) pure bending; (c and d) four-point bending; (e) five-point bending; (f) bar with fixed ends.

			d	e	f
Loading scheme			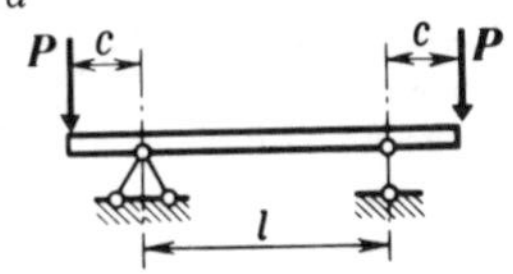 d	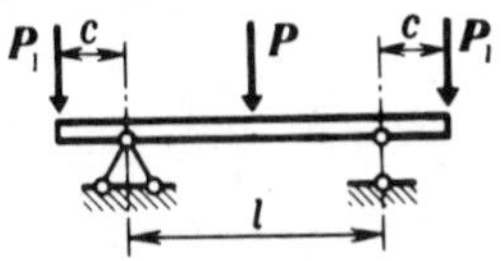 e	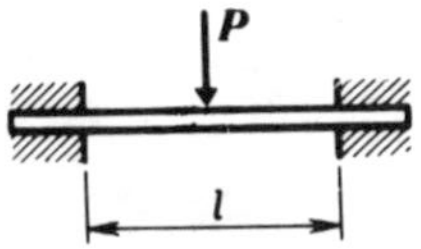 f
Determinable characteristics			E_x^b, σ_x^{bu}	E_x^b, G_{xz}^b, E_x^b / G_{xz}^b	E_x^b and G_{xz}^b
Measurable values			P, w	P, P_1, w	p, w
Geometrical sizes			$l/h, c, b, h$	$l/h, l/c, b, h$	$l/h, b, h$
Limitations	Structural	Layup	0° ; 90° ; 0/90°		
		Orientation	0°, 90°		
	Physical				
	Geometrical		For σ_x^{bu}: $\frac{c}{x} > \frac{\sigma_x^{bu}}{4\tau_{bz}^{bu}}$	$\frac{c}{h} > \sigma_x^{bu}/4\tau_{xz}^{bu}$ $p_{max} = p_{1\,max} < bh^2\,\sigma_x^{bu}/\sigma c$	For E_x^b and G_{xz}^b $\frac{l}{h} \leqslant 1.55\sqrt{\frac{1-K_\tau}{K_\tau}\frac{E_x^b}{G_{xz}^b}}$

Therefore bending tests of ring segments are examined separately in Section 5.5.4.

The need to take into account the characteristic behavior of bars of anisotropic materials in bending has spurred the development of appropriate engineering formulas and the evaluation of their validity. In the precise evaluation of experimental results, the limitations imposed by theoretical considerations must be taken into account. The refined formulas used at the present time for determination of deflection and stress distribution were obtained for bars of a homogeneous anisotropic material with the same tensile and compression characteristics, on the assumption that deflection of a bar is small.

The task is further complicated when the tensile and compression characteristics of the material differ and when the principal axes of elastic symmetry of the material do not coincide with the longitudinal axis of the specimen, as well as in the case where the deflection is large and slipping of the bar from the supports is observed. The appearance of materials with low transverse tension strength has forced an evaluation of the error introduced by the anisotropy of the reinforcing fibers. The majority of these questions have been studied insufficiently and they are not treated in this book. Only when they can significantly affect the results are they pointed out.

The structural peculiarities of reinforced plastics result in the situation that in bending tests essentially a composite structure and not a homogeneous material is studied. Therefore, the experimental characteristics determined in bending can only be provisionally considered to be material characteristics, with due consideration given the conditions of transition to continuous medium. The equations for various methods of loading in bending were obtained for a homogeneous material and they do not take into account the reinforcement layup, the number of layers, the general and local waviness of the reinforcement, interactions of the reinforcement with the polymer matrix, or defects in fabrication (voids, destruction of the adhesive bonds, etc.).

Unlike uniaxial tension, the bending state of stress varies with height and, in some cases, with the length of the specimen. Together with local effects due to the structure of the material and manufacturing variations, this can lead to theoretically unreliable results, and in the majority of cases to lower values of the strength and elastic constants. Experience shows that reliable bending test results which correspond to theoretical assumptions can be obtained only be means of careful selection of loading schemes and conditions and specimen dimensions (primarily the number of reinforcing layers necessary for transition from a laminated, nonhomogeneous material to a continuous, homogeneous medium).

5.1.2. Prismatic Bars

Prismatic bars are used for determination of the elastic characteristics and bending strength of a material. Selection of the loading method depends on the purpose of the investigation. The longitudinal axis of the specimen should coincide with one of the principal axes of elastic symmetry of the material. In the opposite case, the Poisson's ratios* and coefficients of mutual influence of the material should also be taken into account in processing the test results. Formulas which allow for these coefficients have been obtained so far only for the case of pure bending [256]. Special supports are necessary for bending tests of bars which do not exhibit elastic symmetry relative to their axis, since such specimens are twisted and do not lie flat on the surface of standard fixed supports under transverse loads.

ISO recommendations specify the following dimensions of prismatic glass fiber composite bars:

LOADING SCHEME	l/h	HEIGHT h, mm	WIDTH b, mm
Three-point loading	15–17	$1 \leqslant h \leqslant 10$	15 ± 0.5
	15–17	$10 \leqslant h \leqslant 20$	30 ± 0.5
Four-point loading	28	$20 \leqslant h \leqslant 35$	50 ± 0.5
	28	$35 \leqslant h \leqslant 50$	80 ± 0.5

Experience shows that for advanced high-modulus polymeric composites, the ratio l/h for the ISO three-point test is too small for reliable determination of the modulus of elasticity E_x^b and strength from normal stresses σ_x^{bu}; in these cases a value of $l/h \geqslant 40$ must be chosen. Among four-point arrangements, that shown in Fig. 5.1.1(d) is the most practical in determination of the modulus of elasticity E_x^b a value of $l/h = 28$ is applicable to all types of composites.

In selecting the dimensions of nonstandardized specimens, it is necessary to clearly understand how the cross-sectional area (scale effect), relative span l/h, and relative width b/h show up in the test results. In selecting the cross-sectional area, it must be kept in mind that the height and width affect the deflection and bending strength differently as a consequence of the anisotropy of the scale effect. Bending tests of high-modulus and high-strength composites are frequently performed on very thin specimens ($h = 0.5$–1.5 mm), any effects of scale being eliminated by means of an increase in specimen width. The possible errors of this approach are treated in Section 5.2.4.

*Poisson's ratios must be considered also in bending of specimens the longitudinal axis of which coincides with the axis of elastic symmetry of the material.

The effect of specimen cross-sectional area on bending strength has not been sufficiently studied. The effects of specimen thickness and width have been studied in greater detail. Results of investigations for unidirectional boron epoxy composites [131] show that in the experimental determination of strength there is an optimum cross-sectional area below and above which the strength decreases; simultaneous change in the cross-sectional area of the specimen and the relative span l/h results in a small shift of this optimum. However, the available experimental data are not sufficient for general conclusions.

In selecting the specimen thickness it is necessary to fulfill the condition that ensures transition to a continuous medium, i.e., the number of reinforcing layers must be greater than the minimum necessary. The results of calculation by the method presented in [231, p. 90] show that the minimum necessary number of reinforcement layers at which the transition from laminated to continuous homogeneous material is possible (the number of layers $n \to \infty$) depends on the material anisotropy, which is characterized by the parameter

$$\kappa = \pi \frac{h}{2l} \sqrt{\frac{E_x^b}{G_{xz}^b}}.$$

For a simply supported bar loaded with a sinusoidal load, the minimum necessary number of layers of reinforcement is as follows:

κ	1	3	5	7
n	5	15	17	20

The minimum necessary number of layers of reinforcement is determined also by the condition of ensuring structural symmetry of the specimen relative to its neutral plane; the effect of manufacturing defects on symmetry of the material in a specimen is negligible at $n > 8$ [39].

However, it should be noted that for bars reinforced with stiff fibers of large diameter (for example, boron fibers) the formulas obtained for a continuous medium may often be inappropriate.

The specimen width should be selected so that loss of specimen stability under load is eliminated. Most often, the specimen cross section is nearly square. However, in practice tests of high-modulus composites are complicated by the tendency to test thin specimens (1–2 mm thick) in bending. The scale effect in bending has not been systematically studied and, at the present time, the minimum cross-sectional area above which the measured bending strength σ_x^{bu} are sure to be constant cannot be specified. If it is necessary to test specimens, for example, with a 20 mm^2 cross-sectional area and with the thickness specified above, specimens 10–20 mm wide would have to be prepared. This is also done in practice, although for a different purpose—to ensure the reliability of measurement of load or deflection. As a result, $b/h = 10–20$ is obtained, which differs greatly from the value proposed by elementary bending theory ($b/h \approx 1$); such specimens are not bars any more, but plates. In processing test results of such specimens the distribution of stresses over the width of the specimen must also be taken into account.

Specimen length is selected in accordance with the conditions of the experiment and the ratio l/h. Hence two limiting cases should be distinguished here: determination of shear characteristics on relatively short specimens (with a small l/h ratio) and determination of the modulus of elasticity E_x^b on flexible specimens (with a large l/h ratio). In determination of the interlaminar shear strength τ_{xz}^{bu} the relative span is selected so that failure due to tangential stresses is ensured. In this case, $l/h = 5$ is most often selected. However, experience shows that some types of high-strength carbon composites, for example, fail due to shear even at $l/h = 10–12$. The accuracy of determination of the shear modulus G_{xz}^b increases with decreasing l/h, i.e., with increasing deflection by shear, but then diminishes, since the calculation procedure is no longer applicable.

It should be noted that in tests by three-point loading the l/h value cannot be arbitrarily decreased. With decreasing l/h, with the same bending moment, the shearing force and the possibility of damage to the support surfaces of the specimen increase. This is treated in greater detail in Section 5.2.4.

In determination of the modulus of elasticity E_x^b an effort is usually made to eliminate the effect of shearing force and to use specimens with a large l/h ratio (up to 60). The use of such specimens sharply increases the requirements for accuracy of load and deflection measurements. Moreover, in this case large deflections and the undesirable phenomena connected with them fail to be eliminated; these include slipping of specimens from the supports, and the change in actual deflection and bending moment.

5.1.3. Sandwich Beams

Sandwich beams are used for determination of the elastic constants and tensile and compression strengths $E_x^{t(c)}$ and $\sigma_x^{t(c)}$ of materials with a reinforcement layup symmetric about the specimen axes; determination of Poison's ratios ν_{xy} and ν_{yx} is unreliable because of the effect of the filler of the sandwich beam. The development of sandwich beams was mainly due to difficulties in tensile and compression tests of high-modulus and high-strength composites.

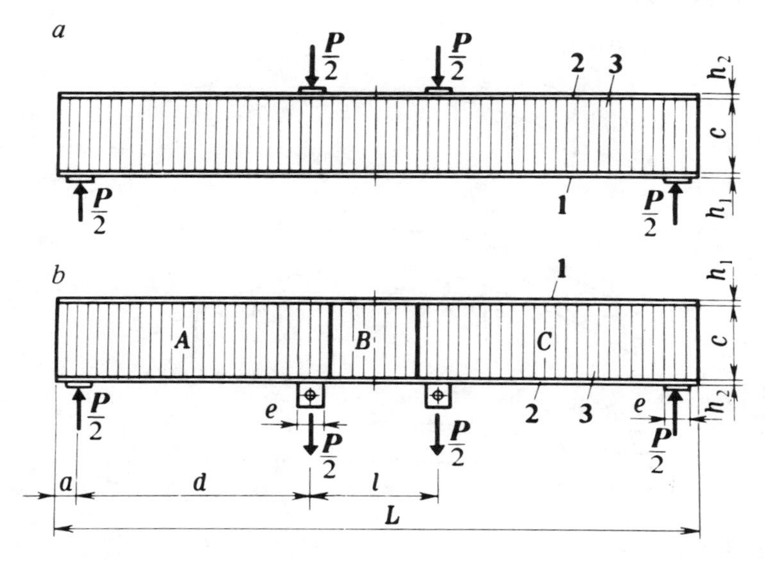

Fig. 5.1.2. Loading schemes of sandwich beams in tension (a) and compression (b) tests: (1) active facing; (2) passive facing; (3) honeycomb core. Basic dimensions in mm [89, 200]:

$$a = 13\text{–}25, \quad c = 25\text{–}38, \quad d = 203,$$

$$e = 25\text{–}38, \quad l = 38\text{–}102, \quad L = 406\text{–}559;$$

Width of a beam 25 mm, thickness of an active facing 0.75–1.1 mm, of a passive facing 1.0–3.2 mm.

At the present time they are rarely used, mainly in research practice. This is not only because the method is laborious and technically complex, but also because reliable ways of testing high-strength unidirectional materials in tension-compression had been found (see Chapters 2 and 3).

In the study of tensile and compression characteristics, sandwich beams are loaded in four-point bending (Fig. 5.1.2). The separate layers of the beam differ both as to purpose and in material. The upper and lower facings absorb only the bending moment; the lower facing usually is loaded in tension and the upper facing in compression. The active facing of the beam is made of the material being tested and the passive one (the opposite side, through which the load is applied) is made of glass cloth composite or a light metal (aluminum or titanium alloy). At the loading and supporting areas the beam has metallic pads.

The inner layer of the beam—the core—absorbs only the shearing force.

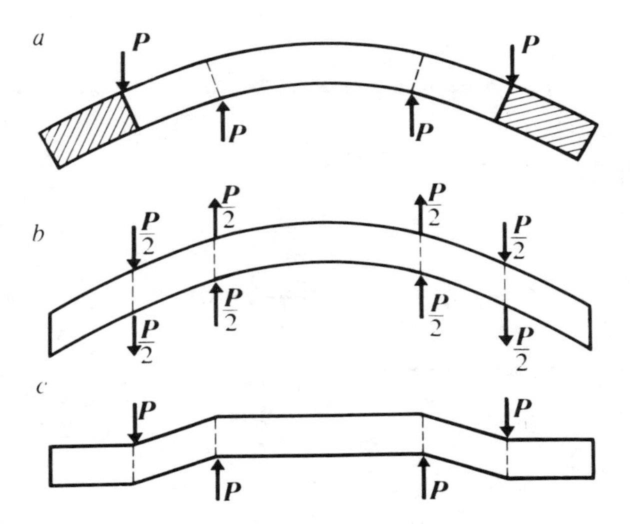

Fig. 5.1.3. Some characteristic types of deformation of a sandwich beam [6]: (a) with rigid inserts (striated); (b) with a compliant core; (c) pure shear.

The core material should have as low a specific weight as possible, negligible rigidity in the midplane of the beam, and high rigidity perpendicular to the midplane. Usually, aluminum honeycomb core is used, with the axis of its cells lying perpendicular to the midplane of the beam. The density of the honeycomb is 70–370 kg/m^3, and it may be constant [Fig. 5.1.2(a)] over the length of the beam or reduced in the gage length [Fig. 5.1.2(b) zone B]. The layers of the beam are bonded together with an epoxy adhesive; thin but sufficiently rigid high-temperature-resistant epoxy films are used.

The profile of a sandwich beam in the deformed state depends on the design and the ratio between the rigidities of the facings and the core. Several characteristic examples are shown in Fig. 5.1.3.

The dimensions of sandwich beams are selected so that at failure of the active facing the load carrying capacity of the core and its bond to the facings of the beam is not exceeded. Furthermore, the shearing force acting on the beam should be small. In this connection, as a consequence of these conditions the dimensions of the sandwich beam turn out to be quite large; length L attains 400–560 mm, with gage length $l = 40$–100 mm [89, 121, 200]. The validity of the equations for sandwich beams (see Section 5.3) has occasionally been questioned [249], since in their derivation the effects of core rigidity and the adhesive joints are not taken into account. The effect of the number of layers in the active facing and of the span and height of the beam have also been insufficiently studied.

The advantages of the use of sandwich beams are the absence of all difficulties connected with fastening of the specimen in the testing machine grips, and the absence of the end effect zone and stress concentrations. As can be seen from Fig. 5.1.2, under compression [Fig. 5.1.2(b)] the active facing is free of external loads, and in tension [Fig. 5.1.2(a)] the bearing reactions are applied to the ends of the facing, far from its gage section. Owing to this, a uniform state of stress is ensured on the gage area of the specimen, and the experimentally determined values prove to be the most reliable [82, 97]. However, the increase in accuracy of the experimental results becomes significant only when the reinforcement is laid at an angle to the beam axis. Fabrication of sandwich beams is much more complicated and costly than that of conventional specimens. Therefore, the use of sandwich beams is advisable mainly for special research (scientific, exploratory) and in fields where items with honeycomb cores are used.

5.1.4. Loading Conditions, Quantities Measured

In bending tests, the specimens are installed on the supports in accordance with the loading method selected. The axis of the specimen should lie precisely along the straight line connecting the centers of the supports with the points of loading. Standard testing machine supports are frequently unsuitable for detailed bending studies. The supports on which the specimen is placed and the testing machine punch or another loading device have cylindrical surfaces. The minimum radius of curvature of a support is $1.5h$ (h is the specimen height), but no less than 3 mm [11]; the numerical values of the radii of curvature given in various standards differ somewhat.

If there are no special conditions (as, for example, in tests by the five-point loading), the load P and bar deflection w are measured in bending, and in individual cases also the tension or compression strains of the outer layers of the specimen $\varepsilon_x^{t(c)}$. The limitations on the maximum load in the determination of the elastic constants are examined in the descriptions of the individual loading schemes. In the determination of the bending strength, the specimen is loaded at a constant rate up to failure and the breaking load P^u is registered. It has been established experimentally that with uniform loading the load-deflection curve is linear almost up to specimen failure. Just before ultimate failure of the specimen a drop in the load is observed because of partial failure of the material. In stepwise loading (by means of weights) the load-deflection curve can become nonlinear as a result of creep of the polymer binder. Therefore, in short-term strength tests uniform machine loading is preferable. However, it should be kept in mind that the load-deflection curve also becomes nonlinear at large deflections (geometric nonlinearity).

In deflection measurement, it should be borne in mind that deflection of small specimens of stiff composites can prove to be very small (several hundredths of a millimeter). Therefore, it is advisable to use indicators with scale divisions no greater than 0.002 mm. In measuring the deflection of specimens with low rigidity, one must allow for the fact that the resistance of the mechanism of a clock-type indicator can be of the same order of magnitude as the external load. For deflection measurement cantilevers with glued-on resistance strain gages are also used; these permit the recording of the load-deflection curve by means of an oscillograph. However, such beams require accurate calibration and careful selection of dimensions, to ensure the necessary sensitivity.

In measurements of the deflection of relatively short specimens, account must be taken of the possible shrinkage of the specimen as a consequence of squeezing of the support surfaces. Measurement of tension strain ε^t or compression strain ε^c of the outer layers of the specimen is carried out by means of resistance strain gages. Resistance strain gages are also used to obtain the distribution of normal stresses over the height of the specimen; for this purpose they are glued to the side faces of the specimen over its height or "molded" into the specimen [58].

The strain rate in bending $\dot{\varepsilon}^b$ is determined by the following relation:

In three-point loading [Fig. 5.1.1(a)]:

$$\dot{\varepsilon}^b = \frac{\Delta P}{\Delta t}\frac{lh}{8E_x^b I} = \frac{6h}{l^2}\frac{v}{1 + 0.486\kappa^2}; \tag{5.1.1}$$

In the four-point loading [Fig. 5.1.1(c)]:

$$\dot{\varepsilon}^b = \frac{\Delta P}{\Delta t}\frac{ah}{2E_x^b I} = \frac{3h}{l^2}\frac{v}{3\dfrac{a}{l} - 4\dfrac{a^2}{l^2} + 0.243\kappa^2}; \tag{5.1.2}$$

In four-point loading [Fig. 5.1.1(d)] and by in pure bending [Fig. 5.1.1(b)]:

$$\dot{\varepsilon}^b = \frac{\Delta M}{\Delta t}\frac{h}{2E_x^b I}. \tag{5.1.3}$$

In the formulas (5.1.1)–(5.1.3) Δt is the time necessary for increasing the load by ΔP or the bending moment by ΔM [for four-point loading according to Fig. 5.1.1(d), moment $\Delta M = c\Delta P$]; v is the speed of displacement of the testing machine punch or some other loading mechanism; $\kappa = \pi(h/2l)\sqrt{E_x^b/G_{xz}^b}$ is the anisotropy parameter.

If $\dot{\varepsilon}^b$ is given, the speed v may be obtained or the rate of increase in the load ΔP or ΔM determined by means of calibration of the testing machine. As can be seen from Equations (5.1.1)–(5.1.3), these values are not constant but depend (with $\dot{\varepsilon}^b = \text{const.}$) on the properties of the material (modulus of elasticity E_x^b, shear modulus G_{xz}^b), the specimen dimensions (height h, moment of inertia I), the loading geometry (span l, distance a or c).

The strain rate $\dot{\varepsilon}^b$ for composites has not been standardized and there is no theoretical basis for selection of $\dot{\varepsilon}^b$ or data on the effect of $\dot{\varepsilon}^b$ on the measured values. We present the loading conditions in three-point bending of rigid nonreinforced plastics recommended in standards as approximate data: according to GOST 4648-71, $v = 2 \pm 0.5$ mm/min for standard size specimens or $v = h/2$ (mm/min) for specimens of other dimensions; according to GOST 9550-71, $v = 1 \pm 0.5$ mm/min; according to ASTM D 790-71 and DIN 53457, $\dot{\varepsilon}^b = 0.01$ min^{-1}. ISO, R 178 recommends a speed $v = h/2$ (h in mm) or $v = 10$ mm/min (for serial tests) for testing glass fiber composite. In testing sandwich beams, the loading rate (movement of the movable testing machine head) is adopted as 0.56 mm/min [120, 121] regardless of the dimensions of the beam, or a loading rate is selected which depends on the purpose of the test and the material tested.

5.1.5. Failure Modes in Bending

In three-point bending of bars of isotropic materials the relation of maximum tangential strength τ_{max} to normal stresses σ_{max} is of the order of $h/2l$, but the strength in terms of normal stresses (σ^u) is close to the strength in terms of tangential stresses (τ^u). Therefore, the design of structures of these materials is based on normal stresses. σ^u and τ^u for composites can differ by an order of magnitude or more, but the tangential stresses due to low shear strength of the material can significantly affect the mode of failure in bending. Therefore, in bending of composites one should strictly differentiate strengths in terms of normal (σ^u) and tangential (τ^u) stresses. The need for such differentiation is enforced by the fact that in bending tests of composites, determination of interlaminar shear strength and, consequently, assurance of specimen failure from tangential stresses through selection of its dimensions, is sometimes the main objective.

Upon failure of the specimen due to normal stresses, failure of the outer stretched or compressed layers [Fig. 5.1.4(a)] is observed; upon failure, due to tangential stresses, splitting on the level of the mid-plane of the specimen

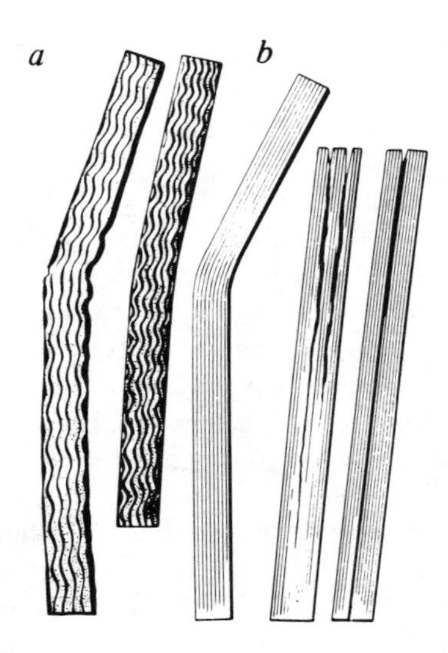

Fig. 5.1.4. Types of failure of fiberglass composites in bending: (a) due to normal stresses; (b) due to tangential stresses.

or in its vicinity is observed [Fig. 5.1.4(b)]. It has been established experimentally that the interlaminar shear strength τ_{xz}^u is not constant, but decreases with an increase in the relative span l/h (for more detail, see Section 5.2.4). In very short bars (with a large h/l) the third mode of failure is observed—due to bearing and shearing, which is accompanied by an apparent growth in material resistance to tangential stresses. Redistribution of stresses in a bar and change in the failure mode, depending on l/h ratio, is shown in Fig. 5.1.5; the mode of failure due to bearing-shearing is shown in Fig. 5.1.6. In the determination of bending strength it is necessary to indicate the failure mode, otherwise the results would not be comparable.

For materials with a laminated structure one more failure mode is possible, as shown in Fig. 5.1.7: the failure of a beam due to normal stresses is preceded by peeling of the outer compressed layer. In three-point bending, the critical stress σ_{cr} at which this failure mode is possible is determined from the following equation [227]:

$$\frac{\sigma_{cr}}{E_x^b} = 16\pi^2 \left(\frac{h_1}{l}\right)^2 \left(1 + \sqrt{1 + \frac{3\gamma l^4}{64\pi^4 E_x^b h_1^5}}\right) \qquad (5.1.4)$$

where l is the span; h_1 is the thickness of delamination, which can be preset by the structure of the material; the parameter γ is the work of fracture

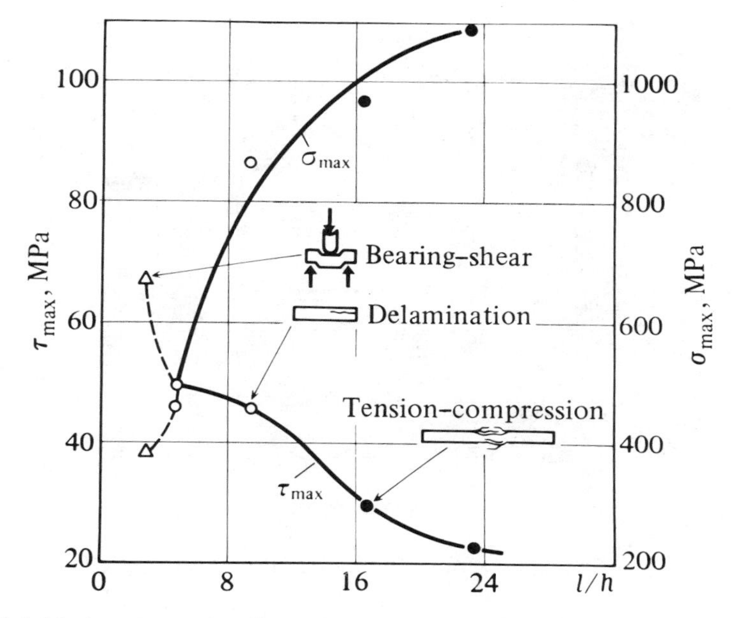

Fig. 5.1.5. Maximum normal and tangential stresses and type of failure in bending versus relative span l/h of roving reinforced polyester-glass fiber composite; specimen thickness $h = 4$ mm [141].

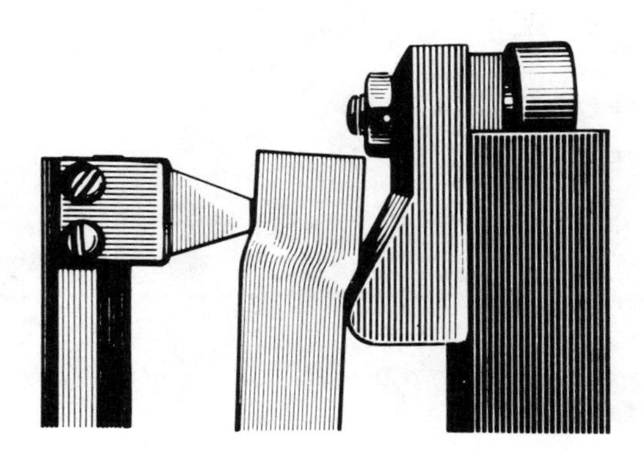

Fig. 5.1.6. Failure of glass fiber composite due to bearing-shear.

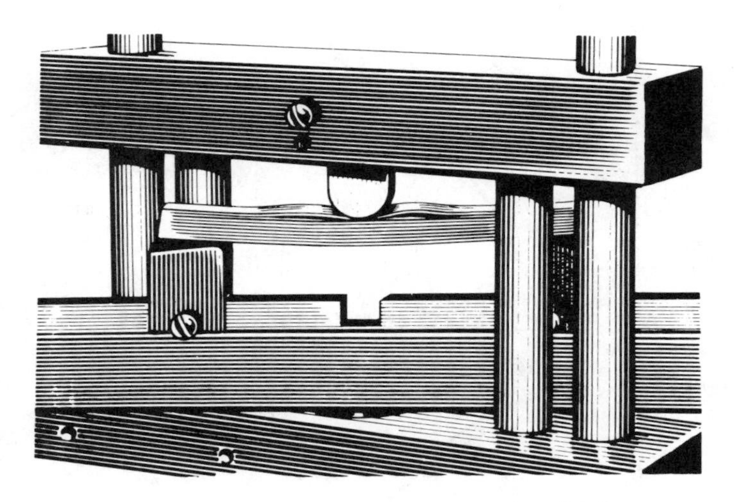

Fig. 5.1.7. Failure of specimen by delamination [228].

according to Griffith (which characterizes delamination strength of composites).

A numerical analysis [227] performed for glass fiber composites with $E_x^b = 5 \times 10^4$ MPa at various span lengths, showed in the majority of cases that the lower boundary of critical stresses calculated according to Equation (5.1.4) lies higher than the strength in terms of normal stresses σ^u; $\sigma_{cr} < \sigma^u$ only at very small γ. The main cause of delamination can be a change in the modulus of elasticity of the peeled layer from that of the rigidity of the bar from which peeling starts, or a sharp change in γ due to introduction of transverse layers in the vicinity of the outer surface for materials with a reinforcement layup that varies through the thickness of a bar.

5.2. THREE-POINT LOADING

5.2.1. Technical Theory of Bending

In the investigation of small deflections of bars of isotropic materials the theory of plane cross sections is used. In this case, the analytical relations turn out to be simple.* For a simply supported bar with a concentrated load P in the middle of span l, i.e., in tests by three-point loading, modulus of elasticity E_x^b may be calculated with high accuracy if the load P and deflection at midspan w_{max} are known:

* A survey of solutions of problems in the bending of composite bars is given in works [123, 205].

$$E_x^b = \frac{Pl^3}{48Iw_{max}} \tag{5.2.1}$$

where $I = bh^3/12$ is the moment of inertia of the cross section of the bar.

The bending strength can be calculated if the ultimate load P^u at which the specimen failed and the corss-sectional dimensions of the specimen are known. For a rectangular specimen with dimensions $b \times h$, the strength under normal stresses is

$$\sigma_x^{bu} = \frac{M^u}{W} = \frac{3}{2}\frac{P^u l}{bh^2} \tag{5.2.2}$$

and under tangential stresses

$$\tau_{xz}^{bu} = \frac{3}{2}\frac{Q^u}{F} = \frac{3}{4}\frac{P^u}{bh} \tag{5.2.3}$$

where $W = bh^2/6$ is moment of resistance; $F = bh$ is cross-sectional area; b is specimen width; h is specimen height.

The applicability of the hypotheses of plane cross sections, of incompressible normals, and of small deflection of bars of anisotropic materials must be evaluated. The assumption of plane cross sections is permissible if shear deformations in the bar are eliminated or are negligibly small. The effect of shear (deflection of the bar is assumed to be small) on the deflection of bars and the type of failure depend on the anisotropy of the bar material E_x^b/G_{xz}^b, its relative height h/l, the support scheme, type of load, and ratio of strength σ_x^u to τ_{xz}^u.

These factors are not all of equal importance, but they are intterrelated and their combined evaluation in bending tests of anisotropic materials is obligatory. This requirement is overlooked in standards for rigid nonreinforced plastics, which for three-point tests fix the relative height of a bar h/l and simultaneously recommend use of Equation (5.2.1) and (5.2.2). Even in the ISO recommendations a ratio $h/l = 1/15–1/17$ is specified. Such a restriction is allowable for isotropic materials [the error in the determination the modulus of elasticity from (5.2.1) for a steel bar of the relative height $h/l = 1/16$ is equal to 3%], but it is completely inapplicable to highly anisotropic materials (at $E_x^b/G_{xz}^b = 100$, the error of processing the results is equal to 32% even before taking into account shear stress). The systematic error in the use of (5.2.1) is characteristic also at a ratio $l/h = 40$; however, in this case it is negligible. It should be noted that the requirements involved in the design of supports and in the consideration of the change in the specimen gage length become more stringent with increasing relative span l/h.

Equations which take into account the effect of shear on bar deflection are more complicated than the simple, single-term formula (5.2.1). As will be shown subsequently, the refined formulas, even in the somewhat simplified approach of S.P. Timoshenko, become equations in two unknowns. This makes it impossible to determine the elastic constants directly from loads and deflection measured experimentally with only one series of specimens. A large error can be introduced in estimation of strength when several failure modes are possible, for example, in the case where a specimen fails due to shear [Equation (5.2.3)] and the Equation (5.2.2) is used for processing the results. The necessity of clear separation of the types of failure of composites due to normal or tangential stresses or due to bearing-shearing imposes very stringent requirements on the selection of dimensions of specimens of anisotropic materials (see Section 5.2.4).

There is no complete theoretical and experimental substantiation for estimation of the error introduced by the hypothesis of incompressible normals. Solutions of two- and three-dimensional problems, which are necessary for theoretical evaluation, exist only for special cases of loading which are rarely used in experimental practice (pure bending, loading with uniformly distributed load, bending of cantilevers). Consequently, there are practically no analytical relations for processing experimental results. There is an approximate solution [231, p. 93] for the simply supported bar loaded with a sinusoidally distributed load:

$$w_m = u_m^* \left[1 + 0.1\lambda_m h^2 \left(\frac{E_x^b}{G_{xz}^b} - \frac{G_{xz}^b}{E_x^b} \right) \right]. \tag{5.2.4}$$

This equation shows that the effect of compressibility of normals increases with decreasing modulus of elasticity E_z^b, i.e., the effect of compressibility of normals is greater in materials reinforced with anisotropic fibers (for example, carbon composites) than in glass fiber composites.

In practice, we frequently come across the concept of "large deflections", i.e., deflections which cannot be determined using elementary bending theory. Clearly formulated limits (in terms of the ratio w/l) of application of the elementary theory are not presented in the literature. The refined formulas for determination of large deflections of bars of isotropic materials are complex and are not applicable for processing of experimental results of anisotropic materials.

For an approximate estimate of the limits of application ofthe elementary bending theory, we present data from [53]. With a fixed ratio $p = w_{\max 1}/w_{\max}$

(where $w_{max\,1}$ is the deflection according to the refined theory and w_{max} is that according to elementary bending theory) or the error $(p - 1) \times 100$ (in %) in three-point loading, the dimensionless parameter $(p/8E_x^bI)(l/h)^2h^2$ should be less than the following values (μ is the friction coefficient of the support surfaces):

	$\frac{p}{8E_x^bJ}\left(\frac{l}{h}\right)^2 h^2$	
p	$\mu = 0$	$\mu = 0.3$
1.05	$\leqslant 0.23$	$\leqslant 0.30$
1.10	$\leqslant 0.60$	$\leqslant 0.65$

These values have been obtained for bars of isotropic materials without taking account of transverse shear. For bars of anisotropic materials, these conditions will be more strict, i.e., the permissible values of the parameter $(p/8E_x^bI)(l/h)^2h^2$ will be less, since these bars are more yielding as a consequence of the effect of transverse shears.

5.2.2. Determination of Elastic Constants

For experimental determination of the elastic constants of a material in bending, the refined formulas for bar deflection are used. The deflection at midspan of a rectangular bar, with allowance for shear, is determined by the formula [231, p. 104; 236, p. 153]*

$$\begin{aligned} w_{max} &= w_\sigma + w_\tau = \frac{Pl^3}{48E_x^bI} + \frac{\alpha}{4}\frac{Pl}{G_{xz}^bF} \\ &= w_{max}^*\left[1 + \alpha\left(\frac{h}{l}\right)^2 \frac{E_x^b}{G_{xz}^b}\right] = w_{max}^*(1 + 0.486\kappa^2) \end{aligned} \tag{5.2.5}$$

where $w_{max}^* = Pl^3/48E_x^bI$ is the maximum deflection of the bar without allowance for shear; $\kappa = \pi(h/2l)\sqrt{E_x^b/G_{xz}^b}$; h is the height of the bar; and α is a coefficient depending on the cross-sectional of the bar (for a rectangular cross section $\alpha = 1.2$).

It is seen from a comparison of Equations (5.2.1) and (5.2.5) that the

*Here and later, it has been assumed that the deformation of the material under tension and compression is linearly elastic. This condition is often not fulfilled for organic composites, the deformation of which under compression may be nonlinear.

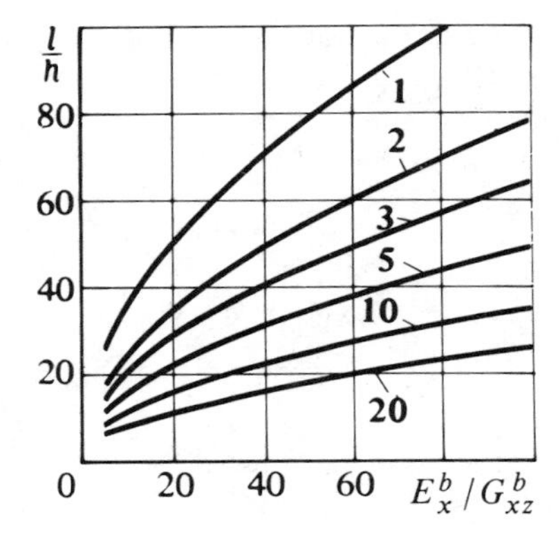

Fig. 5.2.1. Dependence of the value of minimum l/h ratio on the E_x^b/G_{xz}^b relation in determination of the modulus of elasticity without accounting for shears with a given error δ (numbers on the curves denote values of δ in %).

modulus of elasticity E_x^b determined by (5.2.1) is fictitious, since it does not reflect the effect of shear stress. Consequently, let us designate the modulus of elasticity determined by (5.2.1) by E_f. The following relation exists between E_f, E_x^b, and G_{xz}^b:

$$\frac{1}{E_f} = \frac{1}{E_x^b} + \frac{1.2}{G_{xz}^b}\left(\frac{h}{l}\right)^2. \tag{5.2.6}$$

The difference between the fictitious modulus E_f and the actual E_x^b is the more significant, the greater h/l and the higher the anisotropy $\beta^2 = E_x^b/G_{xz}^b$

In order for the error in determination of the modulus E_x^b without allowance for shears not to exceed a fixed value δ, tests should be carried out with l/h values above the corresponding δ-curve (Fig. 5.2.1). As can be seen from the figure, the l/h ratio for anisotropic materials turn out to be sufficiently large. This can lead to certain technical difficulties (measurement of deflections and loads). Therefore it is more appropriate to allow for the effect of shear in determining the modulus of elasticity in bending E_x^b.

For the experimental determination of the elastic constants in bending, several series of specimens with different h/l ratios should be tested. For processing experimental results Equations (5.2.5) or (5.2.6) are used. The curve of (5.2.6) in $(h/l)^2$, $1/E_f$ coordinates has the form of a straight line of slope $1.2/G_{xz}^b$ and $1/E_f$ = intercept $1/E_x^b$ (Fig. 5.2.2). By means of a graph in the $(h/l)^2$, $1/E_f$ coordinate system it is possible to determine not only the elastic constants E_x^b and G_{xz}^b, but also the scatter of the experimental data. If a straight line cannot be fitted to the experimental points by the method

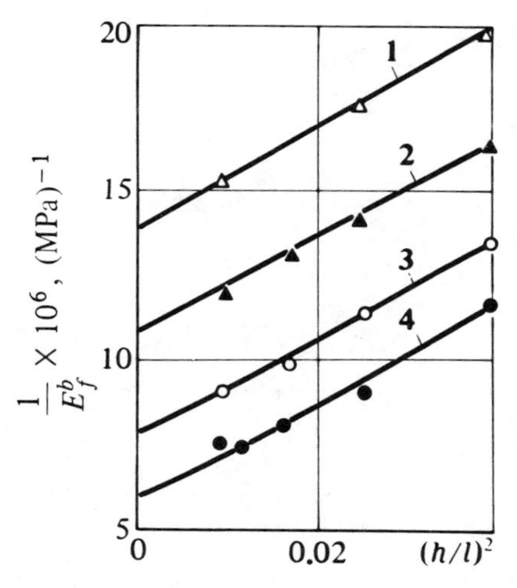

Fig. 5.2.2. Examples of processing of results in the determination of the modulus of elasticity E_x^b and shear modulus G_{xz}^b by Equation (5.2.7) for boron composite. Layups: (1) 1 : 1 : 1; (2) 0/90°; (3) 2 : 1; (4) 0°.

of least squares, an incorrect experiment or incorrect processing of results is indicated.

After an experimentally plotted straight line (5.2.6) is obtained, the elastic constants can be determined directly from the graph or by analysis. In the latter case, by selecting two points (subscripts 1 and 2) from the straight line (5.2.6) (in the analytical determination of the elastic constants, this straight line is used only for verification of the quality of the experiment), the following system of equations can be obtained:

$$\begin{aligned} \frac{1}{E_{f1}} &= \frac{1}{E_x^b} + \frac{1.2}{G_{xz}^b}\left(\frac{h}{l}\right)_1^2, \\ \frac{1}{E_{f2}} &= \frac{1}{E_x^b} + \frac{1.2}{G_{xz}^b}\left(\frac{h}{l}\right)_2^2 \end{aligned} \tag{5.2.7}$$

where

$$\frac{1}{E_{f1}} = \frac{P_1 l_1^3}{48 I_1 w_{\max 1}}, \quad \frac{1}{E_{f2}} = \frac{P_2 l_2^3}{48 I_2 w_{\max 2}}.$$

It is also possible to obtain the following formulas for determination of E_x^b and G_{xz}^b:

$$E_x^b = E_{fi}\left[1 + 1.2\beta^2\left(\frac{h}{l}\right)_i^2\right]$$

$$G_{xz}^b = \frac{E_x^b}{\beta^2}$$

where $i = 1$ or 2;

$$\beta^2 = \frac{E_{f1} - E_{f2}}{1.2\left[E_{f2}\left(\frac{h}{l}\right)_2^2 - E_{f1}\left(\frac{h}{l}\right)_1^2\right]}. \tag{5.2.8}$$

Equation (5.2.8) contains the difference between two large numbers $(E_{f1} - E_{f2})$, so that considerable accuracy in processing of the experimental results is required. Other methods of solution of the system of equations (5.2.7) are also possible.

$$E_x^b = \frac{ms_{11} - s_1^2}{s_{11}s_2 - s_1 s_2} \tag{5.2.9}$$

$$G_{xz}^b = \frac{ms_{11} - s_1^2}{ms_{12} - s_1 s_2}. \tag{5.2.10}$$

Here m is the number of experimental points equal to the sum of all $(h/l)_i n_w$, where n_w is the number of deflection measurements with $(h/l)_i = \text{const}$. Further,

$$s_1 = \sum_{i=1}^{n}\left(\frac{h}{l}\right)_i^2; \qquad s_{11} = \sum_{i=1}^{n}\left(\frac{h}{l}\right)_i^4$$

$$s_2 = \sum_{i=1}^{n}\frac{1}{E_{fi}}; \qquad s_{12} = \sum_{i=1}^{n}\frac{1}{E_{fi}}\left(\frac{h}{l}\right)_i^2.$$

For the determination of E_x^b and G_{xz}^b Equation (5.2.5) can also be used; it has the form of a straight line in l^2, w_{max}/l coordinates, of slope is equal to $1/48E_x^b I$ with a unit load $(P = 1)$, and $(w_{max}/l) =$ intercept $\alpha/4G_{xz}^b F$ (Fig. 5.2.3).

The maximum load P_{max} (with the largest h/l ratio) used in the determination of the elastic constants is selected so that the stresses in the bar do not exceed the level of the first break point in the tension (or compression)

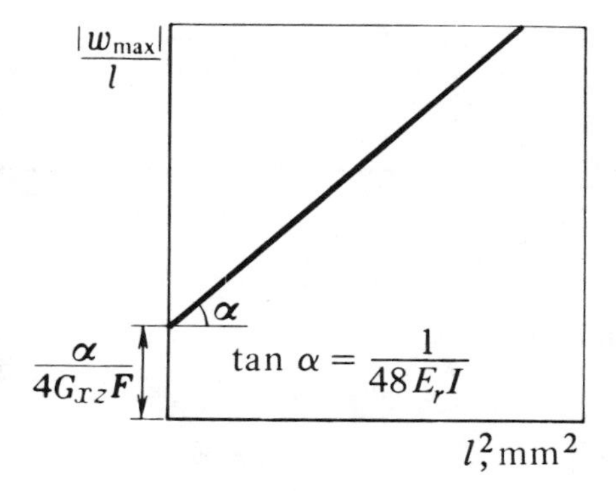

Fig. 5.2.3. Determination of the modulus of elasticity E_x^b and shear modulus G_{xz}^b according to Equation (5.2.5) in bending tests.

curve and the material does not fail due to tangential stresses. The overhang of the specimen ends has practically no effect on the bar deflection.

The fraction of deflection due to shear stresses, w_τ, constitutes only part of what is in the majority of cases a very small specimen deflection w. For reliable determination of the interlaminar shear modulus G_{xz}^b through selection of the relative span l/h, depending on the anisotropy E_x^b/G_{xz}^b of the material tested, sufficient sensitivity of the method, i.e., a sufficiently great deflection ratio $k_\tau = w_\tau/w$ should be ensured. Usually $k_\tau \geqslant 0.3$. For determination of the necessary relative span l/h as a function of the fixed numerical values E_x^b/G_{xz}^b and $k_\tau = w_\tau/w$, the following formula can be used:

$$\frac{l}{h} < 1.095\sqrt{\frac{1-k_\tau}{k_\tau}\frac{E_x^b}{G_{xz}^b}}. \tag{5.2.11}$$

The permissible minimum value of the l/h ratio is in need of refinement.

5.2.3. Sliding at Supports

Large deflections of a bar are related to slippage of the specimen at supports and to changes in span length. This phenomenon should be taken into account in tests of bars with large l/h ratios (flexible bars), i.e., in determination of the modulus of elasticity E_x^b without taking into account the shear stress. Change in span length takes place as a consequence of rotation of the bar on the supports at large deflections, and it depends on the design of the supports; it is not taken into account in elementary bending theory. This can distort the experimental results considerably.

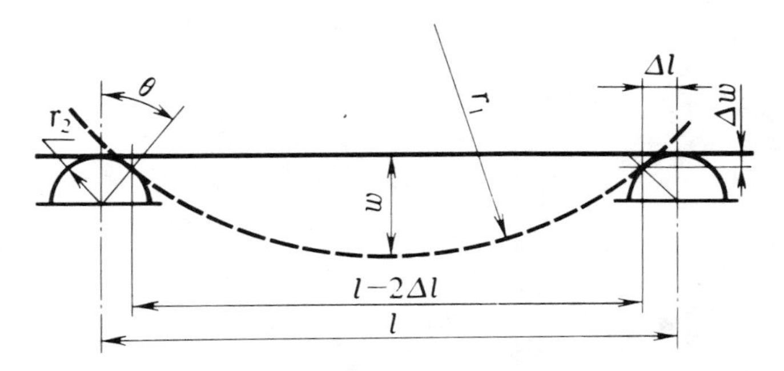

Fig. 5.2.4. Deformation of specimens in bending supported on cylindrical supports.

The effect of the type of supports on measured deflection in tests by the three-point loading has been studied in detail [160]. The effect of six different types of supports on deflection was evaluated. The specimens were made of steel and glass fiber composite. The specimen dimensions were selected so that the effect of shear could be disregarded: span $l = 200$ mm, $l/h = 250$ (for steel) and 64 (for glass fiber), $b/h = 16.5$ and 5.8, respectively. The test results show that, with small loads, the P versus w curve is practically independent of support design and remains rectilinear up to a deflection $w = 0.03l$. With large loads, the support method has an appreciable effect: the scatter of the measured deflection is 6–9%, depending on the type of supports.

The simplest type of support is the knife-edge support, which provides a constant span over the entire range of loads. A deficiency of knife-edge supports is that they cut into the specimen surface, even when the edges of the supports are rounded. This is especially noticeable in relatively short specimens, for which the shearing force in loading is greater than for long specimens (with $M =$ const.). This shows up unfavorably on the stress next to the support area and it introduces an error in deflection measurements. Therefore, when knife-edge supports are used it is advisable to employ hard protective pads, bonded to the specimen.

When specimens are supported by large-radius cylindrical supports, the decrease in span of the specimen must be taken into account and a correction must also be introduced into the deflection measurement (Fig. 5.2.4). Changes in deflection Δw and span Δl are determined by the following equations (for notation, see Fig. 5.2.4):

$$\Delta w = \frac{2R_2}{\dfrac{l^2}{4w} + 1} \qquad (5.2.12)$$

$$\Delta l = \frac{l}{2}\frac{\Delta w}{w} = \frac{R_2}{\dfrac{l}{2w} + \dfrac{w}{l}}. \tag{5.2.13}$$

When medium-radius cylindrical supports are used, a correction for changes in span Δl and deflection Δw have also proved necessary. This makes the use of cylindrical supports inconvenient.

Supports in the form of rotating rollers have been studied for the purpose of determining the effect of friction when fixed supports are used. Within the range of normal scatter of the measurements, the experimental results turn out to be the same, i.e., friction on the supports has practically no effect on the results of bending tests.

Roller supports on swinging frames, where the axis of rotation of the frame coincides with the roller surface, have the advantage that span l does not change in loading. Unlike cylindrical supports, these supports do not require corrections Δl and Δw. A deficiency of swinging roller supports is their instability in unloading of the specimen. This makes them practically unusable for repeated loading of specimens.

When small diameter roller supports are used, it is theoretically necessary to introduce corrections Δl and Δw, but with a sufficiently small roller diameter (2.5–3 mm) these corrections can be disregarded. Test results are independent of the method of attachment of the rollers (whether the rollers have a fixed axle or rotate freely in a groove, which is half the roller diameter deep). Rollers made of silver steel have been investigated. To ensure reliable contact with the specimen, their diameter has to be at least 2.4 mm.

5.2.4. Strength

Mathematical expressions for calculation of the maximum normal and tangential stresses in three-point bending are simple. However, in the case of their use for determination of the strength of the material, a series of factors connected with the type of failure of the specimen are considered. Analytical research shows that the effect of shear on the distribution of stress over the height of a bar is considerably less than its effect on deflection [231, p. 123]. For sufficient practical accuracy, the deviations of the normal stress distributions from linear and the tangential stresses from parabolic can be disregarded, if $\kappa < 1.2$.

To reach an approximate determination of the maximum stresses from experimental results, the following equations can be used [231, p. 127]:

$$\frac{\sigma_{\max}}{\sigma^*_{\max}} = 1 + \frac{1}{15}\kappa^2 - \frac{1}{125}\kappa^4, \tag{5.2.14}$$

$$\frac{\tau_{max}}{\tau^*_{max}} = 1 - \frac{1}{60}\kappa^2 + \frac{1}{12{,}600}\kappa^4. \tag{5.2.15}$$

Stresses σ^*_{max} and τ^*_{max} are determined from the following equations, known from the strength of materials:

$$\sigma^*_{max} = \sigma^u_x = \pm\frac{P^u lh}{8I} \tag{5.2.16}$$

$$\tau^*_{max} = \tau^u_{xz} = \frac{3}{4}\frac{P^u}{F} \tag{5.2.17}$$

where P^u is the breaking load; $I = bh^3/12$ is the moment of inertia; and $F = bh$ is the cross-sectional area of the bar.

Equation (5.2.16) does not take into account the additional bending moment due to the horizontal components of the support reaction. To eliminate this error, with $w > 0.1l$ the following equation can be recommended [11]:

$$\sigma^u_x = \frac{P^u lh}{8I}\left(1 + 6\frac{w^2}{l^2} - 4\frac{wh}{l^2}\right) \tag{5.2.18}$$

where w is the deflection at failure.

The expected nature of failure in bending can be predicted by means of a l/h versus τ_{max} curve [146]. The tangential stresses in the specimen at failure due to normal stresses $\sigma_{max} = \sigma^x_u$ are equal to:

$$\tau_{max} = \frac{h}{2l}\sigma^u_x. \tag{5.2.19}$$

Equation (5.2.19) is a rectangular hyperbola in l/h, τ_{max} coordinates, the asymptotes of which coincide with the coordinate axes (Fig. 5.2.5). If the interlaminar shear strength τ^u_{xz} is plotted together with this curve, the intersection of the straight τ^u_{xz} line with the τ_{max} curve will determine the relative span l/h at which the type of failure (due to normal or tangential stresses) changes.

In practice, because of defects in manufacture a distinct intersection point of the τ_{max} curve with a straight line τ^u_{xz} (Fig. 5.2.5) cannot be obtained (strength τ^u_{xz} is very sensitive to all the factors which determine the strength of the matrix and of the adhesive bond), and there always is some transition region around the theoretical intersection point where failure of the material is possible due to both normal and tangential stresses. The difficulties of

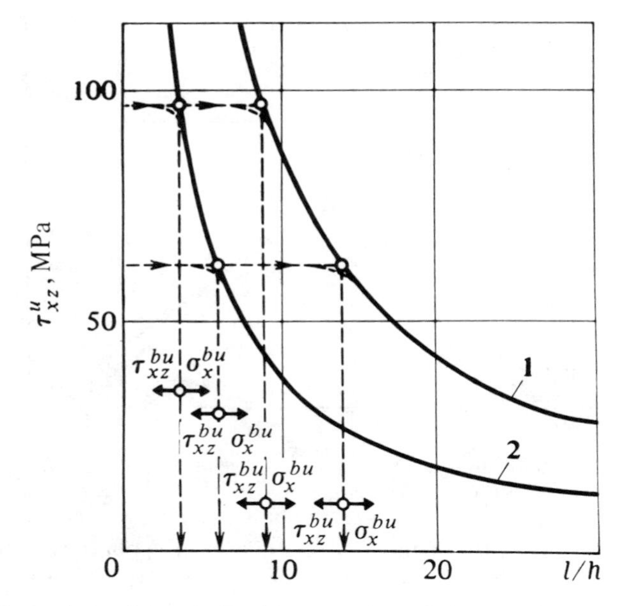

Fig. 5.2.5. Change in maximum tangential stresses (at a given strength σ_x^{tu}) with the relative span l/h in three-point bending [146] (material—unidirectional boron composite): (1) $\tau_{max} = 840\ h/l$; $\sigma_x^{tu} = 1680$ MPa, $\tau_{xz}^{u} = 96$ MPa, $V_f = 70\%$, $V_v = 0\%$; (2) $\tau_{max} = 360\ h/l$, $\sigma_x^{tu} = 720$ MPa, $\tau_{xz}^{u} = 62$ MPa, $V_f = 30\%$, $V_v = 6\%$. V_v is the volume content of voids.

such a preliminary evaluation of the type of failure in bending is that the strengths of the material σ_x^u and τ_{xz}^u must be estimated ahead of time, even if only theoretically.

In isotropic materials, τ_{xz}^u in Fig. 5.2.5 is set high, and failure due to shear is practically impossible. In the testing of polymeric composites, as can be seen from Fig. 5.2.5, failure from tangential stresses will not always be assured in standardized specimens for determination of the interlaminar shear strength ($l/h = 5$); on the other hand, high-strength composites are more sensitive to manufacturing defects which reduce the interlaminar shear strength, and their failure due to tangential stresses can take place at large l/h values.

Increases in interlaminar shear strength have forced experimentalists to use ever smaller l/h values. However, at $l/h < 6$ elementary bending theory becomes invalid. Let us treat this problem in more detail.

The derivation of Equations (5.2.2) and (5.2.3), assumes a linear distribution of the normal stresses σ_x and a parabolic distribution of the tangential stresses τ_{xz} over the height of the bar; the normal stresses τ_x change linearly

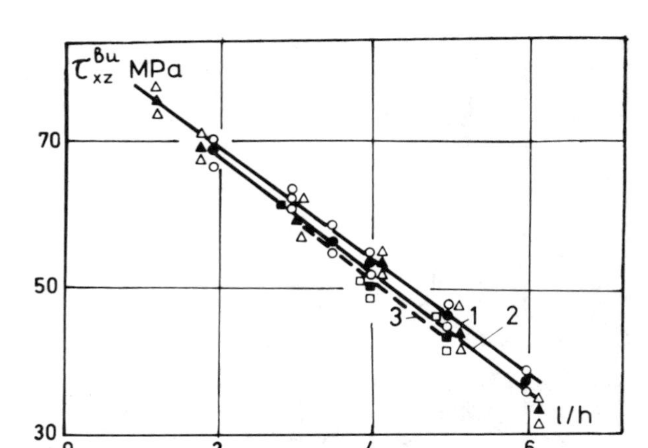

Fig. 5.2.6. Dependence of strength τ_{xz}^{bu} on the relative span l/h in three-point bending [274]. $b = 10$ mm; $h = 10$ (1), 15 (2), and 20 (3) mm.

in direct proportion to the distance from the supports, but the absolute values of tangential stresses τ_{xz} are constant over the entire length of the bar.

Practical experience and theoretical analysis show that for short bars ($l/h < 6$) of anisotropic material such assumptions are unrealistic. It has been established [40, 228] that the experimentally determinable interlaminar shear strength τ_{xz}^{u} is not constant, but decreases with increasing relative span l/h (Fig. 5.2.6). There are several causes of this phenomenon.

Theoretical investigation [22, 228] shows that the distribution of tangential stresses τ_{xz} over the height of a short bar of anisotropic material coincides approximately with a quadratic parabola only in the middle of a half-span [within the range of $(0.25–0.75)l/h$]. At the points of application of concentrated force (the load and the support reaction) the distribution of the tangential stresses over the height of a bar differs from parabolic and it has a distinct maximum near the loaded surface of the bar several times larger than the mean tangential stress $\tau_{\text{mean}} = Q/F$ (Fig. 5.2.7). Furthermore, there is no section with a constant ordinate of maximum tangential stresses $\tau_{xz\,\text{max}}$ for relatively short bars of anisotropic material (Fig. 5.2.8). Owing to the peculiarities indicated, which depend on the characteristics of anisotropy of the bar material E_x^b/E_z^b and E_x^b/G_{xz}^b, the possibility of delamination of the material due to tangential stresses acting on the midplane of a relatively short bar ($l/h < 4$) is considerably limited.

The classical theory of bending does not take into account the effect of transverse stresses σ_z^b. Calculations show [228] that actually the compressive transversal stresses σ_z^c act over the entire length of a relatively short bar;

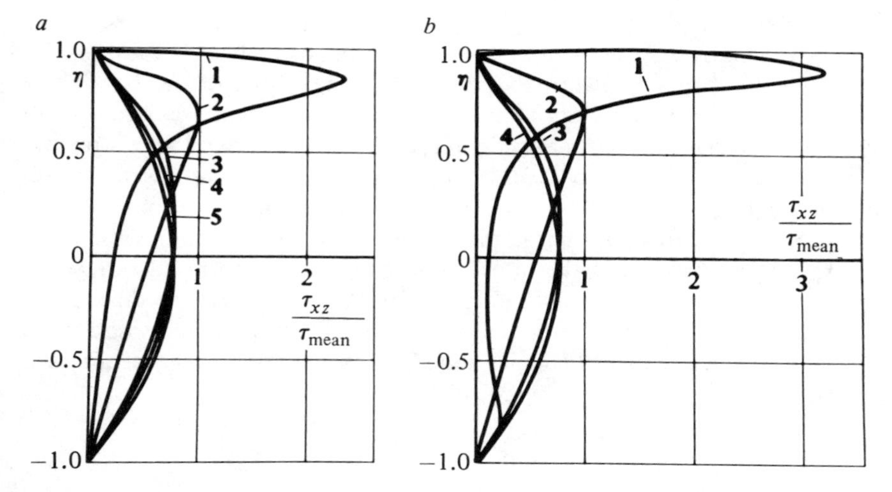

Fig. 5.2.7. Change in τ_{xz}/τ_{mean} over the ordinate η, depending on the parameters $\xi = x/l$ and $\bar{l} = l/h$ [228]: (a) $\bar{l} = 4$; $\xi = 0.05$ (1); 0.15 (2); 0.25 (3); 0.35 (4); 0.50 (5); (b) $\xi = 0.05$; $\bar{l} = 1$ (1); 4 (2); 10 (3); 15 (4) (ξ readings from midspan).

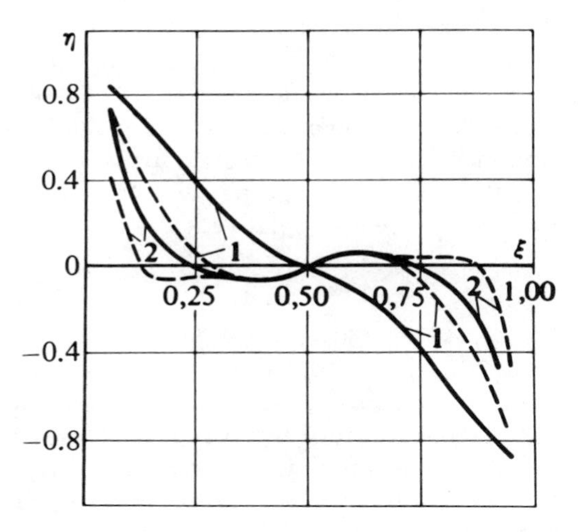

Fig. 5.2.8. Change in coordinates of maximum tangential stresses (ξ and η) for anisotropic (solid line) and isotropic (broken line) materials [228]; $\bar{l} = l/h = 4$ (curve 1); 10 (curve 2); $\xi = x/l$ (from midspan); $\eta = z/h$.

near the points of application of concentrated forces these stresses exceed the mean value of the tangential stresses by a factor of 13–15 (with $l/h = 5$) and sharply decrease toward the middle of a half-span of the bar. These compressive stresses restrict the initiation of cracks during delamination of the bar material and seemingly increase the interlaminar shear strength. In the contact regions of relatively short bars high local compressive stresses σ_x^k are also observed. They are several times higher than the mean tangential stress τ_{mean} and can result in microbuckling of the reinforcing fibers.

Experience shows that the interlaminar shear strength τ_{xz}^u also depends on the reinforcement layup scheme, i.e., on the edge effect. In three-point bending tests of 24-ply Thornel 300/Narmco 5208 carbon composites with a specimen relative span $l/h = 4$, the following results have been obtained [250]:

LAYUP	STRENGTH τ_{xz}^u	
	MPa	%
$[0°]_{24}$ (unidirectional)	108	100
$[0/90°]_s$	84	78
$[0/\pm 45/90°]_{3s}$	71	66

As a consequence of these deviations of stress in a relatively short bar from the theoretically supposed results of determination the interlaminar shear strength τ_{xz}^u by three-point bending, this test can only be used for approximate qualitative comparison of different composites.

It is known that for bars of isotropic materials the distribution of tangential stresses over the width of a specimen with $b/h > 1$ is nonlinear and that the ratio of the maximum value of the tangential stresses τ_{max} (along the side faces of the specimen) to the corresponding value from elementary bending theory τ_{max}^* increases with increasing relative width of the specimen:

b/h	1	2	4
τ_{max}/τ_{max}^*	1.126	1.396	1.986

In the case of isotropic materials, the change in numerical values of the tangential stresses over the width of the bar is not of special significance, since the strength obtained from tangential loads τ_{xz}^u of isotropic materials is close to the strength obtained from normal loads σ_x^u. The situation is different for modern composites in which $\sigma_x^u \gg \tau_{xz}^u$; tangential stresses, despite their low value, can cause failure of these materials. Study of the distribution of tangential stresses over the width of a bar of anisotropic

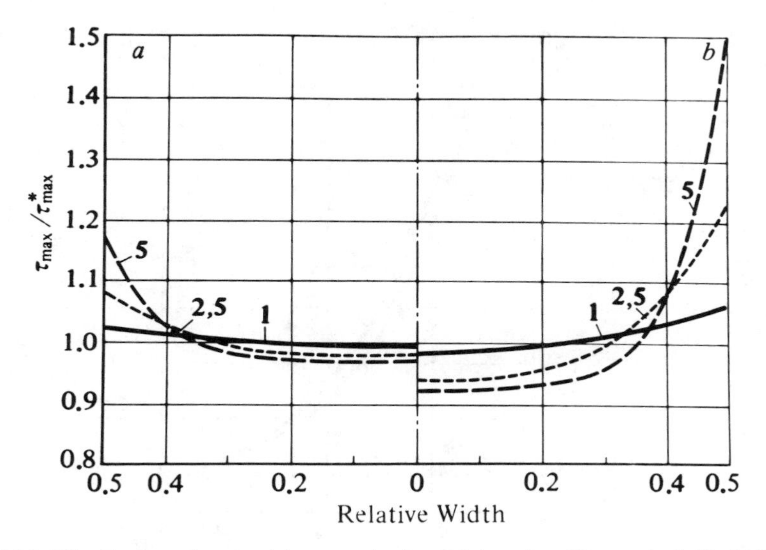

Fig. 5.2.9. Distribution of tangential stresses in the midplane ($z = 0$) over the half-width of a carbon composite (a) and glass fiber composite bar (b) with various b/h ratios (the numbers on the curves). According to the elementary theory, $\tau_{max}/\tau^*_{max} = 1 =$ const. over the entire width of the bar [110].

material has begun recently [110, 207]. A precise analytical estimation of the actual distribution of tangential stresses over the width of a bar of composite is impossible, since it is impossible to take into account all the local irregularities in the polymer interlayer caused by material structure and fabrication.

The results of calculations for unidirectional glass fiber and carbon composites [110] are shown in Fig. 5.2.9. It can be seen from the figure that, with large relative width b/h, the τ_{max}/τ^*_{max} ratio can attain a large value, and actual failure of the material due to tangential stresses begins at higher values of these stresses than are given by calculations using mean tangential stresses. The error is made augmented by the edge effect—the appearance of interlaminar stresses near the specimen side faces, depending on the stacking sequence of the reinforcement. This phenomenon in the bending of composites has not been investigated to any great extent. Considering these deviations from the viewpoint of classical bending theory, estimation of the interlaminar shear strength will be inaccurate unless the distribution of the tangential stresses over the width of the specimen is taken into account. This is confirmed by experience (Fig. 5.2.10). This peculiarity has not taken into account by existing standards.

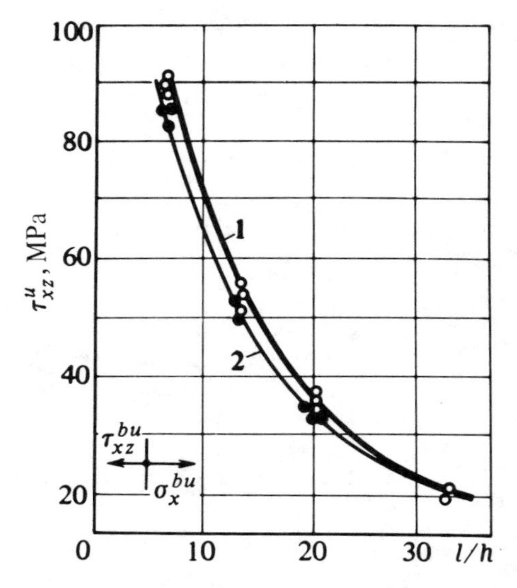

Fig. 5.2.10. Effect of specimen width and relative span l/h on the three-point bending strength of 12-ply unidirectional boron composites [131]. Specimen width: (1) 6.35 mm (open circle); (2) 10.2 mm (filled circle).

The effect on strength of overhang of the bars has not been studied sufficiently. In practice, for glass fiber composites an overhang length 1–2 times greater than height h is sufficient.

5.3. MULTI-POINT LOADING METHODS

5.3.1. Pure Bending

Tests of this type [Fig. 5.1.1(b)] are conducted for the purpose of determining the moduli of elasticity E_x^t and E_x^c and the pure bending strength σ_x^{bu}. Loading in pure bending is accomplished, by applying bending moments to the end faces of the bar. The advantage of the pure bending test method is the uniform state of stress over the entire length of the specimen, the absence of contact stresses at the points of application of concentrated forces (load and support reactions), and the elimination of the overhang effect. In this loading technique, the specimen is accessible for measurement over its entire length. Because of the absence of shear strains in the specimen, the methods of measurement of deflection w and the strains of the outer layers of the bar ε^t and ε^c with proper construction of the loading devices, i.e., with the absence of local distortions of the elastic line of the bar in the loaded sections, are qualitatively equivalent.

The moduli of elasticity of the material under tension (E_x^t) and compression (E_x^c) can be determined by pure bending, using the following strength of materials relationships:

$$E_x^t = \frac{M}{W} \cdot \frac{\varepsilon^t + \varepsilon^c}{2(\varepsilon^t)^2} \tag{5.3.1}$$

$$E_x^c = \frac{M}{W} \cdot \frac{\varepsilon^t + \varepsilon^c}{2(\varepsilon^c)^2} \tag{5.3.2}$$

$$E_x = \frac{Ml^2}{8Iw_{max}} \tag{5.3.3}$$

$$E_x = \frac{Ml}{2I\varphi}. \tag{5.3.4}$$

where M is the bending moment; $I = bh^3/12$ and $W = bh^2/6$ are the moments of inertia and resistance of the cross sections of the bar; ε^t and ε^c are the experimentally measured strains of the outer layers of the bar under tension and compression; w_{max} is the deflection at the mid-span; φ is the rotation angle of the end section of the bar.

Equations (5.3.1) and (5.3.2) are more universal, since they permit evaluation of different moduli in the test material ($E_x^t \neq E_x^c$).

The strength at failure of a specimen in its gage length, i.e., due to the action of normal stresses, is determined by the formula

$$\sigma_x^{bu} = \frac{M^u}{W}. \tag{5.3.5}$$

Pure bending tests also permit the determination of the coefficients of transverse strains a_{13}, a_{23}, a_{21} of orthotropic materials [211]. For this purpose, three series of specimens, must be prepared cut relative to the axes of orthotropic material as shown in Fig. 5.3.1. In the experiment, displacements are measured over the width of specimens between two points A and B (Fig. 5.3.2): Δu_1, Δv_2, and Δv_3 for specimens of series I, II and III, respectively. Coefficients of transverse strains are calculated by the formulas

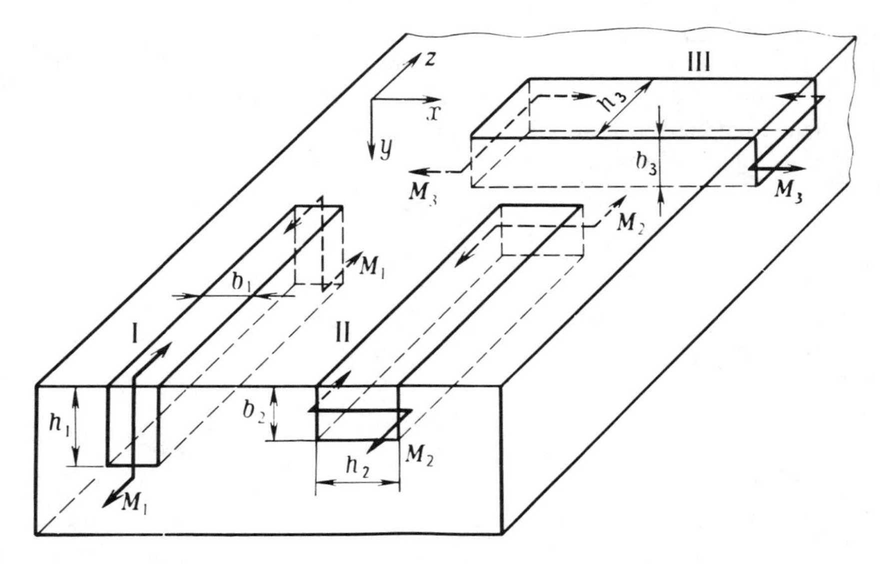

Fig. 5.3.1. A scheme of cutting specimens for determination of Poisson's ratio in pure bending tests [211].

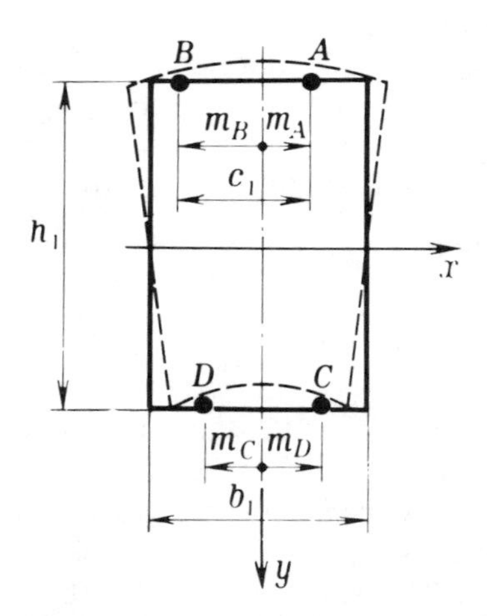

Fig. 5.3.2. Scheme for measurement of transverse displacements in determination of Poisson's ratio in pure bending tests [211].

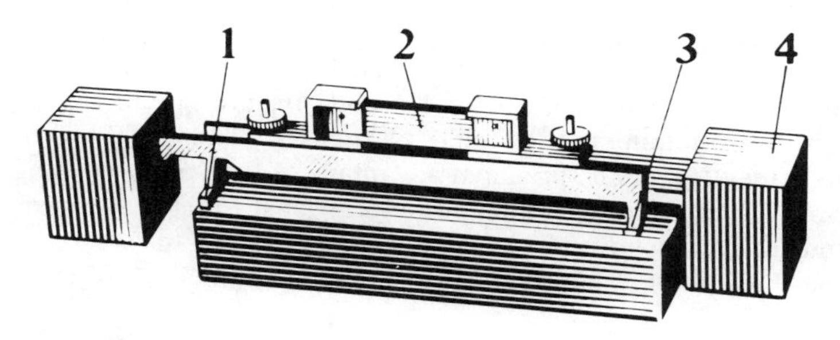

Fig. 5.3.3. Device for pure bending tests [262]: (1) support (knife edge); (2) a specimen; (3) hinge; (4) weight.

$$a_{13} = -\frac{\nu_{zx}}{E_z} \cdot \frac{2I_{x1}\Delta u_1}{M_1 h_1 c_1}$$

$$a_{23} = -\frac{\nu_{zy}}{E_z} \cdot \frac{2I_{y2}\Delta v_2}{M_2 h_2 c_2} \qquad (5.3.6)$$

$$a_{21} = -\frac{\nu_{xy}}{E_x} \cdot \frac{2I_{y3}\Delta v_3}{M_3 h_3 c_3}$$

where c_1, c_2, and c_3 are the distances between points A and B (the basis of a tensometer); I_{x1}, I_{y2}, I_{y3} are moments of inertia of transverse sections of the specimens; h_1, h_2, h_3 are the specimen heights; M_1, M_2, M_3 are bending moments.

The deficiency of this method is the necessity to have specimens with a relatively large cross section (approximately 30 × 30 mm).

In analogous studies of isotropic materials, optical methods of measurement [57] are sometimes used, which ensure high accuracy (the error does not exceed 1%).

Certain difficulties arise in designing loading devices for pure bending, since theoretically the bending moment must be applied on the end faces of the specimen. Because of the necessity of fastening the ends of the specimen in the loading device, this requirement is impracticable. Hence the appearance of an end effect zone in the specimen is unavoidable. Standard devices (Fig. 5.3.3) have been developed for testing small specimens in pure bending. Another well substantiated but rather complicated device for loading straight and curved bars in pure bending has been described in [93–95].

5.3.2. Four-Point Loading

Four-point bending tests (Fig. 5.1.1(c and d)] are conducted with the purpose of determining the moduli of elasticity $E_x^{b(t)}$ and $E_x^{b(c)}$ and the pure bending strength σ_x^{bu}. The principal advantage of four-point bending is the uniform state of stress over the entire gage length of the specimen, as a consequence of which four-point loading is considered to be more reliable than the three-point loading method for the determination of the moduli of elasticity [84].

In principle, the four-point loading method is also a pure bending scheme. However, in this case, the bending moment is created by concentrated forces P acting at a distance [a or c, see Fig. 5.1.1(c and d)] from the bar supports. Consequently, a shearing force Q and shear strains appear on bar sections of length a or c. Efforts to use comparatively short specimens result in increase in the applied force and, consequently, in an increase in deflection from shear and possibility of delamination. This difference in the methods of introduction of the bending moment in bars of anisotropic materials is very significant and must be taken into account in the conduct of the experiment and the processing of results.

Four-point bending can be realized by two methods: by loading inside span l [Fig. 5.1.1(c)] or outside it [Fig. 5.1.1(d)].

In loading inside the span [Fig. 5.1.1(c)] the deflection at the midspan of the bar is determined by the following relation, with the effect of shear taken into account:

$$\begin{aligned} w_{l/2} = w_\sigma + w_\tau &= \frac{Pal^2}{24E_x^bI}\left(3 - 4\frac{a^2}{l^2}\right) + \frac{\alpha Pa}{G_{xz}F} \\ &= \frac{Pal^2}{24E_x^bI}\left(3 - 4\frac{a^2}{l^2}\right)\left[1 + \frac{2\alpha}{3 - 4\dfrac{a^2}{l^2}} \cdot \frac{h^2}{l^2}\frac{E_x^b}{G_{xz}^b}\right]. \end{aligned} \tag{5.3.7}$$

In practice, the four-point loading test is sometimes called a pure bending method. However, in loading by the arrangement of Fig. 5.1.1(c), as can be seen from Equation (5.3.7), there is no state of pure bending, i.e., bending is only due to normal stresses. At the same time, the difference

$$w_{l/2} - w_{x=a} = \frac{Pal^2}{8E_x^bI}\left(1 + 4\frac{a}{l} + 4\frac{a^2}{l^2}\right) \tag{5.3.8}$$

($w_{x=a}$ is the deflection at the loading point) does not depend on shear, since

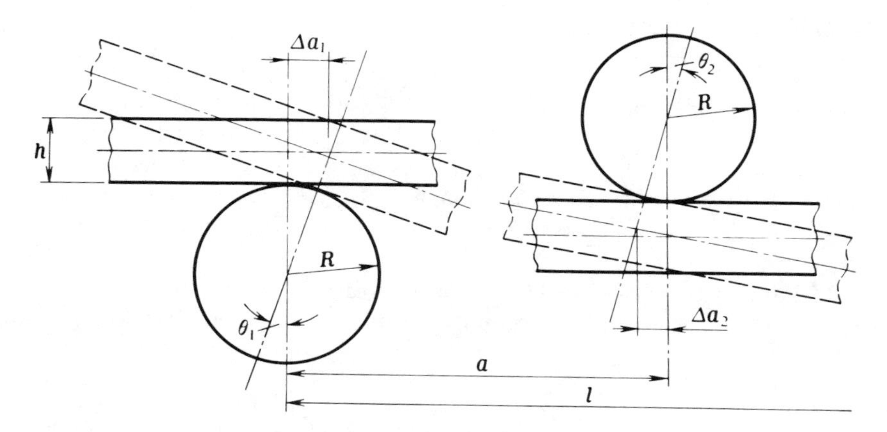

Fig. 5.3.4. The change in span and distance a in four-point bending.

the section of the bar between the loading points is bent only by normal stresses (if the end effect at the loading point is disregarded). This feature of the state of stress in the bar shows up in the technique of the experiment. If Equation (5.3.8) is used for determination of the modulus of elasticity $E_x^{b(t,c)}$, the deflection ought to be measured at two points: at $x = a$ and $x = l/2$. It is more reasonable to measure the strains ε^t and ε^c and determine the modulus of elasticity using Equations (5.3.1) and (5.3.2); in this case, the difference in the moduli in tension and compression of the specimen material can also be estimated.

The strength of the material in the loading method shown in Fig. 5.1.1(c) is determined from Equation (5.3.5) by substitution of $M^u = P^u a$. In this case, the change in the span l and distance a which should be taken into account becomes obvious. For example, in testing glass fiber composites [265] the change in distance a and, consequently, the error in determination of the strength was 14–56%; this error decreases with increase in the E_x^b/σ_x^{bu} ratio. The deflection at specimen failure was $(0.07–0.29)h$, i.e., in the majority of cases, it exceeded the limit assumed by elementary bending theory. In the case of more anisotropic materials (boron and carbon composites) the error will be still greater.

The changes in span l and distance a can be estimated by means of the following equations (for notation see Fig. 5.3.4):

$$l_1 = l - 2\Delta a_1 \tag{5.3.9}$$

$$a_1 = a - (\Delta a_1 + \Delta a_2) \tag{5.3.10}$$

where

$$\Delta a_1 = \left(R_1 + \frac{h}{2}\right)\sin\theta_1 \approx \left(R_1 + \frac{h}{2}\right)\theta_1$$

$$\Delta a_2 = \left(R_2 + \frac{h}{2}\right)\sin\theta_2 \approx \left(R_2 + \frac{h}{2}\right)\theta_2.$$

At large deflections ($w > 0.1l$) the horizontal components of support reactions must be taken into account; in this case, the strength is determined [11] by

$$\sigma_x^{bu} = \frac{2Pl}{\sigma h^2}\left(1 + 4.70\frac{w^2}{l^2} - 7.04\frac{wh}{l^2}\right). \tag{5.3.11}$$

Stresses in the active facing of the sandwich beam, which is also loaded by the scheme of Fig. 5.1.1(c), are calculated by the following formula (for notation see Fig. 5.1.2; the core absorbs only the shearing force):

$$\sigma_x^{t(c)} = \frac{M}{W} = \frac{Pd}{W} \tag{5.3.12}$$

where σ_x are tensile (t) or compressive (c) stresses in the active facing; P is the load; d is the distance between the support and a loading point.

The resistance moment of the cross section W depends on the construction of the beam. In the case of tension [the active facing of thickness h_1; see Fig. 5.1.2(a)] it is equal to

$$W_1 = \frac{I_y}{z_1}.$$

In the case of compression [the active facing of thickness h_2; see Fig. 5.1.2(b)]

$$W_2 = \frac{I_y}{z_2}$$

where

$$I_y = \frac{b}{3}[(z_1^3 - z_2^3) - (z_1 - h_1)^3 - (z_2 - h_2)^3],$$

$$z_1 = \frac{h^3 - c(2h_1 + c)}{2(h - c)}.$$

$$z_2 = h - z_1$$

and b is the width of the beam.

In processing the experimental results on sandwich beams, the calculated stresses $\sigma_x^{t(c)}$ and measured strains $\varepsilon_x^{t(c)}$ and $\varepsilon_y^{t(c)}$, the stress-strain curves σ_x versus ε_x or σ_x versus ε_y are plotted and the modulus of elasticity $E_x^{b(t,c)}$ and Poisson's ratios determined.

The shortcomings of the loading method shown in Fig. 5.1.1(c)—the variable state of stress over the span length of the bar and the change in loading geometry during the test—are completely absent in loading by the method shown in Fig. 5.1.1(d). In this case, a bending moment $M = P_c$ is created by a pair of forces outside span l; in this way the bending moment is made accessible for measurement over the entire length. The deflection at midspan of the bar is determined from

$$w_{\max} = \frac{Ml^2}{8E_x^bI} = \frac{Pcl^2}{8E_x^bI}. \tag{5.3.13}$$

It does not depend on the shear characteristics of the material. It follows from Equation (5.3.13) that

$$E_x^b = \frac{Pcl^2}{8Iw_{\max}}. \tag{5.3.14}$$

At large deflections of the bar a change in span l and distance c is also observed as a consequence of rotation of the bar on the supports. The effect of these changes is smaller than in loading by the method of Fig. 5.1.1(c):

$$l_1 = l + 2\Delta c \tag{5.3.15}$$

$$c_1 = c - \Delta c \tag{5.3.16}$$

where

$$\Delta c = \left(R_1 + \frac{h}{2}\right)\sin\theta_1 \approx \left(R_1 + \frac{h}{2}\right)\theta_1.$$

In processing the results of tests by this loading technique, Equations (5.3.14) and (5.3.1), (5.3.2) for determination of the modulus of elasticity are equivalent, i.e., the deflection $w_{\max}$ (estimation of the different moduli of the material in tension and compression is impossible) or the strains ε^t and ε^c can be measured. The bending strength σ_x^{bu} is determined using Equation (5.3.5). When using (5.3.5) and (5.3.14) in the case of large deflections, the change in bending moment as a consequence of change in distance c should be taken into account.

Failure of specimens loaded by the four-point method can begin in a section of length a or c due to bearing (see Fig. 5.1.7), when these sections are very short, or due to interlaminar splitting (see Fig. 5.1.5). To eliminate failure of the specimen due to splitting the following relationship must be observed

$$\frac{a}{h} = \frac{c}{h} > \frac{\sigma_x^{bu}}{4\tau_{xz}^{bu}}. \tag{5.3.17}$$

5.3.3. Five-Point Loading

Five-point bending tests of prismatic bars [Fig. 5.1.1(e)] are conducted with the purpose of determining the modulus of elasticity E_x^b and shear modulus G_{xz}^b or their ratio $\beta^2 = E_x^b/G_{xz}^b$. The bar is loaded with a force P at the middle of span l and by two equal forces P_1 at the ends of the bar. By varying forces P and P_1 (by not exceeding the material strength), span l, and distance c, any desired deflection w at the middle of span l can be obtained.

Deflection of the bar w is determined by

$$w = w_P + w_{P_1} + w_Q = \frac{Pl^3}{48E_x^bI} - \frac{P_1cl^2}{8E_x^bI} + \frac{\alpha Pl}{4G_{xz}^bF}. \tag{5.3.18}$$

There are three cases of using the five-point loading technique:

1. With $|w_P| = |w_{P_1}|$, the deflection $w = w_Q$, i.e., from the deflection, measured experimentally at midspan, the shear modulus G_{xz}^b can be determined:

$$G_{xz}^b = \frac{\alpha Pl}{4Fw}. \tag{5.3.19}$$

The condition $|w_P| = |w_{P_1}|$ is fulfilled at the following relation of forces P and P_1:

$$\frac{P_1}{P} = \frac{l}{6c}. \qquad (5.3.20)$$

The ultimate values of forces P and P_1 are established in terms of bending strength of the material σ_x^{bu}. At $M_{2\,\max} < 2M_{1\,\max}$ (where $M_{1\,\max} = P_1 c$; $M_{2\,\max} = Pl/4$) limiting value of the moment $M_{1\,\max}$; this condition follows:

$$P_{1\,\max} < \frac{W\sigma_x^u}{c}. \qquad (5.3.21)$$

At $M_{2\,\max} > 2M_{1\,\max}$ limiting value of the moment $M_{2\,\max}$; it follows that

$$P_{\max} < \frac{4W\sigma_x^u}{l} \qquad (5.3.22)$$

where $W = bh^2/6$ is the moment of resistance of the specimen cross section.

Besides, for the purpose of elimination the interlaminar splitting of a cantilever of length c, the following condition should be fulfilled:

$$\frac{c}{h} > \frac{\sigma_x^{bu}}{4\tau_{xz}^{bu}}. \qquad (5.3.23)$$

2. Equation (5.3.18) in $Pl/(4P_1 c)$, $w/(P_1 c)$ coordinate system describes a straight line which intersects the ordinate at the point $-l^2/8E_x^b I$, and at abscissa $Pl/(4P_1 c) = 3/2$ it has ordinate $3\alpha/(2G_{xz}^b F)$ (Fig. 5.3.5). By testing a bar with several arbitrary selected values of P, P_1, and c (span l remains constant; P and P_1 are selected with allowance for the same limitations on the maximum load, as in case 1) this straight line can be plotted graphically (see Section 5.2.2) and from the ordinates $-l^2/8E_x^b I$ and $3\alpha/2G_{xz}^b F$ the moduli E_x^b and G_{xz}^b determined.

3. The $\beta^2 = E_x^b/G_{xz}^b$ ratio can be determined directly from the parameters P, P_1, l, and c. The bar is loaded initially with forces P_1, and force P is then applied so that deflection w at midspan becomes zero. With $w = 0$, it follows from Equation (5.3.18) that

$$\beta^2 = \frac{E_x^b}{G_{xz}^b} = \frac{1}{2\alpha} \cdot \frac{F}{I}\left(\frac{P_1}{P} cl - \frac{l^2}{6}\right) = \frac{1}{\alpha}\left(\frac{l}{h}\right)^2 \left(\frac{6c}{l}\frac{P_1}{P} - 1\right). \qquad (5.3.24)$$

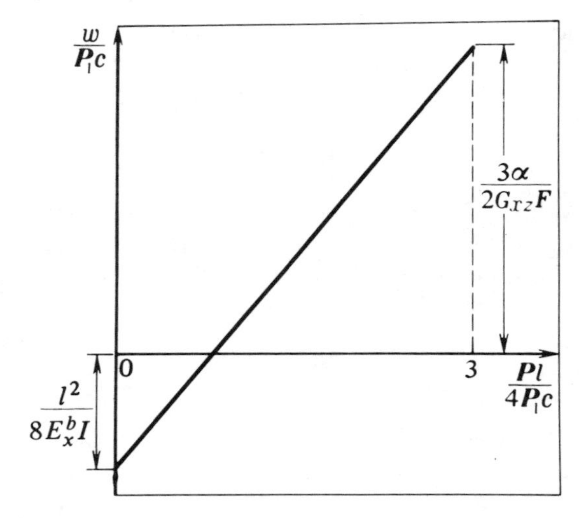

Fig. 5.3.5. Determination of modulus of elasticity E_x^b and shear modulus G_{xz}^b in five-point bending tests.

Further, from the known G_{xz}^b and β^2 the modulus of elasticity E_x^b can be determined

$$E_x^b = \beta^2 G_{xz}^b. \tag{5.3.25}$$

Five-point bending tests avoid the errors involved in the handling of large numbers, and the testing time is some 25% less than in three-point bending [100]. At the present time, the five-point scheme is not sufficiently used.

5.4. BARS WITH FIXED ENDS

In some works (for example, [150]) for the determination of the shear modulus G_{xz}^b use of test bars with fixed ends loaded with force P at midspan is recommended [Fig. 5.1.1(f)]. An advantage of this method is that the fraction of the deflection due to shear in the bar with fixed ends is significantly greater than in a simply supported bar and, consequently, the accuracy of determination of the shear modulus should be higher. Actually, practically always (with $l/h \geqslant 5$, $E_x^b/G_{xz}^b \geqslant 5$) the shear fraction of deflection $k_\tau = w_\tau/w > 0.3$ can be ensured. However, it should be taken into account that in loading of bars having equal span l and the same of cross-sectional (F, I) and material (E_x^b, G_{xz}^b) characteristics with the same load, the deflection at midspan of a bar with fixed ends is considerably less (without shear, four times) than the deflection of a simply supported bar. This puts higher requirements to accuracy of deflection measurements.

Furthermore, in the case of anisotropic materials, the deflection of a bar with fixed ends depends considerably more strongly on the attachment conditions than is true for isotropic materials [231, p. 87]. Two cases of attachment of the bar are usually examined in the theory of elasticity: $dw/dx = 0$ (rigid attachment, the horizontal element of the bar axis fixed) and $(\partial u/\partial z)_{z=0} = 0$ (flexible attachment, the vertical element of the support section fixed at the level of the neutral axis of the bar). Experimental studies show [231, p. 108] that rigid attachment is unrealistic and that the support sections of bars take some middle positions described by the conditions $dw/dx = 0$ and $(\partial u/\partial z)_{z=0} = 0$, which cannot be estimated analytically. Therefore, in determining the shear modulus G_{xz}^b by this scheme, a certain uncertainty always remains.

The calculation relationships in bending of bars with fixed ends are presented below:

At $dw/dx = 0$,

$$w_{\max} = w_{\max}^* \left(1 + 2\alpha \frac{h^2}{l^2} \frac{E_x^b}{G_{xz}^b}\right). \tag{5.4.1}$$

At $(\partial u/\partial z)_{z=0} = 0$,

$$w_{\max} = w_{\max}^* \left(1 + 4.5\alpha \frac{h^2}{l^2} \frac{E_x^b}{G_{xz}^b}\right). \tag{5.4.2}$$

At $dw/dx = \theta$,

$$w_{\max} = w_{\max}^* \left(1 + 2\alpha \frac{h^2}{l^2} \frac{E_x^b}{G_{xz}^b} + 48 \frac{E_x^b I}{Pl^2} \theta\right) \tag{5.4.3}$$

where $w_{\max}^* = Pl^3/(192E_x^b I)$; θ is the angle of rotation of the support sections as a consequence of elasticity of the supports; α is a correction factor which depends on the shape of the cross section. The experimental results are processed by the same methods as in three-point bending tests.

5.5. RINGS AND SEGMENTS

5.5.1. Loading Methods

Bending of intact and split rings in their planes can be realized by loading with concentrated forces. Loading methods of intact and split rings in bending are shown in Figs. 5.5.1 and 5.5.2; the material characteristics, force factors, and variable geometric dimensions are also shown there.

Loading of intact rings in their planes can be realized in two-point [Fig. 5.5.1(a)], three-point [5.5.1(b)], or four-point [Fig. 5.5.1(c)] arrangements. Loading of intact rings with concentrated forces is used both for determination of the moduli E_θ and $G_{\theta r}$ and interlaminar shear strength $\tau^u_{\theta r}$. Efforts to use this method for determination of tensile or compression strengths $\sigma^{t(c)}_\theta$ have proven unsuccessful because of difficulties in processing the experimental results. Two-point loading with two diametrically opposed tensile or compressive forces [Fig. 5.5.1(a)] is the most commonly used of the loading methods shown in Fig. 5.5.1. Three- and four-point loadings [Fig. 5.5.1(b and c)], because of somewhat greater complexity of obtaining and processing results, are mainly used to determine the possible "trough" in the modulus of elasticity as a consequence of local distortions of the reinforcing fibers.

Split rings are loaded in their planes with concentrated forces in the cut [see Fig. 5.5.2(a)] or on the diameter [see Fig. 5.5.2(b)], or with a concentrated moment and forces in the cut [see Fig. 5.5.2(c and d)]. Tests of split rings are considerably less common than tests of intact rings. This would appear to be explained by the more rigid requirements on experimental technique. These involve increased accuracy of installation of the rings and difficulty in loading. However, the possibilities of split ring testing are clearly underestimated. These methods make it possible to determine not only the modulus of elasticity E_θ, but also the transverse (radial) tension strength σ^{tu}_r, which is difficult to determine in practice by other methods and the interlaminar shear strength $\tau^u_{\theta r}$. The capabilities, advantages, and disadvantages of each loading method listed above are evaluated in the subsequent sections.*

5.5.2. Intact Rings

The technique of testing rings with concentrated forces is simple: the load and the displacement across the vertical or horizontal diameter are measured. Difficulties arise in processing the results. The equations given by strength of materials have proved to hold true only in a very narrow range of h/R ratios. There are two causes of error: for thin rings, transition into the geometrically nonlinear region of deformations; for thick-walled rings, the effect of shear stresses and strains.

In loading thin-walled rings with concentrated forces for obtaining reliable

*The method of determination Poisson's ratios on ring specimens has been treated in [157].

Loading scheme			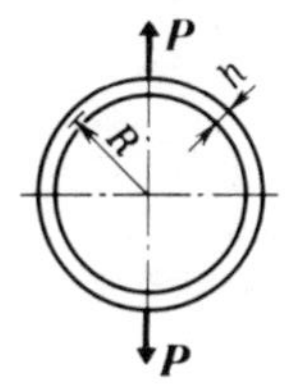 *a*	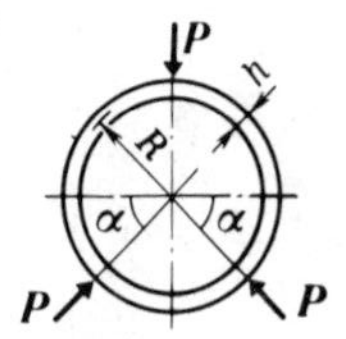 *b*	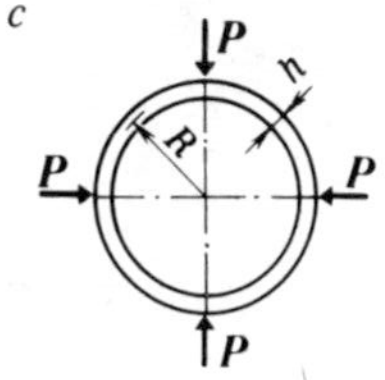 *c*
Determinable characteristics			E_θ^t, $G_{\theta r}$	E_θ^t, $G_{\theta r}$	E_θ^t, $G_{\theta r}$
Measurable values			P, Δw	P, Δw	P, Δw
Geometrical sizes			R/h, b, h	R/h, b, h	R/h, b, h
Limitations	Structural	Layup	0°; 90°; 0/90°		
		Orientation	0°, 90°		
	Physical		For $G_{\theta r}$: linear range of the curve $P \sim \Delta w$		
	Geometrical		For $G_{\theta r}$: $\frac{R}{h} \leqslant 0.72\sqrt{\frac{1-K_\tau}{K_\tau}\frac{E_\theta}{G_{\theta r}}}$	—	—

Fig. 5.5.1. Loading schemes of intact ring specimens with a concentrated force: (a) two-point; (b) three-point; (c) four-point.

		a	b	c	d
Loading scheme		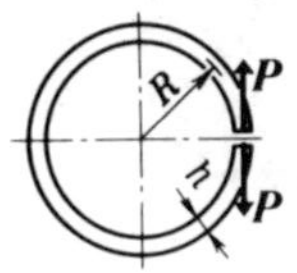	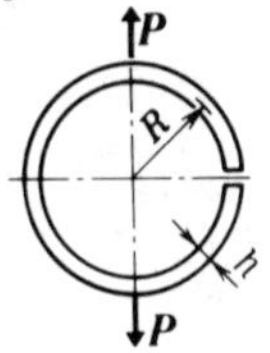	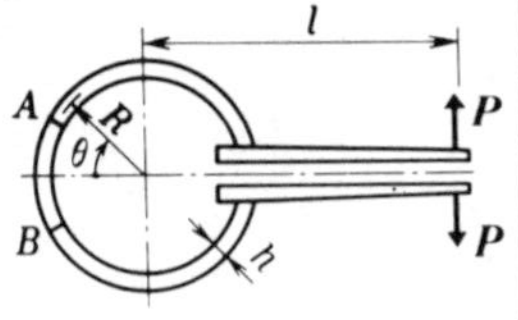	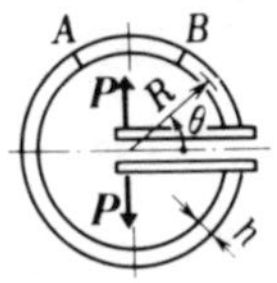
Determinable characteristics		E_θ^t	E_θ^t	σ_r^{tu}	$\tau_{\theta r}^u$
Measurable values		P, u_p	P, W_p	P	P
Geometrical sized		$R/h, b, h$	$R/h, b, h$	$R/h, b, h$	b/h
Limitations: Structural	Layup	0°; 90°; 0/90°			
Limitations: Structural	Orientation	0°, 90°			
Limitations: Physical		For E_θ^t, linear range of the curve $P \sim u_p$ (or w_p)			—
Limitations: Geometrical		$\frac{R}{h} > \frac{5}{3\sqrt{3}}\beta$	$\frac{R}{h} > \frac{5}{3}\beta$	$h/R > 4\,\sigma_r^{tu}/\sigma_\theta^u$ $l/R > \sigma_r^{tu}/\tau_{\theta r}^u$	—

Fig. 5.5.2. Loading of a split ring in its plane. The loading scheme is explained in the text.

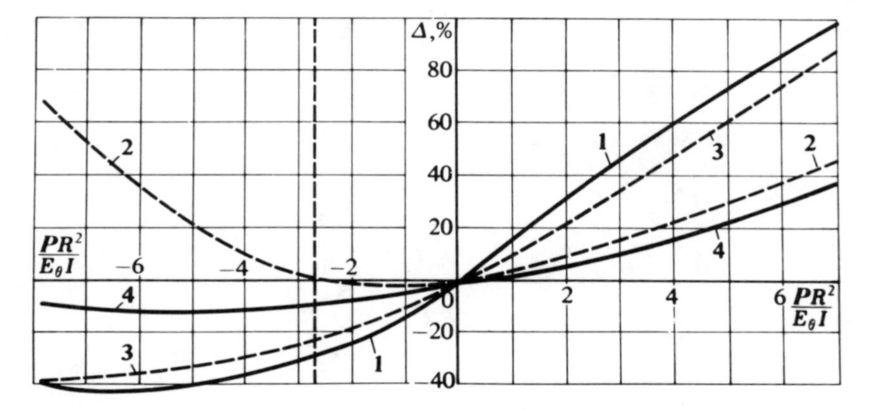

Fig. 5.5.3. The error Δ introduced by not taking account of the geometrical nonlinearity under tension (right side of the diagram) and compression (left side of the diagram) of rings [186]: (1) error in determination of E_θ with measurement of displacement w_v at point of application of load; (2) the same, with measurement of displacement w_h of points on the horizontal diameter; (3) the error in determination of bending stresses in sections along horizontal diameter; (4) the same, in sections under the load.

information it is impossible, as a rule, to restrict oneself to the region of small displacements. The measured radial displacements of the ring become commensurable with the diameter; the load-displacement curve becomes significantly nonlinear. In this case, the analytical relationships are complicated. However, the error, introduced by use of the elementary formulas, can be estimated by means of a graph in which the dimensionless parameter $PR^2/(E_\theta I)$ is plotted on the abscissa and the error in the determination of stiffness and strength (Fig. 5.5.3) on the ordinate. For rings made of composites, frequently $PR^2/(E_\theta I) > 3$. This means that the error in determination of the displacements on the vertical diameter exceeds 50%. At the same time, in the region of deflections up to $0.5R$, the nonlinearity shows up slightly in displacements on the horizontal diameter.

With increase in relative thickness, the low shear resistance plays a larger and larger role. It becomes necessary to take shear stress into account in processing the experimental results for comparatively thin rings (for unidirectional glass fiber composites at $h/R > 1/20$). Such an approach enables not only E_θ and $G_{\theta r}$ to be determined, but also the interlaminar shear strength $\tau^y_{\theta r}$. The expressions which connect the vertical diameter change w_v and horizontal diameter change w_h with the load P have the following form [231, p. 165; 236, p. 320]:

$$w_v = w_v^* \left(1 + 0.528\beta^2 \frac{h^2}{R^2}\right)$$
$$w_h = w_h^* \left(1 + 0.366\beta^2 \frac{h^2}{R^2}\right) \tag{5.5.1}$$

$$w_v^* = 0.149 \frac{PR^3}{E_\theta I}$$
$$w_h^* = 0.137 \frac{PR^3}{E_\theta I} \tag{5.5.2}$$

where the asterisk designates displacements calculated without taking shear stress into account; $\beta^2 = E_\theta/G_{\theta r}$ is the anisotropy of the material of the specimen; h is the ring thickness; and R is the mean radius of the ring.

The connection between the moduli of elasticity determined by strength of materials formulas without taking shear into account* and the quantities $1/E_\theta$ and $(h/R)^2$ is linear:

$$E_{fv} = 0.149 \frac{PR^3}{Iw_v} \tag{5.5.3}$$

$$E_{fh} = 0.137 \frac{PR^3}{Iw_h} \tag{5.5.4}$$

$$\frac{1}{E_{fv}} = \frac{1}{E_\theta} + \frac{0.528}{G_{\theta r}} \left(\frac{h}{R}\right)^2 \tag{5.5.5}$$

$$\frac{1}{E_{fh}} = \frac{1}{E_\theta} + \frac{0.366}{G_{\theta r}} \left(\frac{h}{R}\right)^2. \tag{5.5.6}$$

In the $(h/R)^2$, $1/E_f$ coordinate system, the straight lines (5.5.5) and (5.5.6) intersect the ordinate at the point $1/E_\theta$ and have slopes $0.528/G_{\theta r}$ and $0.366/G_{\theta r}$ (Fig. 5.5.4).

The straight line $1/E_f = f(h/R)^2$ is plotted experimentally from the results of testing several series of rings with various relative thickness h/R; in this

*The moduli determined in this manner are fictitious (subscript f).

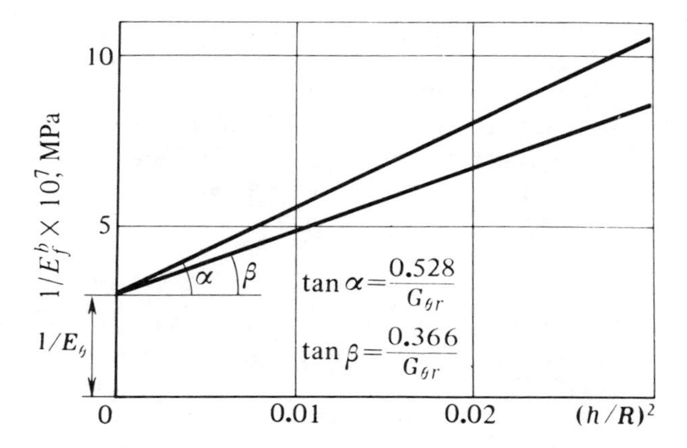

Fig. 5.5.4. Determination of moduli E_θ and $G_{\theta r}$ in tests of rings of differing relative thickness [231].

manner, the values of moduli E_θ and $G_{\theta r}$ are experimentally determined.* The method of least squares (see Section 5.2.2) and also be employed for processing the experimental results.

Bending tests of rings using concentrated forces has experimentally been well substantiated, and in the determination of elastic constants it gives results comparable with those obtained by other methods. The following data are presented as examples:

	$E_\theta \times 10^{-5}$, MPa		$G_{\theta r} \times 10^{-4}$, MPa	
MATERIAL	BENDING WITH CONCENTRATED FORCES	TENSION OF RINGS BY A RUBBER DISK	BENDING WITH CONCENTRATED FORCES	TORSION OF SPLIT RINGS
High-modulus glass fiber composite	0.692	0.689	0.92	1.05
Boron composite	2.120	2.010	0.71	0.75
Carbon composite	1.180	0.880	0.27	0.265

Reliability of the described method of determining the interlaminar shear modulus $G_{\theta r}$ depends on proper selection of the relative thickness of the specimen h/R. Displacements w_v and w_h are usually small and displacements due to transverse shears w_τ is a part of displacements w_v or w_h, depending

*Determination of the modulus of elasticity E_θ from independent tests is undesirable owing to introduction of additional rather notable errors.

on the anisotropy of the material β^2 and the relative thickness of the specimen h/R. To ensure sufficient sensitivity of the method, the displacement w_τ must equal no less than 25–30% of w_v or w_h. On the basis of this condition and approximately estimated anisotropy β^2 of the material, the relative specimen thickness h/R is determined from the formula:

$$0.200 \geqslant \frac{h}{R} \geqslant 1.376\sqrt{\frac{k_\tau}{1-k_\tau}\frac{G_{\theta r}}{E_\theta}} \qquad (5.5.7)$$

where $k_\tau = w_\tau/w$ is the fraction of displacement due to tangential stresses (in experiment of determination the elastic constants, the change in a vertical diameter w_v is usually employed).

The upper limit ($h/R \leqslant 0.200$) in (5.5.7) means that (5.5.1) has been obtained for rings of small curvature.

In tests of thin-walled rings, when shear strains are negligibly small the circumferential modulus of elasticity E_θ can be determined by one more method [4]. By means of resistance strain gages, glued to ring sections along the horizontal diameter, the mean circumferential strain $\varepsilon_\theta = (\varepsilon_{\theta 1} + \varepsilon_{\theta 2})/2$ is determined and the circumferential modulus of elasticity calculated

$$E_\theta = \frac{3PRb}{h^2\varepsilon_\theta}\left(1 - \frac{2}{\pi}\right). \qquad (5.5.8)$$

It has been suggested (see, for example, [191]) that the elastic constants E_θ and $G_{\theta r}$ be determined from an experimentally determined difference in displacements $\Delta w = w_v - w_h$. With the help of Equations (5.5.1) and (5.5.2) it is easy to convince oneself of the fact that in this case the reliability of the method, i.e., the fraction of displacements due to tangential stresses $k_\tau = w_\tau/w$, decreases, compared to measurements only according to horizontal or vertical diameters.

The two-point loading method for rings of sufficient relative thickness, when the material fails due to shear, can be used also for estimation of the shear strength. For fiberglass composites, this range of relative thicknesses is $h/R = 0.08$–0.18. The strength is calculated by the formula

$$\tau_{\theta r}^u = \frac{3}{4}\frac{P^u}{h} \qquad (5.5.9)$$

where P^u is the load at which circular cracks appear.

Mathematical expressions and results of experimental verification of three- and four-point loading schemes [see Fig. 5.5.1(b and c)] are presented in

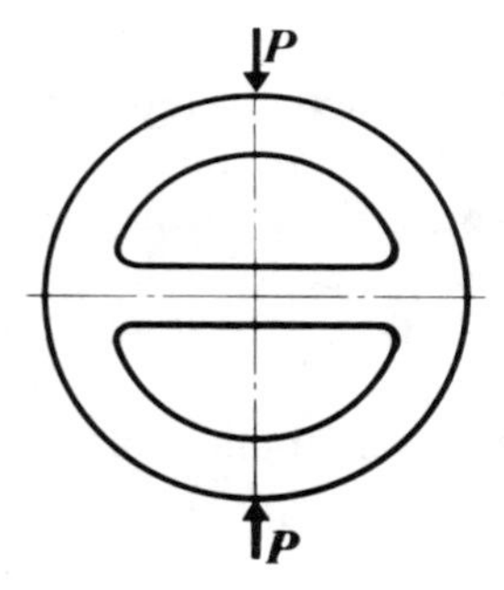

Fig. 5.5.5. Theta-shaped specimen.

[142, 143]. With these schemes, displacement of the loading point is determined from the following formulas:

For a three-point scheme,

$$w = w^* \left(1 + 1.6\beta^2 \frac{h^2}{R^2}\right) \tag{5.5.10}$$

where $w^* = 0.0319PR^3/2E_\theta I$.

For a four-point scheme,

$$w = w^* \left(1 + 4.75\beta^2 \frac{h^2}{R^2}\right) \tag{5.5.11}$$

where $w^* = 0.006PR^3/2E_\theta I$.

Three- and four-point loadings are more complicated than two-point loading: a uniform load distribution and free deformation of the specimen should be ensured. Therefore, use of these methods is only justified for special purposes, for example, investigation of the elastic characteristics and strength along the circumference of the ring.

In [54] a theta-shaped specimen has been proposed (the specimen shape and loading scheme are shown in Fig. 5.5.5), which is a combination of a ring and a prismatic specimen. The circular part of the specimen serves as the load transmitting element, the prismatic part is the gage section. Theta-shaped specimens are made by winding or they are cut from a plate. They are intended for study of the elastic and strength characteristics of composites and for investigation of stress concentrations (at the sites of openings, grooves, etc.). The main advantage of the theta-shaped specimen is a uniform state of stress in the gage section and absence of difficulties connected with fastening of prismatic specimens in the testing machine grips. Disadvantages are difficulties in fabrication, necessity of preliminary selection of the specimen dimensions and their calibration. Theta-shaped specimens are successfully used in the study of isotropic materials by photoelastic methods [67]. The similar experience of application to composites is unknown. Obviously, this method can be employed only for transversely isotropic (in the plane of cutting) materials; in other cases, we have to deal with anisotropy of a general character.

5.5.3. Split Rings

In the determination of the elastic constants two methods of loading split rings with concentrated forces are possible. They were shown previously in Fig. 5.5.2(a and b). Both schemes have been developed for the purpose of obtaining a simple method of determining the circumferential modulus of elasticity of would materials.

In loading a split ring by the method, presented in Fig. 5.5.2(a) when the load is applied at the site of the cut, the mutual displacement of the loading points is determined by the following equation [150]:

$$u_P = \frac{3\pi PR^3}{2E_\theta I}\left(1 + \frac{1}{36}\beta^2\frac{h^2}{R^2}\right). \tag{5.5.12}$$

In this case, the effect of shear stress is considerably less than in tests of intact rings. Thus, with $h/R = 0.1$ and $\beta^2 = 25$, the correction due to shear is only 0.7%; in the case of intact rings with identical parameters, the correction is 13.2%. Therefore, the effect of shear in tests by this scheme is disregarded, and the measured displacement u_P is used directly for determination of the modulus of elasticity:

$$E_\theta = \frac{3\pi PR^3}{2Iu_P}. \tag{5.5.13}$$

Loading by the method shown in Fig. 5.5.2(b), when the forces are applied along the vertical diameter, is simpler to accomplish. Mutual displacement of the points of application of the load is determined by the following equation [150]:

$$w_P = \frac{\pi P R^3}{4E_\theta I}\left(1 + \frac{1}{12}\beta^2\frac{h^2}{R^2}\right). \qquad (5.5.14)$$

With $h/R = 0.1$ and $\beta^2 = 25$, the correction due to shear is about 2%. With correct selection of h^2/R^2 ratio at given β^2 the term $(1/12)\beta^2(h^2/R^2)$ can be disregarded compared to unity. Equation (5.5.14) can be simplified and the modulus of elasticity determined directly from the following relation between the displacement and the load:

$$E_\theta = \frac{\pi P R^3}{4I w_P}. \qquad (5.5.15)$$

It must be noted that Equations (5.5.12) and (5.5.14) were obtained on the basis of the assumption that $G_{\theta r}/E_\theta \ll 1$. As a consequence of this, they give somewhat exaggerated values of the moduli of elasticity.

A method of determination of the transverse (radial) tensile strength σ_r^t was proposed in [154]. The scheme of split ring loading used in that case is shown in Fig. 5.5.2(c). In bending such a specimen, failure can take place in the form of delamination as a result of combined action of the normal radial stresses σ_r^t and tangential stresses $\tau_{\theta r}$. Therefore, in the study of the radial tensile strength, pure bending is a faultless method so long as there are no tangential stresses. However, it is difficult to perform this type of loading on ring specimens. The loading method shown in Fig. 5.5.2(c) permits the maximum tangential stresses $\tau_{\theta r}$ to be reduced to a desirable minimum, compared with radial stresses σ_r^t. According to this scheme, a segment of a ring is loaded with bending moment $M = Pl$, produced by means of two cantilevers to the ends of which the load P is applied, and the concentrated force P. The maximum radial stresses are determined by the following approximate equation [154] (the exact expression is unwieldy and inconvenient for processing the experimental data):

$$\sigma_{r\,\max}^t = \frac{3}{2}\frac{Pl}{bhR}\left(1 + \frac{R}{l}\cos\theta\right). \qquad (5.5.16)$$

The maximum tangential stresses are found by

$$\tau_{\theta r\,\max} = -\frac{\sin\theta}{\dfrac{l}{R} + \cos\theta}\sigma_{r\,\max}. \qquad (5.5.17)$$

It is advisable to select a specimen gage length AB [see Fig. 5.5.2(c)] which is narrower than the width of the specimen at the points of attachment of the cantilevers. The shorter the gage length, the smaller the value $\tau_{\theta r}$ at the boundaries of segment AB (for $\theta = 0$, stresses are $\tau_{\theta r} = 0$). If the gage length is fixed, reduction of the maximum tangential stresses should be achieved by selection of the l/R ratio. In an experimental test of the method in [154], glass fiber composite specimens and a device with the following dimensions were used: specimen inside diameter 150 mm, thickness 20 mm, gage length determined by angle $2\theta = 90°$, width of the gage length $b = 8$ mm, arm $L =$ 600 mm. With these dimensions, the maximum tangential stresses were about $0.1\sigma^{t}_{r\,\max}$. Failure of all specimens took place as a result of formation of circular cracks.

Failure of the specimen can also be caused by tangential normal stresses σ_θ. In order for failure to occur by interlaminar tension, the dimensions of the specimen must satisfy the inequality

$$\frac{h}{R} > \frac{4\sigma_r^{tu}}{\sigma_\theta^u} \tag{5.5.18}$$

where σ_r^{tu} is the interlaminar tension strength and σ_θ^u is the strength under normal circumferential stress.

Failure from interlaminar tension takes place at $h/R \geqslant 0.25$. The relation $(l/R)(b/b_s) \geqslant 10$ must be fulfilled to eliminate shear; hence $\tau_{\theta r} \ll \tau_{\theta r}^u$, but $\sigma_r^t \to \sigma_r^{tu}$. Decrease in $\tau_{\theta r}$ is obtained, first of all, at the expense of selection the l/R ratio.

The method has one technical deficiency—large displacements of the points of application of the load P; as a consequence of this, large cross-head speed rates for testing machine have to be used.

A method for determining the interlaminar shear strength from bending tests of split rings was proposed and evaluated in [153]. By this method, the split rings are loaded with concentrated forces, which are applied to rigid cantilevers in such a manner that their line of action passed through the center of the ring [Fig. 5.5.2(d)]. The tangential stresses of the ring are determined by the formula

$$\tau_{\theta r} = \frac{3}{2}\frac{Q}{bh},$$

the normal radial stresses by

$$\sigma_r = \frac{3}{2}\frac{M}{bhR}$$

where $Q = P\sin\theta$; $M = PR\cos\theta$.

In the part of the ring where $\theta = \pi/2$, coinciding with the lines of action of the load, tangential stresses have the highest value and radial normal stresses are zero. In this case a gage length AB [see, Fig. 5.5.2(d)] is selected which is narrower than the width of the ring at the point of attachment of the cantilevers. At the ends of the gage length, the radial stresses are $\sigma_r = (s/2R)\tau_{\theta r}$, where s is the gage length along the midline of the specimen. In the part of the ring where $\theta = \pi$, in which the bending moment reaches a maximum, a steel clamp is placed to prevent the failure of the specimen in this region. For verification of the method, glass fiber composite specimens of the following dimensions have been used: inside diameter of the ring 150 mm, thickness 20 mm, width 15 mm, gage length 50 mm, width 8 mm [153].

5.5.4. Segments

The loading methods for ring segments are shown in Fig. 5.5.6; the experimentally determinable values, force factors, and the variable specimen dimensions are also indicated in the figure. The loading technique shown in Fig. 5.5.6(c) reflects the first efforts at testing wound materials. In tension, a bending moment is also induced in the segment. Therefore, special devices have been developed to eliminate bending. This loading method has not been widely used, therefore it is not treated in this book.

The interlaminar shear strength $\tau_{\theta r}^u$, the strength due to circumferential stresses σ_θ^u, interlaminar tension strength σ_r^{tu}, and modulus of elasticity E_θ can be determined by tests of ring segments. However, a number of features of the testing of composite ring segments impose stringent requirements on the capabilities of this method.

The method of determination of the interlaminar shear strength $\tau_{\theta r}^u$ on ring segments has been standardized. According to ASTM Standard D 2344-76, segments are cut from NOL rings with inside diameter 146.05 ± 0.05 mm, thickness $h = 3.18 \pm 0.05$ mm, and width $b = 6.35 + 0.13$ mm by means of a special device. The shape and dimensions of the specimen are shown in Fig. 5.5.7. ASTM D 2344-76 specifies the following relative dimensions of the specimen (l is the distance between supports; L is the entire length of the specimen; h is thickness):

Reinforcement material	l/h	L/h
Boron, graphite, silica	4	6
Carbon, glass (filament or cloth), steel	5	7

Selection of l/h and L/h according to ASTM is based on the following

		a 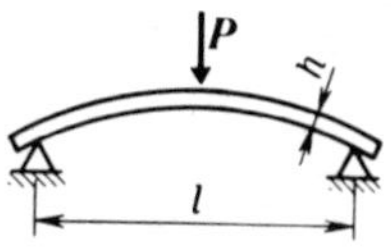	*b* 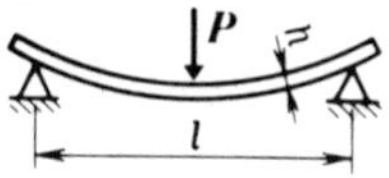	*c* 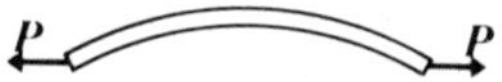
Loading scheme				
Determinable characteristics		$\tau_{\theta r}^{u}$	$\tau_{\theta r}^{u}$	—
Measurable values		P	P	—
Geometrical sized		l/h	l/h	—
Limitations: Structural	Layup	0° ; 90° ; 0/90°		—
	Orientation	0°, 90°		—
Limitations: Physical				—
Limitations: Geometrical		$\frac{l}{2R} < \frac{\sigma_r^{tu}}{\tau_{\theta r}^{u}}$; $\frac{h}{2l} > \tau_{\theta r}^{u} / \sigma_\theta^u$	$\frac{l}{2R} < \sigma_r^{cu}/\tau_{\theta r}^{u}$; $\frac{h}{2l} > \tau_{\theta r}^{u} / \sigma_\theta^u$	—

Fig. 5.5.6. Loading schemes of ring segments: (a) bending with concave side down; (b) bending with concave side up; (c) tension.

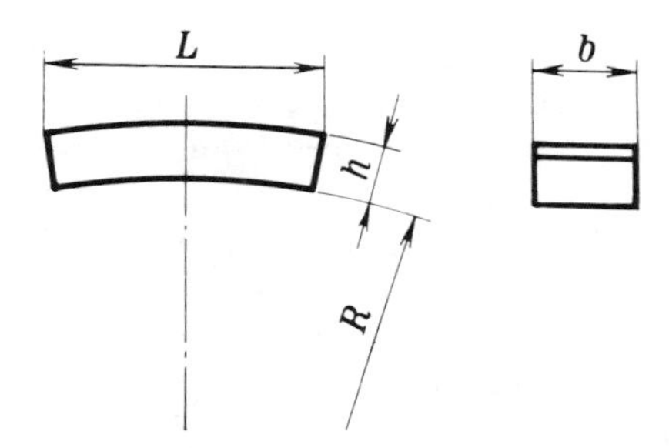

Fig. 5.5.7. A specimen-segment of a NOL ring for determination of the interlaminar shear strength according to ASTM D 2344–67 [11].

conditions: if the modulus of elasticity of reinforcement fibers $E_f \leqslant 100$ GPa, the ratio $l/h = 5$; if $E_f < 100$ GPa, then $l/h = 4$; $L = l + 2h$. The small sizes of specimens require high accuracy in machining. Therefore, ASTM D 2344-76 allows specimens of larger thickness h, but maintenance of the prescribed l/h and L/h ratios is obligatory.

In determination of the interlaminar shear strength $\tau_{\theta r}^{u}$, ring segments are subjected to bending tests by the three-point scheme. Here, the specimen can be installed in the supports with its concave side down, as the ASTM standard requires [Fig. 5.5.6(a)] or with its concave side up [Fig. 5.5.6(b)]. According to ASTM D 2344-76, the specimen is loaded with a constant rate of cross-head movement of 1.3 mm/min until it fails; the strength is determined by the formula for prismatic bars:

$$\tau_{\theta r}^{u} = \frac{3}{2}\frac{Q^u}{F} = \frac{3}{4}\frac{P^u}{bh} \tag{5.5.19}$$

ASTM D 2344-76 does not take into account the main difference between ring segments and prismatic bars: in ring segments, besides normal stresses σ_θ and interlaminar tangential stresses $\tau_{\theta r}$, the normal interlaminar stresses σ_r also act over the entire length; these are approximately determined by the following equation:

$$\sigma_{r\,\max} = \frac{3}{2}\frac{M}{bhR}. \tag{5.5.20}$$

The direction of stresses σ_r depends on the loading method: in loading of segments with the concave side down, the stresses σ_r will be tensile (σ_r^t); in loading of segments with the concave side up they will be compressive

(σ_r^c). Depending on the specimen geometry and interlaminar shear ($\tau_{\theta r}^u$) and transverse tension (σ_r^{tu}) strengths of the material, the normal stresses $\sigma_r^{t(c)}$ can heavily affect the strength and type of failure of the segments.

Due to the interlaminar tangential stresses $\tau_{\theta r}$ and radial normal stresses σ_r, the segments placed with the concave side down fail by means of delamination, in which the circumferential crack begins not far from the loading point. The identical nature of failure due to stresses $\tau_{\theta r}$ and σ_r hampers evaluation of the cause of failure. The segments, placed with their concave side up, fail due to normal circumferential stresses σ_θ and local stresses directly under the concentrated load. In prismatic bars, normal stresses σ_r act only at the points of application of the load and support reactions and they are compressive; therefore, in this case, their effect can be disregarded.

In the determination of strength $\tau_{\theta r}^u$, an l/R ratio should be selected such that at specimen failure the normal circumferential stresses σ_θ and normal radial stresses σ_r are negligibly small compared with tangential stresses $\sigma_{\theta r}$. If the permissible limit of stresses σ_r is considered to be 1/10 of their maximum value (i.e., of strength σ_r^u), the condition of selection of the l/R ratio is written as follows:

$$\sin\theta = \frac{l}{2R} \leqslant 0.1 \frac{\sigma_r^{t(c)u}}{\tau_{\theta r}^u} \tag{5.5.21}$$

where (t) indicates segments with their concave side down, and (c) indicates segments with their concave side up.

In order for segments not to fail due to normal circumferential stresses σ_θ, the following condition must be satisfied:

$$\frac{h}{2l} > \frac{\tau_{\theta r}^u}{\sigma_\theta^u}. \tag{5.5.22}$$

The approximate values of $\sigma_r^{tu}/\tau_{\theta r}^u$, $\sigma_r^{cu}/\tau_{\theta r}^u$, $\tau_{\theta r}^u/\sigma_\theta^u$, $l/2R$ and $h/2l$ for glass fiber, boron, and carbon composites are presented in Table 5.5.1.

In the determination of the strength of ring segments, it should be taken into account that the shearing force Q depends on the specimen supports. In loading with the supports shown as solid lines in Fig. 5.5.8 (plates, knife edges) the shearing force is determined by the formula

$$Q = \frac{P}{2}\cos(\theta - \alpha). \tag{5.5.23}$$

It reaches a maximum at $\alpha = \theta$, i.e., under the load: $Q_{max} = P/2$. In this

Table 5.5.1. Approximate Values of $\sigma_r^{tu}/\tau_{\theta r}^u$, $\sigma_r^{cu}/\tau_{\theta r}^u$, $\tau_{\theta r}^u/\sigma_\theta^u$, $l/2R$, and $h/2l$ for glass fiber, boron, and carbon composites.

				$l/2R$ FOR SEGMENTS WITH		
MATERIAL	$\frac{\sigma_r^{tu}}{\tau_{\theta r}^u}$	$\frac{\sigma_r^{cu}}{\tau_{\theta r}^u}$	$\frac{\tau_{\theta r}^u}{\sigma_\theta^u}$	CONCAVE SIDE DOWN	CONCAVE SIDE UP	$\frac{h}{2l}$
Glass fiber composite	0.3–2.0	3–7	1/20–1/40	⩽(0.03–0.20)	⩽(0.3–0.7)	⩾(1/4–1/10)
Boron composite	0.5–0.8	2–5	1/20–1/50	⩽(0.05–0.08)	⩽(0.2–0.5)	⩾(1/4–1/10)
Carbon composite	0.2–0.8	1.5–7	1/15–1/40	⩽(0.02–0.08)	⩽(0.15–0.7)	⩾(1/3–1/8)

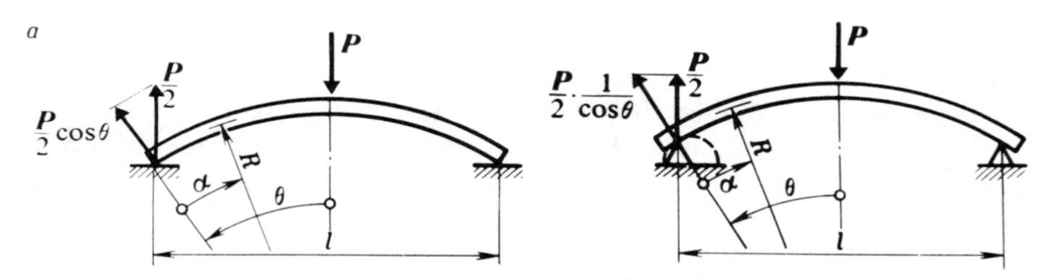

Fig. 5.5.8. Schemes for supporting of ring segments: (a) on a plate; (b) on a cylindrical support.

case, the strength $\tau_{\theta r}^u$ is determined by the same formula as the strength of a prismatic rod:

$$\tau_{\theta r}^u = \frac{3}{4}\frac{P^u}{F}. \tag{5.5.24}$$

In loading with the support method shown in broken lines in Fig. 5.5.8(b) (cylindrical supports) the shearing force is determined by the formula

$$Q = \frac{P}{2}\frac{1}{\cos(\theta - \alpha)}. \tag{5.5.25}$$

It reaches a maximum at $\alpha = 0$, i.e., above the support: $Q_{\max} = P/2\cos\theta$. Consequently, in this case, the strength $\tau_{\theta r}^u$ is determined from the relation

$$\tau_{\theta r}^u = \frac{3}{4}\frac{P}{F}\frac{1}{\cos\theta}. \tag{5.5.26}$$

In determination of the shearing force Q the friction on the supports is not taken into account in Equations (5.5.23) and (5.5.24).

Beside the characteristics of stress in ring segments, the same deviations from the theoretical stress distribution take place as in prismatic bars—high stress concentration around the points of application of the load and of support reactions, and shift of the maximum stress (see Section 5.2). At the present time, there are no analytical studies of these phenomena in ring segments made of composites.

It is reasonable to determine the strength due to circumferential stresses σ_θ^u and interlaminar tension strength σ_r^{tu} from pure bending tests. However, difficulties arise in realizing of this loading method. The four-point technique used in this case is suitable only with small displacements, and this is difficult to achieve in the case of ring segments without producing axial loads in the specimen. Therefore, it is preferable to load the segments with moments applied to the end faces of the specimen. One of the devices used for this purpose has been described in [93–95].

The circumferential and radial stresses in pure bending of a curved bar with cylindrical anisotropy are determined by the following equations [124, p. 9]:

$$\sigma_\theta = -\frac{M}{R_0^2 hg}\left[1 - \frac{1-c^{k+1}}{1-c^{2k}}k\left(\frac{r}{R_0}\right)^{k-1} + \frac{1-c^{k-1}}{1-c^{2k}}kc^{k+1}\left(\frac{R_0}{r}\right)^{k+1}\right] \tag{5.5.27}$$

$$\sigma_r = -\frac{M}{R_0^2 hg}\left[1 - \frac{1-c^{k+1}}{1-c^{2k}}\left(\frac{r}{R_0}\right)^{k-1} - \frac{1-c^{k-1}}{1-c^{2k}}c^{k+1}\left(\frac{R_0}{r}\right)^{k+1}\right] \tag{5.5.28}$$

where M is the bending moment; r is the current radius; R_0 is outside radius; R_i is inside radius; and

$$g = \frac{1-c^2}{2} - \frac{k}{k+1}\frac{(1-c^{k+1})^2}{1-c^{2k}} + \frac{kc^2}{k-1}\frac{(1-c^{k-1})^2}{1-c^{2k}}$$

$$c = R_i/R_0, \qquad k = \sqrt{E_\theta/E_r}.$$

The circumferential stresses will be at a maximum on the outer or inner surfaces of the specimen, depending on the numerical value of coefficient k. With $r = R_i$ and $r = R_0$, the radial stresses $\sigma_r = 0$; their maximum value is observed around the neutral axis of the bar (the corresponding value of radius r depends on coefficients c and k). The structure of Equations (5.5.27) and (5.5.28) does not permit an analytical evaluation of the change from one type of failure to another, i.e., selection of the segment dimensions. This must be accomplished experimentally.

In determining the modulus of elasticity E_θ, the load P and the rotation angle φ of the end sections of the specimen are measured. The modulus of elasticity E_θ is calculated by the formula

$$E_\theta = \frac{Ml}{2I\varphi} \tag{5.5.29}$$

where M is the bending moment; l is the span; $I = bh^3/12$.

* * *

In summary Table III specimen shapes and loading conditions are presented. Bending tests of prismatic bars, rings and their segments are shown.

Summary Table III. Bending Tes

METHOD	STANDARDS	CHARACTERISTICS DETERMINED: ELASTIC CONSTANTS	CHARACTERISTICS DETERMINED: STRENGTH	EXPERIMENTALLY MEASURED V: IN ELASTIC CONSTANT DETERMINATION	EXPERIMENTALLY MEASURED V: IN ST DETER
BARS WITH A STRAIGHT AXIS					
Three-point bending	BS 2782, Part 10, Method 1005; ISO R 178	E_x^b, E_y^b, G_{xz}, G_{yz}	σ_x^{bu}, σ_y^{bu}, τ_{xz}^{bu}	P (load), w (deflection at the midspan)	P^u (load a failure)
Pure bending	—	$E_x^{t(c)}$, $E_y^{t(c)}$, ν_{xy}	σ_x^{bu}, σ_y^{bu}	M (bending moment at end faces of bars), $\varepsilon_x^{t(c)}$, $\varepsilon_y^{t(c)}$ (strains of outer layers)	M^u (bendi at specim
Four-point bending	ISO R 178	$E_x^{t(c)}$, $E_y^{t(c)}$	$\sigma_x^{tu(cu)}$	P (load), $\varepsilon_x^{t(c)}$, $\varepsilon_y^{t(c)}$ (strains of outer layers)	P^u (load a failure)
Five-point bending	—	E_x^b, E_y^b, G_{xz}, G_{yz}	—	P (load at midspan); P_1 (load at specimen ends); w (deflection at midspan)	—
BARS WITH A CIRCULAR AXIS					
Intact Rings Bending with a concentrated force (two-point method)	—	E_θ^t, $G_{\theta r}$	$\tau_{\theta r}^{tu}$	P (load), w_0, w_1 (changes in vertical and horizontal diameters of rings)	P^u (load a failure)
Split Rings Loading at the point of split in the plane of the ring	—	E_θ^t	—	P (load), u_p (displacement at the point of loading)	—
Loading over vertical diameter in the plane of the ring	—	E_θ^t	—	P (load), w_p (displacement along the vertical diameter)	
Ring Segments Three-point bending	ASTM D 2344-76	—	$\tau_{\theta r}^u$	—	P^u (load a failure)

Fibrous Polymeric Composites.

SPECIMEN SHAPE				
DETERMINATION ASTIC CONSTANTS	FOR DETERMINATION OF STRENGTH	TESTING EQUIPMENT	DEFICIENCIES OF THE METHOD	SECTION OF THIS BOOK
ngular	Parallelepiped	Tensile testing machine with fixture for bending tests	For determination of elastic constants, several series of specimens with different h/l ratios are necessary; failure mechanism of the specimen depends on h/l ratio; only qualitative comparison of interlaminar shear strength of different materials is possible	5.2
ame		Special device	Complexity of producing pure bending	5.3.1
wich beam, angular parallel-ed	Sandwich beam	Tensile testing machine with fixture for bending tests or special device	With incorrect selection of a/l ratio, failure of a specimen from interlaminar shear or bearing-shearing is possible; allowance for large deflections is necessary; in the case of testing sandwich beams, complicated specimen design; complicated loading scheme	5.3.2
angular parallel-ped		Special device		5.3.3
n-walled ring or g of medium ative thickness	Ring of medium relative thickness	Tensile testing machine	In determination of elastic constants, high measurement accuracy is required (maximum displacements limited)	5.5
t ring	—	The same	—	5.5.3
same	—	The same	—	5.5.3
	Ring segments	Tensile testing machine with fixture for bending tests	Method is applicable only for qualitative comparison of materials	5.5.4

References

1. Abramchuk, S.S., and V.P. Buldakov. Allowable Values of the Poisson's Coefficients of Anisotropic Materials. *Mechanics of Composite Materials*, 1979, Vol. 15, No. 2, pp. 174–178.
2. Adams, D.F., and R.L. Thomas. Test Methods for the Determination of Unidirectional Composite Shear Properties. *Advances in Structural Composites*. SAMPE 12th National Symposium and Exhibition. Anaheim, California, 1967, AC-5.
3. Adsit, N.R., H. McCutchen, and J.D. Forest. Shear Testing of Advanced Composites. *Composite Materials in Engineering Design*. Proc. Sixth St. Louis Symposium. Metals Park, Ohio, 1973, pp. 448–460.
4. Akasaka, T., Y. Sakai, and K. Kabe. Torsional Buckling of FWP Circular Cylindrical Shell under Inner Pressure. *Composite Materials and Structures* (Japan), 1974, Vol. 3, No. 3, pp. 29–35.
5. Algra, E.A.H., and M.H.B. van der Beek. Standards and Test Methods for Filament-Wound Reinforced Plastics. *Conference on Filament Winding*, Paper 12, London, 1967.
6. Allen, H.G. *Analysis and Design of Structural Sandwich Panels*. Pergamon Press, Oxford, 1969.
7. Ambartsumyan, S.A. *Theory of Anisotropic Plates*. Nauka Press, Moscow, 1967. (in Russian)
8. Amijima, S., and T. Adachi. Linear and Nonlinear Stress-Strain Response of Laminated Composite Materials. Part I. Linear Stress-Strain Response. *Sci. and Eng. Rev. Doshisha Univ.*, 1975, Vol. 16, No. 1, pp. 31–51.
9. Amijima, S., and T. Adachi. Linear and Nonlinear Stress-Strain Response of Laminated Composite Materials. Part II. Nonlinear Stress-Strain Response. *Sci. and Eng. Rev. Doshisha Univ.*, 1975, Vol. 16, Nos. 3–4, pp. 147–167.
10. *Analysis of the Test Methods for High Modulus Fibers and Composites*. ASTM STP No. 521. Philadelphia, Pa., 1973.
11. *1978 Annual Book of ASTM Standards*. Parts 34 (766 p.), 35 (1058 p.), and 36 (1015 p.). Philadelphia, Pa., 1978.
12. Arcan, M., Z. Hashin, and A. Voloshin. A Method to Produce Uniform Plane-Stress States with Applications to Fiber-Reinforced Materials. *Experimental Mechanics*, 1978, Vol. 18, No. 4, pp. 141–146.
13. Ashkenazi, Ye. K., and E. Ya. Abashin. Author's certificate 368, 575, January 19, 1970.—Otkritiya. Izobreteniya. Promyšlenniye obraztsi. Tovarniye znaki, 1973, No. 9 (in Russian).
14. Ashkenazi, Ye.K., and E.V. Ganov. *Anisotropy of Structural Materials: A Handbook*. Mashinostroyeniye Press, Moscow-Leningrad, 1972 (in Russian).

15. Ashkenazi, Ye.K., and F.P. Pekker. Evaluation of the Effect of Specimen Width On Delamination Strength of Glass Fiber Composites. *Zavodskaya Laboratoriya*, 1970, Vol. 36, No. 7, pp. 860–864 (in Russian).
16. Ashton J.E., J.C. Halpin, and P.H. Petit. *Primer on Composite Materials: Analysis*. Technomic Publishing Co., Stamford, Conn., 1969.
17. Auzukalns, Ya.V., A.N. Birze, and F.Ya. Bulavs. Strength Properties of Reinforced Plastics in Compression at an Angle to the Reinforcement Direction. *Methods of Nondestructive Testing for Building Materials*, No. 2, Riga, Polytechnic, Riga, 1976, pp. 86–95 (in Russian).
18. Azzi, V.D., and S.W. Tsai. Anisotropic Strength of Composites. *Experimental Mechanics*, 1965, Vol. 5, No. 9, pp. 283–288.
19. Barnes, P.J. Anticlastic Curvature in Anisotropic Beams. *Aeronautical Journal*, 1974, Vol. 78, No. 767, pp. 525–528.
20. Beckett, P.E., R.J. Dohrmann, and K.D. Ives. An Experimental Method for Determining the Elastic Constants of Orthogonally Stiffened Plates. In: *Experimental Mechanics*, New York, 1963, pp. 129–148.
21. Bel'ankin, F.P. *Wood Strength at Splitting along the Fibers*. U.S.S.R. Academy of Science Press, Kiev, 1955 (in Russian).
22. Berg, C.A., J. Tirosh, and M. Israeli. Analysis of Short Beam of Fiber Reinforced Composites. *Composite Materials: Testing and Design (Second Conference)*. ASTM STP No. 497. Philadelphia, Pa., 1972, pp. 206–218.
23. Bergsträsser, M. Bestimmung der beiden elastischen Konstanten von plattenförmigen Körpern. *Z. Techn. Phys.*, 1927, Vol. 8, No. 9, pp. 355–359.
24. Bert, C.W. Static Testing Techniques for Filament-Wound Composite Materials. *Composites*, 1974, Vol. 5, No. 1, pp. 20–26.
25. Biderman, V.L. Elasticity and Strength of Anisotropic Glass Fiber Composites. *Raschoti na Prochnost*, No. 11, Mashinostroyeniye Press, Moscow, 1965, pp. 3–30 (in Russian).
26. Biderman, V.L., and V.A. Knyazeva. Anti-Planar Deformation of an Anisotropic Beam of Rectangular Cross Section in Bending. *Izvestiya Vuzov*. Mashinostroyeniye Press, Moscow, 1976, No. 6, pp. 5–11 (in Russian).
27. Blagonadyozhin, V.L., G.H. Murzakhanov, and V.P. Nikolaev. *Methods of Experimental Investigation of Composite Materials and Structures*. MEI, Moscow, 1976 (in Russian).
28. Blagonadyozhin, V.L., V.P. Nilolaev, and V.G. Perevozchikov. Investigation of Transverse Compliance of Wound Glass Fiber Items. *Works of MEI. Dynamics and Strength of Machines*, No. 101, MEI, Moscow, 1972, pp. 36–40 (in Russian).
29. Bolotin, V.V., I.I. Gol'denblat, and A.F. Smirnov. *Structural Mechanics: Contemporary State and Prospects of Development*. Stroyizdat Press, Moscow, 1972. (in Russian)
30. Bulmanis, V.N. Compression Tests on Composite Tubes. *Polymer Mechanics*, 1976, Vol. 10, No. 5, pp. 702–706. (in Russian)
31. Chamis, C.C. Failure Criteria for Filamentary Composites. *Composite Materials: Testing and Design*. ASTM STP No. 460. Philadelphia, Pa., 1969, pp. 336–351.
32. Chamis, C.C. Impetus of Composite Mechanics in Test Methods for Fiber Composites. *Fracture of Composite Materials*. Ed. by G.C. Sih and V.P. Tamuzs. Sijthoff & Nordhoff International Publishers, Alphen aan den Rijn, 1979, pp. 329–348.
33. Chamis, C.C., and J.H. Sinclair. Ten-Deg. Off-Axis Test for Shear Properties in Fiber Composites. *Experimental Mechanics*, 1977, Vol. 17, No. 9, pp. 339–346.
34. Chamis, C.C., and J.H. Sinclair. The Effects of Eccentricities in the Fracture of Off-Axis Fiber Composites. *Polym. Eng. and Sci.*, 1979, Vol. 19, No. 5, pp. 337–341.
35. Chandra, R. On Twisting of Orthotropic Plates in Large Deflection Regime. *AIAA*

Journal, 1976, Vol. 14, No. 8, pp. 1130–1131.
36. Chattarji, P.P. Torsion of Curved Beams of Rectangular Corss-Section Having Transverse Isotropy. *Z. angew. Math. u. Mech.*, 1958, Vol. 38, No. 3/4, pp. 157–159.
37. Chiao, C.C., and T.T. Chiao. Aramid Fibers and Composites. Preprint UCRL-80400. Livermore, California, 1977.
38. Chiao, C.C., R.L. Moore, and T.T. Chiao. Measurement of Shear Properties of Fibre Composites. Part I. Evaluation of Test Methods. *Composites*, 1977, Vol. 8, No. 3, pp. 161–169.
39. Chiao, T.T., and M.A. Hamstad. Testing of Fiber Composite Materials. *Proc. of the 1975 International Conference on Composite Materials*. Vol. 2, New York, 1976, pp. 884–915.
40. Christiansen, A.W., J. Lilley, and J.B. Shortall. A Three-Point Bend Test for Fiber-Reinforced Composites. *Fibre Science and Technology*, 1974, Vol. 7, No. 1, pp. 1–13.
41. *Composite Materials*. Ed. by L.J. Broutman and R.H. Crock. *Volume 2. Mechanics of Composite Materials*. Ed. by G.P. Sendeckyj. Academic Press, New York and London, 1974.
42. *Composite Materials*. Ed. by L.J. Broutman and R.H. Crock. *Volume 3. Engineering Applications of Composites*. Ed. by B.R. Noton. Academic Press, New York and London, 1974.
43. *Composite Materials*. Ed. by L.J. Broutman and R.H. Crock. *Volume 5. Fracture and Fatigue*. Ed. by L.J. Broutman. Academic Press, New York and London, 1974.
44. *Composite Materials*. Ed. by L.J. Broutman and R.H. Crock. *Volume 7. Structural Design and Analysis*. Part 1. Ed. by C.C. Chamis. Academic Press, New York and London, 1974.
45. *Composite Materials*. Ed. by L.J. Broutman and R.H. Crock. *Volume 8. Structural Design and Analysis. Part 2*. Ed. by C.C. Chamis. Academic Press, New York and London, 1974.
46. *Composite Materials: Testing and Design*. ASTM STP No. 460. Philadelphia, Pa., 1969.
47. *Composte Materials: Testing and Design (Second Conference)*. ASTM STP No. 497. Philadelphia, Pa., 1972.
48. *Composite Materials: Testing and Design (Third Conference)*. ASTM STP No. 546. Philadelphia, Pa., 1974.
49. *Composite Materials: Testing and Design (Fourth Conference)*. ASTM STP No. 617. Philadelphia, Pa., 1977.
50. *Composite Materials: Testing and Design (Fifth Conference)*. ASTM STP No. 674, Philadelphia, Pa., 1979.
51. *Composite Reliability*. ASTM STP No. 580. Philadelphia, Pa., 1975.
52. *Composite Materials Workshop*. Ed. by S.W. Tsai, J.C. Halpin, and N.J. Pagano. Technomic Publishing Co., Stamford, Conn., 1968.
53. Conway, H.D. The Large Deflection of Simply Supported Beams. *Phil. Mag.*, 1947, Ser. 7, Vol. 38, No. 287, pp. 905–911.
54. Conway, J.C., Jr. Theta-Shaped Test Specimen for Composite Materials. U.S. Patent 3, 842, 664, October 22, 1974.
55. Corten, H.T. Reinforced Plastics. Chapter 14 in: *Engineering Design for Plastics*. Ed. by E. Bear. Reinhold Publishing Co., New York; Chapman and Hall, Ltd., London, 1964, pp. 869–994.
56. Crossman, F.W. Analysis of Free Edge Induced Failure of Composite Laminates. *Fracture of Composite Materials*. Ed. by G.C. Sih and V.P. Tamuzs. Sijthoff & Noordhoff International Publishers, Alphen aan den Rijn, 1979, pp. 291–302.
57. Crow, S.C. A New Method for Determining Poisson's Ratio. *Material Research and Standards*, 1963, Vol. 3, No. 12, pp. 996–1002.

58. Daniel, J.M., J.L. Mullineaux, F.J. Ahimaz, and T. Liber. The Embedded Strain Gage Technique for Testing Boron-Epoxy Composites. In: *Composite Materials: Testing and Design (Second Conference)*. ASTM STP No. 497, Philadelphia, Pa., 1972, pp. 257–272.
59. Dastin, S., G. Lubin, J. Munyak, and A. Slobodzinski. Mechanical Properties and Test Techniques for Reinforced Plastic Laminates. *Composte Materials: Testing and Design*. ASTM STP No. 460. Philadelphia, Pa., 1969, pp. 13–26.
60. Dawihl, W., E. Dörre, W. Eicke, G. Elssner, F.J. Esper, H.W. Grünling, K.H. Grünthaler, J. Hartwig, W. Heinrich, E. Hillnhagen, G. Ibe, S. Janes, A. Kump, H. Leis, B. Liebmann, and J. Nixdorf. Vorschläge zur Klassifizierung von Verbundwerkstoffen und Grundlagen für die technologische Prüfung. In: *Verbundwerkstoffe*, Deutsche Gesellschaft für Metallkunde eV, Oberursel (Taunus bei Frankfurt/M.), 1974, S. 217–252.
61. Dedyukhin, V.G., and V.P. Stavrov. *Molding Technology and Strength of Glass Fiber Products*. Khimiya Press, Moscow, 1968. (in Russian)
62. Dickerson, E.O., and B. Di Martino. Off-Axis Strength and Testing of Filamentary Materials for Aircraft Applications. *Advanced Fibrous Reinforced Composites*. SAMPE 10th National Symposium and Exhibition, San Diego, California, 1966, H-23/H-50.
63. Dickerson, E.O., and L.M. Lackman. Effects of Prestressing in Filamentary Composite Materials. In: *Advances in Structural Composites*. SAMPE 12th National Symposium and Exhibition, Anaheim, California, 1967, D-2.
64. Dingle, L.E., R.G. Williams, and N.J. Parratt. The Disproportionate Weakening of Composites by Sub-Millimetre Defects. In: *Composites—Standards, Testing and Design*. JPC Science and Technology Press Ltd., Guildford, U.K., 1974, pp. 51–53.
65. Dreumel, W.H.M. van, and J.L.M. Kamp. Non-Hookean Behaviour in the Fibre Direction of Carbon-Fibre Composites and the Influence of Fibre Waviness on the Tensile Properties. *J. Composite Materials*, 1977, October, pp. 461–469.
66. Duggan, M.F. An Experimental Evaluation of the Slotted-Tension Shear Test for Composite Materials. 1979 SESA Spring Meeting, San Francisco, Cali., Paper No. R 79-117.
67. Durelli, A.J., S. Morse, and V.J. Parks. The Theta Specimen for Determining Tensile Strength of Brittle Materials. *Materials Research and Standards*, 1962, Vol. 2, No. 2, pp. 114–117.
68. Ewins, P.D. A Compressive Test Specimen for Unidirectional Carbon Fibre Reinforced Plastics. Aeronautical Research Council Current Paper No. 1132, HMSO, London, 1970.
69. Ewins, P.D. Techniques for Measuring the Mechanical Properties of Composite Materials. In: *Composites—Standards, Testing and Design*. JPC Science and Technology Press Ltd., Guildford, U.K., 1974, pp. 144–154.
70. Faupel, J.H. *Engineering Design*. John Wiley and Sons, New York, 1964.
71. Feodosyev, V.I. *Ten Lectures: Talks on Strength of Materials*. Nauka Press, Moscow, 1969. (in Russian)
72. Foye, R.L. Deflection Limits on the Plate-Twisting Test. *J. Composite Materials*, April 1967, pp. 194–198.
73. Fukui, S., et al. Some Theoretical and Experimental Studies on the Width Variation Effects for the Filament-Wound Cylinders. In: *Proceedings 6th International Symposium on Space Technology and Science*, Tokyo, 1965.
74. Garg, S.K., V. Svalbones, and G.A. Gurtman. *Analysis of Structural Composite Materials*. Marcel Dekker, Inc., New York, 1973.
75. Gol'denblat, I.I., and V.A. Kopnov. *Criteria of Strength and Plasticity of Structural Materials*. Mashinostroyeniye Press, Moscow, 1968. (in Russian)
76. Goldman, A.Ya., N.F. Savelyeva, and V.I. Smirnov, Investigation of Mechanical Properties of Fabric Glass-Reinforced Plastics under Tension and Compression Normally

to the Reinforcement Plane. *Polymer Mechanics*, 1968, No. 5, pp. 803–809 (in Russian).

77. Gordon, J. *The New Science of Strong Materials or Why You Don't Fall Through the Floor*, Penguin Book, Harmondsworth, 1968.
78. Greszczuk, L.B. Effect of Material Orthotropy in the Directions of Principal Stresses and Strains. *Orientation Effects in the Mechanical Behaviour of Anisotropic Structural Materials*. ASTM STP No. 405. Philadelphia, Pa., 1966, pp. 1–13.
79. Greszczuk, L.B. Shear-Modulus Determination of Isotropic and Composite Materials. *Composite Materials: Testing and Design*. ASTM STP No. 460. Philadelphia, Pa., 1969, pp. 140–149.
80. Greszczuk, L.B. Compressive Strength and Failure Modes of Unidirectional Composites. *Analysis of the Test Methods for High Modulus Fibers and Composites*. ASTM STP No. 521. Philadelphia, Pa., 1973, pp. 192–217.
81. Grimes, G.C., and J.M. Whitney. Degradation of Graphite/Epoxy Composite Materials Because of Load Induced Micromechanical Damage. *SAMPLE Quarterly*, 1974, Vol. 5, No. 4, pp. 1–13.
82. Hadcock, R.N., and J.B. Whiteside. Special Problems Associated with Boron-Epoxy Mechanical Test Specimens. *Composite Materials: Testing and Design*. ASTM STP No. 460. Philadelphia, Pa., 1969, pp. 27–36.
83. Hahn, H.T. A Note on Determination of the Shear Stress-Strain Response of Unidirectional Composites. *J. Composite Materials*, July 1973, pp. 383–386.
84. Hammant, B. Use of 4-Point Loading Tests to Determine Mechanical Properties. *Composites*, 1971, Vol. 2, No. 4, pp. 246–249.
85. Hancock, J.R., and G.D. Swanson. A Tension Test for Filamentary Composites. *J. Composite Materials*, July 1971, pp. 414–416.
86. Hancox, N.L. The Use of a Torsion Machine to Measure the Shear Strength and Modulus of Unidirectional Carbon Fibre Reinforced Plastic Composites. *J. Mater. Sci.*, 1972, Vol. 7, No. 7, pp. 1030–1036.
87. Hancox, N.L. The Compression Strength of Unidirectional CFRP. In: *Composites—Stanbards, Testing and Design*. Guildford, U.K., 1974, pp. 158–159.
88. Hancox, N.L. The Compression Strength of Unidirectional Carbon Fibre Reinforced Plastic. *J. Mater. Sci.*, 1975, Vol. 10, No. 2, pp. 234–242.
89. *Handbook of Fiberglass and Advanced Plastics Composites*. Ed. by G. Lubin. Van Nostrand Reinhold Company, New York, 1969.
90. Harris, A., and O. Orringer. Investigation of Angle-Ply Delamination Specimen for Interlaminar Strength Test. *J. Composite Materials*, July 1978, pp. 285–289.
91. Hearmon, R.F.S. *An Introduction to Applied Anisotropic Elasticity*. Oxford Univ. Press, London, 1961.
92. Hearmon, R.F.S., and E.H. Adams. The Bending and Twisting of Anisotropic Plates. *Brit. J. Appl. Phys.*, 1952, Vol. 3, No. 5, pp. 150–156.
93. Hill, R.G. Evaluation of Elastic Moduli of Bilaminate Filament-Wound Composites. *Experimental Mechanics*, 1968, Vol. 8, No. 2, pp. 75–81.
94. Hill, R.G., and W.E. Anderson. The Engineering Properties of Polymer Matrix Composite Materials by a Pure Moment Test. *Composite Materials: Testing and Design (Second Conference)*. ASTM STP No. 497. Philadelphia, Pa., 1972, pp. 219–236.
95. Hill, R.G., and E.J. Zapel. Pure Bending Test Machine. U.S. Patent 3,026,720, March 27, 1962.
96. Hofer, K.E., Jr., and P.N. Rao. A New Static Compression Fixture for Advanced Composite Materials. *J. of Testing and Evaluation*, 1977, Vol. 5, No. 4, pp. 278–283.
97. Hoggatt, J.T. Test Methods for High-Modulus Carbon Yarn and Composites. *Composite*

Materials: Testing and Design. ASTM STP No. 460. Philadelphia, Pa., 1969, pp. 48–61.

98. Holmes, R.D., and D.W. Wright. Creep and Fatigue Characteristics of Graphite/Epoxy Composites. ASME Paper 70-DE-32.
99. Hörig, H. Über die unmittelbare Messung der Gleitzahlen S_{44}, S_{55}, S_{66} bei Stoffen von rhombischer Symmetrie und geringer Starrheit. *Ann. Phys.*, 1943, Vol. 43, No. 4, pp. 285–295.
100. Howard, H.B. The Five-Point Loading Shear Stiffness Test. *J. Roy. Aeron. Soc.*, 1962, Vol. 66, No. 621, p. 591.
101. *Inelastic Behavior of Composite Materials*. Ed. by Carl T. Herakovich. ASME, New York, N.Y., 1975.
102. Iosipescu, N. New Accurate Procedure for Single Shear Testing of Metals. *J. Materials*, 1967, Vol. 2, No. 3, pp. 537–566.
103. Ishai, O., and R.E. Lavengood. Characterizing Strength of Unidirectional Composites. *Composite Materials: Testing and Design*. ASTM STP No. 460. Philadelphia, Pa., 1969, pp. 271–281.
104. Jones, R.M. *Mechanics of Composite Materials*. Scripta Book Company, Washington, D.C., 1975.
105. Kalnin, I.L. Carbon Fiber Surfaces—Characterization, Modification and Effect on the Fracture of High Modulus Fiber-Polymer Composites. *Fracture of Composite Materials*. Ed. by G.C. Sih and V.P. Tamuzs. Sijthoff & Noordhoff International Publishers, Alphen aan den Rijn, 1979, pp. 373–384.
106. Kaminski, B.E. Effects of Specimen Geometry on the Strength of Composite Materials. *Analysis of the Test Methods for High Modulus Fibers and Composites*. ASTM STP No. 521. Philadelphia, Pa., 1973, pp. 181–191.
107. Kaminski, B.E., and R.B. Lantz. Strength Theories of Failure for Anisotropic Materials. *Composite Materials: Testing and Design*. ASTM STP No. 460. Philadelphia, Pa., 1969, pp. 160–169.
108. Kargin, V.A. *Contemporary Problems of Polymer Science*, Moscow State University Press, Moscow, 1962. (in Russian)
109. Kassandrova, Yu.N., and V.V. Lebedev. *Processing of Measurement Results*. Nauka Press, Moscow, 1970. (in Russian)
110. Kedward, K.T. On the Short Beam Test Method. *Fibre Science and Technology*, 1972, Vol. 5, No. 2, pp. 85–95.
111. Kelly, A. *Strong Solids*. Second Edition. Clarendon Press, Oxford, 1973.
112. Kelly, A. Composites—A Decade of Progress. *Composites*, 1979, Vol. 10, No. 1, pp. 2–3.
113. Khitrov, V.V., and Yu.I. Katarzhov. Effect of Reinforcement Angle on Load-Carrying Capacity of Compressed Bars. *Mechanics of Composite Materials*, 1979, No. 4, pp. 611–616 (in Russian).
114. Kintsis, T.Ya., and R.P. Shlitsa. Determination of the Shear Modulus of Composites from Experiments on the Twisting of Circular Specimens. *Polymer Mechanics*, 1978, Vol. 14, No. 5, pp. 764–767. (in Russian)
115. Kleiner, W., und W. Lüssmann. Ermittlung charakteristischer Werkstoffkenngrössen von GFK-Wickelstrukturen unter betriebsgerechten Bedingungen. *Kunststoffe*, 1959, Vol. 59, No. 12, pp. 941–947.
116. Klosowska-Wolkowicz, Z., W. Królikowski, and P. Penczek. *Żywice i Laminaty Poliestrowe*. WNT, Warszawa, 1969.
117. Knight, C.E., Jr. Failure Analysis of the Split-D Test Method. In: *Composite Materials: Testing and Design. (Fourth Conference)*. ASTM STP No. 617. Philadelphia, Pa., 1977, pp. 201–214.

118. Kunukkasseril, V.X., R.A. Chaudhuri, and K. Balaraman. A Method to Determine Eighteen Rigidities of Layered Anisotropic Plates. *Fibre Science and Technology*, 1975, Vol. 8, No. 4, pp. 303–318.
119. Langhaar, H.L. Torsion of Curved Beams of Rectangular Cross Section. *J. Appl. Mech.*, March 1952, Vol. 19, pp. 49–53.
120. Lantz, R.B. Boron Epoxy Laminate Test Methods. *J. Composite Materials*, October 1969, pp. 642–650.
121. Lantz, R.B., and K.G. Baldridge. Angle-Plied Boron-Epoxy Test Methods. A Comparison of Beam-Tension and Axial Tension Coupon Testing. *Composite Materials: Testing and Design*. ASTM STP No. 460. Philadelphia, Pa., 1969, pp. 94–97.
122. Lauraitis, K.N. Failure Modes and Strength of Angle-Ply Laminates. *Composite Materials in Engineering Design*. Proc. Sixth St. Louis Symposium, Metals Park, Ohio, 1973, pp. 541–561.
123. Lazaryan, V.A. *Technical Theory of Bending*. Naukova Dumka Press, Kiev, 1976 (in Russian).
124. Lekhnitskii, S.G. *Anisotropic Plates*. Gostekhizdat Press, Moscow, 1957 (in Russian).
125. Lekhnitskii, S.G. *Torsion of Anisotropic and Nonhomogeneous Bars*. Nauka Press, Moscow, 1971 (in Russian).
126. Lekhnitskii, S.G. *Theory of Elasticity of an Anisotropic Body. Second Edition*. Nauka Press, Moscow, 1977 (in Russian).
127. Lempriere, B.M. Poisson's Ratio in Orthotropic Materials. *AIAA Journal*, 1968, Vol. 6, No. 11, pp. 2220–2221.
128. Lempriere, B.M., R.W. Fenn, Jr., D.D. Crooks, and W.C. Kinder. Torsion Testing for Shear Modulus of Thin Orthotropic Sheet. *AIAA Journal*, 1969, Vol. 7, No. 12, pp. 2341–2342.
129. Lenk, R.S. *Plastics Rheology*. MacLaren and Sons, London, 1968.
130. Lenoe, E.M. Testing and Design of Advanced Composite Materials. *J. Eng. Mech. Div., Proc. Amer. Soc. Civ. Eng.*, December 1970, pp. 809–823.
131. Lenoe, E.M., M. Knight, and C. Schoene. Preliminary Evaluation of Test Standards for Boron Epoxy Laminates. *Composite Materials: Testing and Design*. ASTM STP No. 460. Philadelphia, Pa., 1969, pp. 122–139.
132. Leontiev, V.A., B.D. Oleinik, V.G. Perevozchikov, and Yu. V. Sokolkin. Assembly for Strength Testing Rings of Composite Materials in One-Sided Heating. *Polymer Mechanics*, 1977, Vol. 13, No. 6, pp. 1119–1120. (in Russian)
133. Lomakin, V.A. Problems in the Mechanics of Structurally Nonhomogeneous Solids. *Mekhanika Tverdovo Tela*, 1978, No. 6, pp. 45–52 (in Russian).
134. Malmeister, A.K. Geometry of Theories of Strength. *Polymer Mechanics*, 1966, Vol. 2, No. 4, pp. 324–331. (in Russian)
135. Malmeisters, A.K., V.P. Tamuzh, and G.A. Teters. *Strength of Rigid Polymeric Materials. Second Edition*. Zinātne Press, Riga, 1972 (in Russian). German Edition: *Mechanik der Polymerwerkstoffen*. Revised and Bearbeitet edited by A. Duda. Akademie-Verlag, 1977.
136. Markham, M.F., and D. Dawson. Interlaminar Shear Strength of Fiber-Reinforced Composites. *Composites*, 1975, Vol. 6, No. 4, pp. 173–176.
137. Martin, J.W. *The Southseaman*. Blackwoods, 1928.
138. Masuda, Y., and J. Yamasaki. On the Measurement of Shear Modulus of an Orthotropic Plate by Torsional Testing. *J. Society of Materials Science Japan*, 1974, Vol. 23, No. 254, pp. 973–979.
139. McAbee, E., and M. Chmura. Effects of Rate on the Mechanical Properties of Glass Reinforced Polyester. *SPE Journal*, April 1963, Vol. 19, pp. 375–378.

140. McKenna, G.B., J.F. Mandell, and F.J. McGarry. Interlaminar Strength and Toughness of Fiberglass Laminates. *SPI Reinforced Plastic/Composites Institute*. Proc. 29th Annual Conference. Washington, D.C., 1974, Section 13-C, pp. 1–8.
141. Menges, G., und R. Kleinholz. Vergleich verschiedener Verfahren zum Bestimmen der interlaminaren Scherfestigkeit. *Kunststoffe*, 1969, Vol. 59, No. 12, pp. 959–966.
142. Merkushev, A.A., A.P. Mishchenko, and E.I. Kertuzov. Analytic Determination of Stiffness of Circular Elastic Fiberglass Elements. *Collected Scientific Works of the Perm Polytechnic Institute*, No. 127, *Polymers in Machine Building*, PPI, Perm, 1973, pp. 146–156 (in Russian).
143. Merkushev, A.A., and O.G. Tsyplakov, Experimental Evaluation of the Effect of Structural Factors on Stiffness of Circular Elastic Fiberglass Elements. *Collected Scientific works of the Perm Polytechnic Institute*, No. 127, *Polymers in Machine Building*, PPI, Perm, 1973, pp. 157–161.
144. Miller, R.J. End Plugs for External Pressure Tests of Composite Cylinders. In: *Composite Materials: Testing and Design*. ASTM STP No. 460. Philadelphia, Pa., 1969, pp. 150–159.
145. Mitynskii, A.N. Torsion and Shear Moduli of Wood as an Anisotropic Material. In: *Works of the S.M. Kirov FTA*, No. 65, Leingrad, 1949, pp. 49–57. (in Russian)
146. Mullin, J.V., and A.C. Knoell. Basic Concepts in Composite Beam Testing. *Materials Research and Standards*, 1970, Vol. 10, No. 12, pp. 16–20, 33.
147. Nadai, A. *Elastische Platten*. Springer-Verlag, Berlin, 1925.
148. Nederveen, C.J., and J.F. Tilstra. Clamping Corrections for Torsional Stiffness of Prismatic Bars. *J. Physics*, 1971, Ser. D, Vol. 4, No. 11, pp. 1661–1667.
149. Niederstadt, G. Druckprüfung an glassfaserverstärkten Kunststoffen. *Kunststoffe*, 1963, Vol. 53, No. 4, pp. 217–219.
150. Nikolaev, V.P. Candidate's Dissertation. Moscow, 1968 (in Russian).
151. Nikolaev, V.P. On the Method of Testing Wound Glass Fiber Products by the Split Disk Method. In: *Reports of the Scientific and Technical Conference on the Results of 1968–69 Scientific Research Work*, *Dynamics and Strength of Machines*, MEI, Moscow, 1969, pp. 113–122 (in Russian).
152. Nikolaev, V.P. Determination of the Shear Moduli of Glass-Reinforced Plastics Using Ring Specimens. *Polymer Mechanics*, 1971, Vol. 7, No. 6, pp. 984–986. (in Russian)
153. Nikolaev, V.P. Method of Interlaminar Shear Strength Determination on Rings Made of Reinforced Materials. In: *Works of MEI*, No. 164, *Dynamics and Strength of Machines*, MEI, Moscow, 1973, pp. 92–96 (in Russian).
154. Nikolaev, V.P. A Method of Testing the Transverse Tensile Strength of Wound Articles. *Polymer Mechanics*, 1973, Vol. 9, No. 4, pp. 675–677. (in Russian)
155. Nilolaev, V.P. Testing Glass-Reinforced Plastic Rings by Means of Rigid Sectors. *Polymer Mechanics*, 1973, Vol. 9, No. 6, pp. 997–999. (in Russian)
156. Nikolaev, V.P., S.G. Abramov, and V.D. Popov. On the Failure of Segments of a Circular Ring in Bending. *Works of MEI*, No. 164, *Dynamics and Strength of Machines*, MEI, Moscow, 1973, pp. 86–92 (in Russian).
157. Nikolaev, V.P., A.G. Vasil'chenko, and I.P. Levin. On the Determination of Poisson's Ratios on Ring Specimens. *Works of MEI*, No. 353, *Mechanics of Deformable Solids and Reliability Theory*, MEI, Moscow, 1978, pp. 49–52 (in Russian).
158. Nikolaev, V.P., and Yu.N. Novichkov. On the Experimental Determination of Shear Moduli of Glass Fiber Composites. *Raschoti na Prochnost*, No. 13, Mashinostroyeniye Press, Moscow, 1968, pp. 355–369 (in Russian).
159. Nikolaev, V.P., and V.G. Perevozchikov. Strength of Fiberglasses Under Combined Action of Interlaminar Shear and Transversal Tension Stresses. *Problemy Prochnosti*, 1974,

No. 11, pp. 62–64 (in Russian).
160. Ogorkiewicz, R.M., and P.E.R. Mucci. Testing of Fibre-Plastics Composites in Three-Point Bending. *Composites*, 1971, Vol. 2, No. 3, pp. 139–145.
161. Ogorkiewicz, R.M., and A.A.M. Sayigh. Torsional Stiffness of Plastic Tubes Reinforced with Glass Fibers. *J. Strain Analysis*, 1971, Vol. 6, No. 4, pp. 226–230.
162. Oken, S., and J.T. Hoggatt. Behavior of Graphite Composites in a Marine Environment. *SAMPE Quarterly*, 1978, Vol. 9, No. 2, pp. 21–27.
163. Pagano, N.J., and P.C. Chou. The Importance of Signs of Shear Stress and Shear Strain in Composites. *J. Composite Materials*, January 1969, pp. 166–173.
164. Pagano, N.J., and J.C. Halpin. Influence of End Constraint in the Testing of Anisotropic Bodies. *J. Composite Materials*, January 1968, pp. 18–31.
165. Pagano, N.J., and R.B. Pipes. Some Observations on the Interlaminar Strength of Composite Laminates. *Int. J. Mech. Sci.*, 1973, Vol. 15, No. 8, pp. 679–688.
166. Pagano, N.J., and J.M. Whitney. Geometric Design of Composite Cylindrical Characterization Specimens. *J. Composite Materials*, July 1970, pp. 360–378.
167. Pan, H.H. Non-Linear Deformation of a Flexible Ring. *Quart. J. Mech. Appl. Math.*, 1962, Part 4, Vol. 15, pp. 401–412.
168. Panshin, B.I., G.M. Bartenev, and G.N. Finogenov. Strength of Plastic under Repeated Loading. *Plastics*, 1960, No. 11, pp. 47–54 (in Russian).
169. Panshin, B. I., L.P. Kotova, and O.V. Kolchev. Author's Certificate 268,731, December 2, 1968. Otkritiya Izobreteniya. Promišlenniye obrazci. Tovarniye znaki. 1970, No. 14 (in Russian).
170. Panshin, B.I., L.P. Kotova, and O.V. Kolchev. Methods of Determination of Mechanical Characteristics of Sheet Materials under In-Plane Shear. *Zavodskaya Laboratoriya*, 1970, Vol. 36, No. 11, pp. 1371–1374 (in Russian).
171. Park, I.K. Tensile and Compressive Test Methods for High-Modulus Graphite-Fibre-Reinforced Composites. *International Conference on Carbon Fibres, their Composites and Applications*, London, 1971, Paper No. 23.
172. Partsevskii, V.V. On Tension of Anisotropic Ring by Rigid Split Disks. *Polymer Mechanics*, 1970, No. 6, pp. 1113–1116 (in Russian).
173. Partsevskii, V.V. Stresses in an Anisotropic Ring under Tension by Rigid Sectors. *Izvestiya Akademii Nauk SSSR, Mekhanika Tverdogo Tela*, 1971, No. 1, pp. 90–92 (in Russian).
174. Partsevskii, V.V. Basis of a Test Procedure for Anisotropic Annular Samples. *Polymer Mechanics*, 1972, Vol. 8, No. 1, pp. 157–159. (in Russian)
175. Partsevskii, V.V. Stability of a Ring under Compression by a Rubber Ring. *Works of MEI, Dynamics and Strength of Machines*, No. 164, MEI, Moscow, 1973, pp. 74–80 (in Russian).
176. Partsevskii, V.V., and A. Ya. Gol'dman. Mechanical Testing of Glass Fiber Ring Specimens. *Works of MEI, Dynamics and Strength of Machines*, No. 74, MEI, Moscow, 1970, pp. 125–128 (in Russian).
177. Peterson, R.E. *Stress Concentration Factors*. John Wiley and Sons, New York, 1974.
178. Petit, P.H. A Simplified Method of Determining In-Plane Shear Stress-Strain Response of Unidirectional Composites. *Composite Materials: Testing and Design*. ASTM STP No. 460. Philadelphia, Pa., 1969, pp. 83–93.
179. Petker, I. The Status of Organic Materials in Advanced Composites. *SAMPE Quarterly*, 1972, Vol. 3, No. 2, pp. 7–21.
180. Pipes, R.B., B.E. Kaminski, and N.J. Pagano. Influence of the Free Edge upon the Strength of Angle-Ply Laminates. *Analysis of the Test Methods for High Modulus Fibers and Composites*. ASTM STP No. 521. Philadelphia, Pa., 1973, pp. 218–228.

181. Polilov, A.N., and V.K. Khokhlov. Design Strength Criterion of Composite Beams under Bending. *Mashinovedeniye*, 1972, No. 2, pp. 53–57 (in Russian).
182. Polilov, A.N., and V.K. Khokhlov. Interlaminar Strength Criterion of Composites under Transverse Bending. *Mashinovedeniye*, 1977, No. 3, pp. 56–59 (in Russian).
183. Polyakov, V.A., and I.G. Zhigun. Contact Problem for Beams Made of Composite Materials. *Polymer Mechanics*, 1977, Vol. 13, No. 1, pp. 59–69. (in Russian)
184. Polyakov, V.A., and I.G. Zhigun. Estimation of the Zone of Disturbed Stressed State during the Stretching of Composites. *Polymer Mechanics*, 1978, Vol. 14, No. 6, pp. 883–888. (in Russian)
185. Polyakov, V.A., and I.G. Zhigun. Assessment of the Zone of Disturbed Stressed States in the Expansion of Composite Materials. 2. Analysis of Stress Distributions. *Mechanics of Composite Materials*, 1979, Vol. 15, No. 1, pp. 109–116. (in Russian)
186. Popov, Ye.P. *Theory and Design of Flexible Elastic Parts*. LKVVIA Press, Leningrad, 1947 (in Russian).
187. *Properties of Polyester Glass Fiber Composites and Methods of Controlling Them*. Ed. by V.I. Smirnov. Sudostroyeniye Press, Leningrad, 1967 (in Russian).
188. *Properties of Polyester Glass Fiber Composites and Methods of Controlling Them*. No. 2. Ed. by V.V. Meshcheryakov. Sudostroyeniye Press, Leingrad, 1970 (in Russian).
189. *Properties of Shipbuilding Glass Fiber Composites and Methods of Controlling Them* No. 3. Ed. by V.V. Meshcheryakov. Sudostroyeniye Press, Leingrad, 1974 (in Russian).
190. Prosen, S. Destructive and Non-Destructive Test Methods. *Fiber Composite Materials*. American Society for Metals, Metals Park, Ohio, 1965.
191. Protopopov, K.G., and N.V. Piskunov. Determination of the Modulus of Elasticity of Reinforced Plastics under Interlaminar Shear. *Zavodskaya Laboratoriya*, 1974, Vol. 40, No. 10, pp. 1269–1271 (in Russian).
192. Puck, A. Festigkeitsberechnung an Glasfaser-Kunststoff-Laminaten bei zusammengesetzter Beanspruchung. Bruchhypothesen und schichtenweise Bruchanalyse. *Kunststoffe*, 1969, Vol. 59, No. 11, pp. 780–787.
193. Puck, A., and W. Schneider. On Failure Mechanisms and Failure Criteria of Filament-Wound Glass Fibre-Resin Composites *Plastics and Polymers*, 1969, Vol. 37, No. 127, pp. 33–43.
194. Purslow, D. The Shear Properties of Unidirectional Carbon Fibre Reinforced Plastics and Their Experimental Determination. Aeronautical Research Current Paper No. 1381, HMSO, London, 1977.
195. Rabinovich, A.L. *Introduction to the Mechanics of Reinforced Polymers*. Nauka Press, Moscow, 1970. (in Russian)
196. Rabinovich, A.L., and Ya.D. Avrasin. Mechanical Characteristics of Some Laminated Plastics in Connection with the Strength of Bolted and Riveted Joints. In: *Glass Textolites and Other Structural Plastics*. Oborongiz, Moscow, 1960, pp. 78–107. (in Russian)
197. Rabotnov, Yu.N. Mechanics of Composites. *Vestnyk Akademii Nauk SSSR*, 1979, No. 5, pp. 50–58. (in Russian)
198. Rabotnov, Yu.N. *Mechanics of a Deformable Solid Body*. Nauka Press, Moscow, 1979 (in Russian).
199. Rabotnov, Yu.N., S.A. Kolesnikov, V.S. Matitsyn, I.M. Makhmutov, V.A. Rudenko, and E.I. Stepanychev. Mechanical Properties of a Composite with a Carbonized Matrix. *Polymer Mechanics*, 1976, Vol. 12, No. 2, pp. 202–206. (in Russian)
200. Ray, J.D. Mechanical Properties of High-Performance Plastics Composites. *International Conference on Carbon Fibres, Their Composites and Application*. London, 1971, Paper No. 29.

201. Rosato, D.V., and C.S. Grove, Jr. *Filament Winding: Its Development, Manufacture, Applications and Design*. Wiley-Interscience, New York, 1964.
202. Rosen, B.W. Simple Procedure for Experimental Determination of the Longitudinal Shear Modulus of Unidirectional Composites. *J. Composite Materials*, October 1972, pp. 552–554.
203. Rithmann, E.A., and G.E. Molter. Characterization of the Mechanical Properties of a Unidirectional Carbon Reinforced Epoxy Matrix Composite. *Composite Materials: Testing and Design*. ASTM STP No. 460. Philadelphia, Pa., 1969, pp. 72–82.
204. Roze, A.V. Effect of Interlaminar Stiffness and Strength on Planar Loading of Fiber Reinforced Materials. *Polymer Mechanics*, 1970, No. 5, pp. 876–883 (in Russian).
205. Rudenko, V. A. Development of Methods for Solution Problems on Static Bending of Beams of Reinforced Polymer Materials. *Mashinovedeniye*, 1977, No. 5, pp. 101–109 (in Russian).
206. Ryder, J. T., and E. D. Black. Compression Testing of Large Gage Length Composite Coupons. *Composite Materials: Testing and Design (Fourth Conference)*. ASTM STP No. 617. Philadelphia, Pa., 1977, pp. 170–189.
207. Sattar, S. A., and D. H. Kellog. The Effect of Geometry on the Mode of Failure of Composites in the Short-Beam Shear Test. *Composite Materials: Testing and Design*. ASTM STP No. 460. Philadelphia, Pa., 1969, pp. 62–71.
208. Sborovskii, A. K., V. A. Popov, N. F. Savelyeva, and A. V. Lavrov. Modern Methods of Mechanical Compression Testing of Wound Shipbuilding Glass Fiber Composites. *Tekhnologiya Sudostroyeniya*, 1971, No. 6, pp. 116–122 (in Russian).
209. Schmitt, F. Einspannungseinfluss bei Zug- und Biegestäben. *Konstruktion*, 1975, Vol. 27, No. 2, pp. 48–54.
210. Semenov, P. I. Determination of Shear Moduli of Orthotropic Materials by Torsion Tests. *Polymer Mechanics*, 1966, No. 1, pp. 27–33 (in Russian).
211. Semenov, P. I. Determination of the Transverse Strain Coefficients of Orthotropic Materials from Bending Tests. *Polymer Mechanics*, 1969, Vol. 5, No. 6, pp. 1001–1003. (in Russian)
212. Sendeckyj, G. P. A Brief Survey of Empirical Multiaxial Strength Criteria for Composites. *Composite Materials: Testing and Design (Second Conference)*. ASTM STP No. 497. Philadelphia, Pa., 1972, pp. 41–51.
213. Serensen, S. V., and V. S. Strelyaev. Statistical Regularities in Failure of Oriented Glass Fiber Composite by Unwinding. *Problemy Prochnosti*, 1970, No. 1, pp. 8–18 (in Russian).
214. Shlitsa, R. P. Deformation Characteristics of Glass-Reinforced Plastics in Tension. 1. Discontinunity in Stress-Strain Curve. *Polymer Mechanics*, 1966, Vol. 2, No. 2, pp. 194–196. (in Russian)
2. Relation between Longitudinal and Transverse Strains. *Polymer Mechanics*, 1969, Vol. 5, No. 2, pp. 315–317. (in Russian)
215. Sidorin, Ya. S. Experimental Study of Glass Fiber Composite Shear Anisotropy. *Zavodskaya Laboratoriya*, 1966, No. 5, pp. 594–597 (in Russian).
216. Sidorin, Ya. S. On the Possibility of Determination of the Shear Modulus of Glass-Reinforced Plastics by Means of the Four-Arm Frame. *Polymer Mechanics*, 1968, No. 5, pp. 799–802 (in Russian).
217. Sims, D. F. In-Plane Shear Stress-Strain Response of Unidirectional Composite Materials. *J. Composite Materials*, January 1973, pp. 124–128.
218. Sims, D. F., and J. C. Halpin. Methods for Determining the Elastic and Viscoelastic Response of Composite Materials. *Composite Materials: Testing and Design (Third Conference)*. ASTM STP No. 546. Philadelphia, Pa., 1974, pp. 46–66.

219. Skudra, A. M., E. Ya. Bulavs, and K. A. Rocens. *Creep and Static Fatigue of Reinforced Plastics*. Zinātne Press, Riga, 1971 (in Russian). German edition: *Kriechen und Zeitstandverhalten von verstärkten Plasten*. Translated and revised by B. Knauer. VEB Deutscher Verlag für Grundstoffindustrie, Leipzig, 1975.
220. Stepanov, M. N. *Statistical Processing of Mechanical Test Results*. Mashinostroyeniye, Moscow, 1972 (in Russian).
221. Strelyaev, V. S., and V. A. Konstantinov. An Engineering Method of Determination Shear Modulus and Coefficients of Mutual Effect of Oriented Plastics. In: *Mechanics of Composite Materials*, No. 2, Riga Polytechnic Institute, Riga, 1979, pp. 26–32 (in Russian).
222. Strength of a Glass Fiber Ship Hull. Ed. by I. K. Smirnova. Sudostroyeniye Press, Leningrad, 1965 (in Russian).
223. Sullivan, T. L., and C. C. Chamis. Some Important Aspects in Testing High-Modulus Fiber Composite Tubes in Axial Tension. *Analysis of the Test Methods for High Modulus Fibers and Composites*. ASTM STP No. 521. Philadelphia, Pa., 1973, pp. 277–292.
224. Sumsion, H.T., and Y.D.S. Rajapakse. Simple Torsion Test for Shear Moduli Determination of Orthotropic Composites. *ICCM/2*. Proc. of the 1978 International Conference on Composite Materials. New York, 1978, pp. 994–1002.
225. N. Fried, Survey of Methods of Test for Parallel Filament Reinforced Plastics. In: *Symposium on Standards for Filament-Wound Reinforced Plastics*. ASTM STP No. 327. Philadelphia, Pa., 1963, pp. 13–39.
226. Tarnopol'skii, Yu. M. Modern Trends in the Development of Fibrous Composites. *Polymer Mechanics*, 1972, Vol. 8, No. 3, pp. 473–481. (in Russian)
227. Tarnopol'skii, Yu, M. Delamination of Compressed Rods of Composites. *Mechanics of Mechanics of Composite Materials*, 1977, Vol. 15, No. 2, pp. 225–231. (in Russian)
228. Tarnopol'skii, Yu. M., I. G. Zhigun, and V. A. Polyakov. Distribution of Shearing Stresses under Three-Point Flexure in Beams Made of Composite Materials. *Polymer Mechanics*, 1977, Vol. 13, No. 1, pp. 52–58. (in Russian)
229. Tarnopol'skii, Yu. M., and T. Ya. Kincis. Features of Mechanical Testing of High-Modulus Reinforced Plastics. *Zavodskaya Laboratoriya*, 1973, No. 11, pp. 1368–1374 (in Russian).
230. Tarnopol'skii, Yu. M., G. G. Portnov, Yu. B. Spridzans, and V. N. Bulmanis. Carrying Capacity of Rings Formed by Winding Composites with High-Modulus Anisotropic Fiber Reinforcement. *Polymer Mechanics*, 1973, Vol. 9, No. 4, pp. 592–599. (in Russian)
231. Tarnopol'skii, Yu. M., and A. V. Roze. *Characteristics of Calculation of Reinforced Plastic Parts*. Zinātne Press, Riga, 1969 (in Russian).
232. Tarnopol'skii, Yu. M., A. V. Roze, I. G. Zhigun, and G. M. Gunyaev. Structural Characteristics of Materials Reinforced with High-Modulus Fibers. *Polymer Mechanics*, 1971, Vol. 7, No. 4, pp. 600–609. (in Russian)
233. Tarnopol'skii, Yu. M., and A. M. Skudra. *Structural Strength and Deformability of Glass Fiber Composites*. Zinātne Press, Riga, 1966 (in Russian).
234. Terry, G. A Comparative Investigation of Some Methods of Unidirectional, In-Plane Shear Characterization of Composite Materials. *Composites*, 1979, Vol. 10. No. 4, pp. 233–237.
235. Timoshenko, S. *History of the Strength of Materials*. McGraw-Hill Book Co., New York, 1953.
236. Timoshenko, S. *Strength of Materials*. Part I. Third edition. Van Nostrand Company, Inc., Princeton, N. J., 1955.
237. Timoshenko, S. *Strength of Materials*. Part II. Third edition. Van Nostrand Company, Inc., Princeton, N. J., 1956.

238. Timoshenko, S. P. *Course in the Theory of Elasticity*. Ed. by E. I. Grigolyuk. Naukova Dumka Press, Kiev, 1972. (in Russian).
239. Tsai, S. W. Experimental Determination of the Elastic Behavior of Orthotropic Plates. *Trans. ASME*, 1965, Ser. B, Vol. 87, No. 3, pp. 315–318.
240. Tsai, S. W. *Strength Characteristics of Composite Materials*. NASA CR-224. Washington, D. C., 1965.
241. Tsai, S. W., and Azzi V. D. Strength of Laminated Composite Materials. *AIAA Journal*, 1966, Vol. 4, No. 2, pp. 296–301.
242. Tsai, S. W., and G. S. Springer. The Determination of Moduli of Anisotropic Plates. *Trans. ASME*, 1963, Ser. E, Vol. 30, No. 3, pp. 467–468.
243. Tsyplakov, O. G. *Principles of Molding of Glass Fiber Shells*. Mashinostroyeniye, Moscow-Leningrad, 1968 (in Russian).
244. Uemura, M. Mechanical Testing Methods and Design Criterion for Fiber Reinforced Plastic Composites. I–III. *J. Japan Society for Composite Materials*, 1981, Vol. 7, No. 1, pp. 32–37; No. 2, pp. 74–81; No. 4, pp. 154–161 (in Japanese).
245. Uemura, M., and H. Iyama. Bearing Strength of Bolted Joints in Carbon Fiber Reinforced Plastics. *Composite Materials and Structures (Japan)*, 1973, Vol. 2, No. 2, pp. 8–14.
246. Uemura, M., and M. Murata. Evaluation of Tensile Method by Use of Ring Specimens for Filament-Wound Composites. *J. Society of Materials Science Japan*, October 1978, Vol. 27, pp. 1001–1007 (in Japanese).
247. Van Fo Fi, G. A. *Reinforced Plastic Structures*. Tekhnika Press, Kiev, 1971 (in Russian).
248. Varushkin, E. M. Investigation of the Temperature Residual Stresses and Strains. *Polymer Mechanics*, 1971, Vol. 7, No. 6, pp. 925–930. (in Russian)
249. Waddoups, M. E. Characterization and Design of Composite Materials. *Composite Materials Workshop*. Technomic Publishing Co., Stamford, Conn., 1968, pp. 254–308.
250. Walter, R. W., R. W. Johnson, R. R. June, and J. E. McCarty. Designing for Integrity in Long-Life Composite Aircraft Structures. *Fatigue of Filamentary Composite Materials*. ASTM STP No. 636. Philadelphia, Pa., 1977, pp. 228–247.
251. Wedemeyer, E. A. Festigkeitsverhalten von glasfaserverstärkten Kunststoffen. *Maschinenmarkt*, 1964, Vol. 70, No. 16, pp. 17–20.
252. Wende, A., W. Moebes, and H. Marten. *Glasfaserverstärkte Plaste*. VEB Deutscher Verlag für Grundstoffindustrie, Leipzig, 1963.
253. Whitney, J. M. Analytical and Experimental Methods in Composite Mechanics. *J. Struct. Div., Proc. Am. Soc. Civ. Eng.*, 1973, Vol. 99, No. ST 1 (January), pp. 113–129.
254. Whitney, J. M. Free-Edge Effects in the Characterization of Composite Materials. In: *Analysis of the Test Methods for High Modulus Fibers and Composites*. ASTM STP No. 521. Philadelphia, Pa., 1973, pp. 167–180.
255. Whitney, J. M., C. E. Browning, and A. Mair. Analysis of the Flexure Test for Laminated Composite Materials. *Composite Materials: Testing and Design (Third Conference)*. ASTM STP No. 546. Philadelphia, Pa., 1974, pp. 30–45.
256. Whitney, J. M., and R. J. Dauksys. Flexure Experiments of Off-Axis Composites. *J. Composite Materials*, January 1970, pp. 135–137.
257. Whitney, J. M., G. C. Grimes, and P. H. Francis. Effect of End Attachment on the Strength of Fiber-Reinforced Composite Cylinders. *Experimental Mechanics*, 1973, Vol. 13, No. 5, pp. 185–192.
258. Whitney, J. M., and G. E. Husman. Use of the Flexure Test for Determining Environmental Behavior of Fibrous Composites. *Experimental Mechanics*, 1978, Vol. 18, No. 5, pp. 185–190.
259. Whitney, J. M., N. J. Pagano, and R. B. Pipes. Design and Fabrication of Tubular

Specimens for Composite Characterization. *Composite Materials: Testing and Design (Second Conference)*. ASTM STP No. 497. Philadelphia, Pa., 1972, pp. 52–67.

260. Whitney, J. M., D. L. Stansberger, and H. B. Howell. Analysis of the Rail Shear Test—Applications and Limitations. *J. Composite Materials*, January 1971, pp. 24–34.
261. Witt, R. K., W. H. Hoppmann, and R. S. Buxbaum. Determination of Elastic Constants of Orthotropic Materials with Special Reference to Laminates. *ASTM Bull.*, 1953, No. 194, pp. 53–57.
262. Wolf, H. *Spannungsoptik*. Springer-Verlag, Berlin, 1961.
263. Wong, H. Y., and R. D. Gordon. Strain Measurement on Filament-Wound Structural Members. *Strain*, July 1970, pp. 109–110.
264. Wu, E. M., and R. L. Thomas. Off-Axis Test of a Composite. *J. Composite Materials*, October 1968, pp. 523–526.
265. Yamamoto, C. A. Evaluation Techniques for Simple Mechanical Property Tests on Composite Materials. *Advanced Techniques for Material Investigation and Fabrication.* SAMPE 14th National Symposium and Exhibition, Cocoa Beach, Florida, 1968, 1-2-3.
266. Yeow, Y. T., and H. F. Brinson. A Comparison of Simple Shear Characterization Methods for Composite Laminates. *Composites*, 1978, Vol. 9, No. 1, pp. 49–55.
267. Yoshida, H., M. Uemura, and Y. Yamaguchi. On Interlaminar Shear and Flexural Strengths of Composite Beam. Part I. Loacal Stresses in the Vicinity of Loading Point. *J. Japan Society for Composite Materials*, 1979, Vol. 5, No. 2, pp. 62–67 (in Japanese).
268. Zaidel', A. N. *Elementary Estimates of Errors in Measurements.* Nauka Press, Moscow-Leningrad, 1967 (in Russian).
269. Zaitsev, G. P., and L. S. Makhov. Stiffness and Strength of Glass-Reinforced Plastics EF-32-307 in Torsion. *Polymer Mechanics*, 1974, No. 3, pp. 555–558 (in Russian).
270. Zhigun, I. G., and V. V. Mikhailov. Tensile Testing of High-Strength Unidirectional Composites. *Polymer Mechanics*, 1978, Vol. 14, No. 4, pp. 586–591. (in Russian)
271. Zhigun, I. G., and V. A. Polyakov. *Properties of Spatially Reinforced Plastics.* Ed. by Yu. M. Tarnopol'skii. Zinātne Press, Riga, 1978 (in Russian).
272. Zhigun, I. G., V. A. Polyakov, and V. V. Mikhailov. Peculiarities of Composite Testing in Compression. *Mekhanika Kompozitnykh Materialov*, 1979, No. 6, pp. 1111–1118 (in Russian).
273. Zhigun, I. G., V. A. Yakushin, V. V. Tanevskii, and V. V. Mikhailov. Analysis of Certain Methods of Determining Shear Moduli. I. Testing of Composites Uniform over the Thickness. *Polymer Mechanics*, 1976, Vol. 12, No. 1, pp. 112–118. (in Russian)
274. Zhigun, I. G., V. A. Yakushin, and Yu. N. Ivonin. Analysis of Methods of Determining of the Interlaminar Shear Strength of Composite Materials. *Polymer Mechanics*, 1976, Vol. 12, No. 4, pp. 573–580. (in Russian)
275. Zinchenko, V. F. Candidate's Dissertation. Riga, 1972 (in Russian).
276. Zinchenko, V. F., and V. A. Latishenko. Connection between Ratio of the Heat Conductivity and the Interlaminar Shear Modulus of Oriented Glass Reinforced Plastics. *Polymer Mechanics*, 1970, No. 6, pp. 985–989. (in Russian)
277. Zweben, C. The Flexure Strength of Aramid Fiber Composites. *J. Composite Materials*, October 1978, pp. 422–430.
278. Zweben, C., W. S. Smith, and M. W. Wardle. Test Methods for Fiber Tensile Strength, Composite Flexural Modulus, and Properties of Fabric-Reinforced Laminates. *Composite Materials: Testing and Design (Fifth Conference)*. ASTM STP No. 674. Philadelphia, Pa., 1979, pp. 228–262.

Index